Hybrid Computational Intelligent Systems

Hybrid Computational Intelligent Systems – Modeling, Simulation and Optimization unearths the latest advances in evolving hybrid intelligent modeling and simulation of human-centric data-intensive applications optimized for real-time use, thereby enabling researchers to come up with novel breakthroughs in this ever-growing field.

Salient features include the fundamentals of modeling and simulation with recourse to knowledge-based simulation, interaction paradigms, and human factors, along with the enhancement of the existing state of art in a high-performance computing setup. In addition, this book presents optimization strategies to evolve robust and failsafe intelligent system modeling and simulation.

The volume also highlights novel applications for different engineering problems including signal and data processing, speech, image, sensor data processing, innovative intelligent systems, and swarm intelligent manufacturing systems.

Features:

- A self-contained approach to integrating the principles of hybrid computational intelligence with system modeling and simulation.
- Well-versed foundation of computational intelligence and its application to real life engineering problems.
- Elucidates essential background, concepts, definitions, and theories thereby putting forward a complete treatment on the subject.
- Effective modeling of hybrid intelligent systems forms the backbone of almost every operative system in real-life.
- Proper simulation of real-time hybrid intelligent systems is a prerequisite for deriving any real-life system solution.
- Optimized system modeling and simulation enable real-time and failsafe operations of the existing hybrid intelligent system solutions.
- Information presented in an accessible way for researchers, engineers, developers, and practitioners from academia and industry working in all major areas and interdisciplinary areas of hybrid computational intelligence and communication systems to evolve human-centered modeling and simulations of real-time data-intensive intelligent systems.

Quantum Machine Intelligence
Series Editors
Siddhartha Bhattacharyya, Rajnagar Mahavidyalaya, Birbhum, India
Elizabeth C. Behrman, Wichita State University, USA

Quantum Machine Intelligence
Siddhartha Bhattacharyya, Mario Koppen, Elizabeth Behrman, Ivan Cruz

Hybrid Computational Intelligent Systems
Modeling, Simulation and Optimization
Siddhartha Bhattacharyya

For more information about this series, please visit: https://www.routledge.com/Quantum-Machine-Intelligence/book-series/QMI

Hybrid Computational Intelligent Systems

Modeling, Simulation and Optimization

Edited by
Siddhartha Bhattacharyya

CRC Press is an imprint of the
Taylor & Francis Group, an **informa** business

First edition published 2023
by CRC Press
6000 Broken Sound Parkway NW, Suite 300, Boca Raton, FL 33487-2742

and by CRC Press
4 Park Square, Milton Park, Abingdon, Oxon, OX14 4RN

CRC Press is an imprint of Taylor & Francis Group, LLC

ISBN: 9781032393025 (hbk)
ISBN: 9781032463292 (pbk)
ISBN: 9781003381167 (ebk)

DOI: 10.1201/9781003381167

Typeset in Sabon
by codeMantra

The editor would like to dedicate this volume to all those who have inherited the unconscious habit of belittling almost every sincere effort, small or big.

Contents

Preface

Almost every technological innovation in the present times is being driven by intelligence in one form or the other with the advent of computational intelligence. Computational intelligence has made its presence felt in every nook and corner of the world, thanks to the rapid exploration of research in this direction. Computational intelligence is now not limited to only specific computational fields; it has made giant strides into several interdisciplinary fields of science, engineering, medical science, business, and finance including signal processing, smart manufacturing, predictive control, robot navigation, smart cities, sensor design to name a few. Latest advances in evolving soft computing concepts and algorithms toward enabling intelligent solutions to real-life problems have enriched this field of computational intelligence. To add to this, researchers have conjoined different intelligent tools and techniques to evolve hybrid intelligent systems, which results in more efficient alternatives to stand-alone intelligent systems.

Of late, human-centered computing has evolved rapidly to add human-centric components to the hybrid intelligent applications in vogue, thereby aiming at a more human-like realistic approach. Furthermore, in order to envisage efficient, robust, and failsafe applications for intelligent data processing, proper modeling of these data-intensive systems and applications has become a prerequisite. Effective modeling of these hybrid intelligent systems would entail knowledge-based simulation of the systems induced by human-centric factors, thereby making them interactive and responsive. In addition, intelligent agents can also help automate the objective. Although traditional modeling practices are already in vogue, the resultant models need to be optimized for enabling real-time and failsafe operations. Hence, optimization of the modeling and simulation process is also a much-needed affair.

This volume targets to unearth the latest advances in evolving hybrid intelligent modeling and simulation of data-intensive applications for real-time applications.

This volume comprises 23 well-versed contributory chapters entailing different facets of intelligent system modeling and their applications to a wide variety of data- and information-intensive frameworks.

Chapter 1 discusses the issues of generating ratings of the Russian agricultural universities, modeled using their passive, active, and semi-active digital footprints. The results of creating ratings show the areas for increasing the efficiency in implementing the instructions of the President of Russia back in 2012 on the achievement of the indicators for improving the activities of the Russian universities. The method for calculating the passive digital footprint is based on website metrics techniques. The active digital footprint includes information on scientific and educational resources posted on universities' websites in the form of developments, publications, consultations, legal and statutory materials, remote training, custom software, databases, and information about electronic employment agencies and digital marketplaces. The semi-active footprint includes information collected from subordinate educational institutions and processed by the Ministry of Science and Higher Education of the Russian Federation in the form of a self-examination report. The evaluated ratings are compared with an integrated metric, which serves as the indicator of a comprehensive evaluation of the effectiveness of agricultural universities' use of information resources.

Chapter 2 presents a hardware and software model for a mechatronic complex (MC) designed to automate technological processes for performing several technological operations (TO), which include actuating mechanisms (AM), detection, fixation, and storage of moving objects. The manual control mode of the MC implies that the human operator controls the TO using commands on the interactive touch panel (ITP) screen. The automatic control mode differs in that the human operator starts the MC using real and/or virtual buttons, and the rest of the actions occur due to the program embedded in the programmable logic controller (PLC). When performing a working cycle in this mode of operation of the MC, the most difficult technological operation is fixing a moving object. The successful implementation of this operation depends on the prediction and control of the trajectory of movement of the pneumatic AM with nonlinear characteristics. If, as a result of the movements, the central axes of the movable object and the pneumatic gripper are not aligned, then the object with a high degree of probability will not be fixed in the gripper jaws. In this case, the further course of the working cycle is disrupted, which affects the operability of the MC. Thus, for the automatic control mode of the MC, problems arise in finding a balance between the minimum time required for aligning the axes of the moving object and the pneumatic gripper which is controlled by the system timers. This chapter attempts to solve these associated problems by developing a program code for controlling the MC in automatic mode; creating a human-machine interface model and verification of its performance followed by an analysis of the kinematic model, comparison of the obtained experimental results for two control modes, and the assessment of forecasting the moment in time when the axes of the pneumatic gripper and the moving object are aligned.

Soil moisture monitoring is highly significant from the agricultural perspective. The efficient use of water plays an important role in cultivation. Proper water usage can be possible by soil moisture prediction. For better soil moisture content estimation, various parameters have to be considered such as soil moisture, turbidity, pH, temperature, and humidity. The objective of Chapter 3 is to monitor the soil moisture considering these parameters. To accomplish the objective, four different sensors are used in our work, and a comparative analysis between a few machine learning methods is performed for detecting the moisture content in the soil.

As a result of the universality of individual attacks and attacked phenomena in nature, the stability analysis of the predator-prey system has always been one of the research topics in the field of population dynamics. Time delay is a necessary factor to be considered when studying the population change law in the process of species evolution, which has a great impact on the stability of the system. Also, diffusion can represent the distribution state of the population living in the biological space. So, a diffusive ratio-dependent predator-prey system involving two delays is discussed in Chapter 4. The distribution of the eigenvalues corresponding to the linear system is analyzed by taking delay as a Hopf bifurcation parameter. By means of the representation of sets, the conditions for the stability of the system along with the further occurrence of Hopf bifurcation are taken into consideration. The theoretical results are supported by some particular systems and simulations. Besides, the spatial motion behavior of two populations in the food chain is discussed by taking diffusion as the object. Results indicate that the time delay and diffusion have significant effects on the system and spatial motion states of predator and prey populations.

College student health is a social issue and a key to national prosperity and development. Based on the grey GM (1,1) prediction model theory, Chapter 5 conducts a statistical analysis of the physical health test data of a certain grade of a college student, establishes a prediction model of the physical health test scores of the group of college students and the development trend of physical fitness, and discusses the relative residuals. Q test, variance ratio C test, and small error probability P test are used to test the accuracy of the model. Judging from the prediction results, with the strong support of the national labor education system, the overall physical fitness of students is on the rise, and there is room for improvement in the performance of some sports events. Studying college student fitness test data helps to monitor student health, improve fitness, and test the implementation effects of labor education, which has certain practical significance.

The University library is an important channel for students to solve problems and obtain knowledge. Due to different borrowing habits and interests, each student has obvious differences in book borrowing behavior and has a certain borrowing tendency. In order to respond to the call of "national reading" of the two sessions and help library managers carry out book activities better and efficiently, taking university as an example, Chapter 6 adopts the Apriori algorithm of association rules, further discusses the factors affecting students' borrowing through the analysis of the factors affecting the borrowing volume of university libraries and finds out the correlation between book borrowing information. In addition, it also helps to understand students' reading preferences, improves reading literacy, and enables library administrators to make better decisions on book procurement and book recommendation.

The objective of Chapter 7 is to analyze the inspection systems for the recognition of materials by the dual-energy method. The focus of the study is the methods and algorithms to compare bremsstrahlung detectors in inspection systems operating in dual-energy mode. This chapter develops criteria and algorithms for comparing inspection systems with a material recognition option for various bremsstrahlung detectors and experimentally tests them for a betatron with maximum radiation energies of 4 and 7.5 MeV. Based on the analysis of the results of theoretical studies, a set of criteria and algorithms has been developed to compare inspection systems with a material recognition option. To evaluate the effectiveness of the developed criteria and algorithms, a series of experimental studies were conducted to compare the bremsstrahlung detectors manufactured by TSNK (Moscow, Russia) and Detection Technologies (Finland). The test objects consisted of fragments of organics, aluminum, steel, and lead with a mass thickness of 20–120 g/cm^2.

Computer-aided diagnosis and medical imaging systems have evolved in the past decade to a point where they can partially mimic radiologists and doctors. These systems can learn and differentiate the features and abnormalities in medical images and provide objective evidence with higher diagnostic confidence and faster inference. In Chapter 8, the authors focus on generating medical reports on chest X-ray images, which can be adapted later to work with other diagnostic tools such as ultrasounds and mammograms. The Indiana University dataset provides the CXR images corresponding to various lung and heart ailments, along with well-defined reports and findings. The generation of medical reports mainly consists of two broad tasks. The first task is to treat the problem as a multi-label classification task to obtain accurate tags for a particular image from the visual features. The second task is to generate the reports using these aforementioned tags, which require the use of recurrent neural networks such as hierarchical LSTMs, and improve the accuracy using a co-attention mechanism.

Presently, Internet of Things (IoT) is one of the ways of reducing manual intervention in different domains such as smart appliances and smart detection systems. Smart appliances are widely used technology that can be used in many areas such as in our home or office for providing comfort, energy consumption, security, etc. Such kind of system has become an important and integral part of the modern home automation system. However, there exist some problems such as dedicated interfacing, user authentication, high security, and other related factors. To address these problems, the authors develop a low-power smart home automation system in Chapter 9, which not only controls the home appliances but also increases the security of the entire system. In this work, smart appliances can be easily controlled by using a smartphone. The proposed system and its hardware are based on Arduino with its interface and communication via Bluetooth with peripheral devices and the Android device system. The security and authentication of the system are done through RFID. The efficacy of the proposed system has been judged with respect to reduced time delay, power consumption, and security.

The COVID-19 outbreak has caused a sizable number of fatalities and poses an unprecedented threat to public health. Vaccination may be an effective weapon in halting the pandemic, delaying the spread of the disease, and reducing the severity of the sickness. Although most Indians have been fully vaccinated, a substantial number of Indians still have not received precautionary, even second doses of the COVID-19 vaccine. After huge income/job losses due to this pandemic, most Indians cannot afford the expensive COVID-19 vaccines from a private vaccination clinic. As a result, free Government vaccination center slots are always in great demand. It can be challenging for ordinary Indians to find free-of-cost slots at convenient places using the Indian government's vaccination web portal, 'CoWIN'. Chapter 10 presents an Android application that can check every 15 minutes for available vaccination slots at the user's desired location and send notifications to the user if slots match with user-preferred criteria. The novel feature of this app is that it may send SMS to several ordinary phone users according to their desired vaccination criteria from a single installed instance. The app collected users' feedback, and a usability study is included in this chapter.

Feature selection is a state-of-art research area for machine learning classification where extraneous attributes are removed to reduce the data processing load and increase classification accuracy. The search space for combinations of large dimensions of data becomes complex making the feature selection an optimization problem. Evolutionary algorithms are being increasingly applied to various optimization problems including feature selection as they have proven to be advantageous over traditional search due to their limited requirement of domain-specific information. A solution for the feature selection problem is binary in nature (selection or omission of a feature) and hence is represented as a binary vector of dimension n. Several binary variants of evolutionary algorithms have been proposed to deal with the available discrete solution space. Chapter 11 presents a binary variant of MMBAIS (Multi-Modal Bat Algorithm with Improved Search) which is an enhancement over the Bat algorithm. Each prospective solution (feature subset), a bat, in Binary MMBAIS is represented as a binary vector rather than the traditional continuous solution space. The proposed algorithm is compared with some of the recent binary evolutionary algorithms; Binary Bat, BP-SOGSA, Binary Crow Search Algorithm, Binary Bat and Differential Evolution, and Binary Gray Wolf Optimization Algorithm over 15 benchmark datasets. Comparative studies illustrate the computational efficiency of Binary MMBAIS for feature selection with respect to the relative weighted consistency measure and classification accuracy over seven machine learning classifiers.

There are millions of people in this world, who have hearing disabilities due to various reasons. Sign languages (SLs) are used to communicate with these people. However, it is not easy to learn or use sign language. So, people face difficulty to interact with them properly. As hearing-impaired people are mostly familiar with SL, it is a visual language that is utilized by hard-of-hearing individuals to convey. Because of the extensive time needed in learning SL, individuals think that it is hard to speak with these specially-abled individuals, creating a communication gap. Chapter 12 attempts to resolve this issue by presenting "Audio to Indian Sign Language Interpreter (AISLI)", which translates English speech to Indian sign language (ISL) in form of Graphics Interchange Format (GIF) and yields letters as the output according to the phrase or word recognized. The Google Application Program Interface (API) is being utilized as the audio acknowledgment engine for the AISLI. The ISL structure is used for the interpretation of the results. It is expected that it will make the process for hearing-impaired individuals to comprehend simpler and speak with others, who do not know the language, using Natural Language Processing (NLP) and Machine Translation as the principal approach.

Pathological image reports are the primary basis of the diagnosis process and a little modification may mislead a doctor. Thus, tamper detection is essential for any medical image. In Chapter 13, a spatial domain-based fragile image watermarking scheme has been proposed for medical color images. This technique is not only useful for tamper detection but also allows to embed the electronic patient reports in the cover image without affecting the regions of interest. These personal reports are encrypted through an auto-generated key and watermark to provide immense privacy and reliability. The size of the watermark and the key are adaptive in nature and this adds a salient feature to this scheme. Moreover, the fragile watermarking process is performed in a reversible way. The proficiency of this proposed methodology is analyzed by means of imperceptibility, robustness, and hiding capacity and finally compared with some state-of-the-art techniques. Hereafter embedding, more than 60 dB average peak signal-to-noise ratio is obtained for applying 0.6 bits per pixel payload on an average. The overall assessment reflects that this scheme is able to provide enhanced hiding capacity, security, and reliability without affecting the region of medical interest of the images.

Worldwide climate change is significantly influenced by global warming as one of the important factors. The surface temperature of the earth has experienced a steady increase of 0.20 centigrade every decade of the last 30 years. Climate change is happening due to this. United Nations Framework Convention on Climate Change (UNFCCC) concurred in Paris that the earth's temperature increase needs to be limited below 20. The most common contributors or features to global warming need to be identified in order to find the reason behind this issue. Establishing a solution model for global warming and climate change as an act of urgency and working toward mitigation by limiting greenhouse gases emission is one of the primary concerns. Chapter 14 provides the trend and correlation of temperature rise and CO_2 emission and explains the different mechanisms driving optimal mitigation. Integrating the climate data and identifying the problem area is one of the objectives of this research. This will help to get a reliable control system that can deal with the global warming issue. This work suggests Green Machine Automation Model (GMAD) architecture and defines how global warming can be controlled using data engineering and analytics. Therefore, environment protection using machine learning (ML) and data engineering is the main contribution of this research work.

Chapter 15 presents an application-oriented methodology to estimate biological age (Abi) with the prediction of the life expectancy of an individual. Segregating collected data based on their different health parameters such as height, weight, blood pressure, body temperature, and Electromyogram (EMG) is the effective gateway to obtaining the biological age compared with the chronological age (Ach) after several trial-and-error methods. Estimation is performed to check the biological age of human beings of different chronological age groups from 21 to 50 years using a neural network algorithm. The decision-making of health status observation of different levels for biologically aged people is another important outcome of this research work. The error percentage for biological age varies from approximately 0.09% to 22.41%, and this estimation is the best fit for adults and middle-aged people. The mean error values of some particular predicted biological age compared to chronological age are 0.08, 0.32, 0.69, and 1.40.

Suicide risk assessment usually includes a conversation between a doctor and a patient. However, due to restricted access to mental health care facilities, clinician availability, absence of understanding, shame, abandonment, and discriminatory treatment associated with mental disorders, a substantial percentage of mentally ill people do not obtain medical help for their condition. Internet access and social media usage, on the other hand, have grown dramatically, allowing specialists and patients to communicate in ways that could contribute to the development of tools for detecting mental health disorders among people on social media. The primary purpose of Chapter 16 is to analyze online Twitter tweets and see what characteristics may indicate suicidal ideation in individuals. Machine learning and natural language processing techniques are used to train our data and test the efficacy of the proposed technique. Several features including linguistic, topic, sentiment, temporal and statistical are retrieved and merged to obtain a high performance in classification. It is observed that there exist significant statistical differences in the datasets for suicidal ideation and non-suicidal users with respect to the different features. Results indicate that the combination of these different features outperforms the accuracy obtained by individual features separately.

Newspapers play a very significant role in society as it keeps the citizens informed about various events happening around them. The news might have a direct or indirect impact on the life of people, especially during a pandemic like COVID. India, in particular, saw a devastating second wave during March-May 2021 where there were concerns like oxygen and medicine shortages, more casualties, and maladministration. In this scenario, analysis of the news articles to mine the important topics becomes pertinent to understanding the role of media during the pandemic. In Chapter 17, 4,902 articles published from 15/3/2021 to 31/5/2021 in five prominent newspapers including *The Times of India, The Hindu, Hindustan Times, The Indian Express*, and *The Deccan Chronicle* are extracted and Latent Dirichlet Allocation (LDA) algorithm is applied to discover the important topics. Total nine topics are discovered focussing on resource constraints, bed scams, and various aspects of vaccination. This demonstrates that the media was only focused on presenting the ground reality prevailing during the period rather than discussing other important aspects like the well-being and mental health of the citizens. This was important as the second wave saw a lot more casualties and anxiety among the citizens of India.

In Chapter 18, a Fast Converging Flower Pollination Algorithm (FFPA) has been cast off to achieve optimized coefficients for implementing finite impulse response filters. Optimization is performed to minimize the fitness value that measures the weighted mean square error present in the passband and stopband of the designed filters. The hardware efficiency of the filter is calculated by approximating the number of structural adders and multiplier adders. Adder costs of the designed filters are obtained by computing the number of signed power two (SPT) terms present in filter coefficients after quantization stalked by common sub expression (CSE) elimination. Simulation results prove that the desired frequency response is achieved by lessening the word length of coefficients. The proposed algorithm outpaces the Parks McClellan Algorithm and traditional meta-heuristic algorithms like genetic algorithm, particle swarm optimization, differential evolution, cuckoo search algorithm, and flower pollination algorithm. Performance analysis of the designed filter has been shown by filtering a noisy phonocardiogram signal.

Deep learning has significantly helped better results for visual-based speech recognition. However, the majority of the research works focus on improving only the training model's architecture. In Chapter 19, the authors identify that enhancing the pre-processing techniques also causes a surge in the model's accuracy. This chapter presents architecture-level suggestions to improve accuracy at every stage of the problem. First, the object recognition phase (for the lips) is improved by replacing the generic Haar cascade with BlazeFace. This process helps to cut down unnecessary information from being fed as input to the model. Since most real-time prediction demands the model to differentiate words of different lengths, enhancements to the window-size selection algorithm (pre-processing phase) are proposed. This algorithm dynamically calculates frames required for a specific word using silence detection techniques. In the final training phase, ResNets, GRUs, and LSTM models are employed to compare the results. The datasets obtained are from the Lombard-GRID corpus which consists of over 5,000 videos with 51 unique words. The proposed model improves the accuracy by around 15% more than standard techniques.

A modified Harris Hawks Optimization Algorithm is proposed in Chapter 20 that aims at determining the maximal levels of thresholding for grayscale images. In the recent past, the natural habits and mannerisms of a specific variety of hawks, namely, Harris Hawks, have been extensively observed with keen scientific interest, by researchers. This thorough observation has eventually led to the inception of a novel meta-heuristic algorithm regarded as the Harris Hawks Optimization Algorithm. The HHO algorithm has demonstrated capabilities of handling a wide array of optimization problems with startling efficiency. This chapter aims to suggest certain adaptive and intrinsic modifications that may accentuate and optimize the performance of the algorithm. A comparative analysis of the performance improvements based on real-world statistical benchmarks, such as the Kruskal-Wallis test, confirms the imminent superiority of the proposed algorithm in terms of performance and efficiency.

A precise and efficient technique for tumor segmentation in brain MRI is obligatory to acquire and analyze indispensable details for accurate diagnosis and required treatment planning. Deviations in tumor structure and size are fundamental challenges and choosing an appropriate segmentation technique shows considerable outcomes. Analyzing MRI images is a difficult and time-intensive process with human assessment of tumors in routine clinical applications. The accurate outcome of an automatic technique is crucial for the valuation of MRI images. The nonexistence of automatic tumor segmentation techniques presents a prime challenge since large amounts of MRI Image data are generated and consistently required for the segmentation process. Chapter 21 presents a unique framework for tumor detection with the EMGM model to rationalize the segmentation technique. Typically presented MRI sequences are used for experimental evaluation and obtained results with higher accuracy.

Industry 4.0 has highlighted remaining useful life (RUL) estimation as an important component for predictive maintenance. Prediction of remaining useful life (RUL) provides the estimate of residual time for system breakdown. By knowing the time estimate for upcoming failure, downtime caused due to unscheduled maintenance can be reduced. In real-time scenario, there is a dearth of training records for RUL estimation. Using advanced computer vision techniques, the time series data can be encoded into images to allow machines to visually recognize and classify and learn structures and patterns. The authors have proposed two novel techniques in Chapter 22 for data generation using Auxiliary Classifier Generative Adversarial Network (ACGAN)s for improving the RUL classification. Two ACGAN-based RUL techniques viz., time series-based ACGAN, and image-based ACGAN are generated using recurrence plots. Data generated with image ACGAN have disclosed higher similarity to original sensor data while compared to time ACGAN. Image data generated using ACGANs have revealed better generalization of features for improved RUL classification accuracy compared to that of the time series ACGAN. This chapter implements both the time and image-based ACGANs in RUL estimation and classification domain for the first time. The experimental results have showcased higher classification accuracy for RUL prediction using image ACGAN.

Computing algorithms have been developed for solving difficult optimization problems inspired by mimicking natural processes and the application of quantum theory. In the last two decades, Quantum Inspired Evolutionary Algorithms (QIEA) have shown promising results to solve hard problems using classical computers. In Chapter 23, the authors introduce the implementation algorithm of QIEA on a quantum computing machine that is a 5-qubit IBM Noisy Intermediate Scale Quantum (NISQ) system to solve a problem of 100 items knapsack. This chapter describes a practical implementation of the algorithm and compares the results obtained on a quantum machine vis-à-vis results on classical computers. Testing is done using four types of knapsack problems to prove the hypothesis that results on both the quantum and classical computers are comparable in fitness for the test cases.

This volume is a novel attempt to enrich the existing knowledge base on hybrid intelligent system modeling and simulation practices optimized for real-time applications. The editor would feel rewarded if the volume comes to the benefit of budding researchers in exploring the field further in order to unearth indigenous intelligent models and frameworks for the future.

Siddhartha Bhattacharyya
Birbhum, India

Editors

Dr. Siddhartha Bhattacharyya did his Bachelors in Physics, Bachelors in Optics and Optoelectronics, and Masters in Optics and Optoelectronics from the University of Calcutta, India in 1995, 1998, and 2000 respectively. He completed his Ph.D. in Computer Science and Engineering from Jadavpur University, India in 2008. He is the recipient of the University Gold Medal from the University of Calcutta for his Masters. He is the recipient of several coveted awards including the Distinguished HoD Award and Distinguished Professor Award conferred by the Computer Society of India, Mumbai Chapter, India in 2017, the Honorary Doctorate Award (D. Litt.) from The University of South America, and the South East Asian Regional Computing Confederation (SEARCC) International Digital Award ICT Educator of the Year in 2017. He has been appointed as the ACM Distinguished Speaker for the tenure of 2018–2020. He has been inducted into the People of ACM hall of fame by ACM, the USA in 2020. He has been appointed as the IEEE Computer Society Distinguished Visitor for the tenure of 2021–2023. He has been elected as a full foreign member of the Russian Academy of Natural Sciences and the Russian Academy of Engineering. He has been elected a full fellow of The Royal Society for Arts, Manufacturers and Commerce (RSA), London, UK.

He is currently serving as the Principal of Rajnagar Mahavidyalaya, Rajnagar, Birbhum. He served as a Professor in the Department of Computer Science and Engineering of Christ University, Bangalore. He served as the Principal of RCC Institute of Information Technology, Kolkata, India during 2017–2019. He has also served as a Senior Research Scientist in the Faculty of Electrical Engineering and Computer Science of VSB Technical University of Ostrava, Czech Republic (2018–2019). Prior to this, he was the Professor of Information Technology at RCC Institute of Information

Technology, Kolkata, India. He served as the Head of the Department from March 2014 to December 2016. Prior to this, he was an Associate Professor of Information Technology at RCC Institute of Information Technology, Kolkata, India from 2011 to 2014. Before that, he served as an Assistant Professor in Computer Science and Information Technology at the University Institute of Technology, The University of Burdwan, India from 2005 to 2011. He was a Lecturer in Information Technology at Kalyani Government Engineering College, India during 2001–2005. He is a co-author of 6 books and the co-editor of 94 books and has more than 400 research publications in international journals and conference proceedings to his credit. He has got two PCTs and 19 patents to his credit. He has been a member of the organizing and technical program committees of several national and international conferences. He is the founding Chair of ICCICN 2014, ICRCICN (2015, 2016, 2017, 2018), and ISSIP (2017, 2018) (Kolkata, India). He was the General Chair of several international conferences like WCNSSP 2016 (Chiang Mai, Thailand), ICACCP (2017, 2019) (Sikkim, India) and (ICICC 2018 (New Delhi, India), and ICICC 2019 (Ostrava, Czech Republic).

He is the Associate Editor of several reputed journals including *Applied Soft Computing*, *IEEE Access*, *Evolutionary Intelligence*, and *IET Quantum Communications*. He is the editor of the *International Journal of Pattern Recognition Research* and the founding Editor in Chief of the *International Journal of Hybrid Intelligence, Inderscience*. He has guest-edited several issues with several international journals. He is serving as the Series Editor of IGI Global Book Series Advances in Information Quality and Management (AIQM), De Gruyter Book Series Frontiers in Computational Intelligence (FCI), CRC Press Book Series(s) Computational Intelligence and Applications & Quantum Machine Intelligence, Wiley Book Series Intelligent Signal and Data Processing, Elsevier Book Series Hybrid Computational Intelligence for Pattern Analysis and Understanding and Springer Tracts on Human Centered Computing.

His research interests include hybrid intelligence, pattern recognition, multimedia data processing, social networks, and quantum computing.

He is a life fellow of the Optical Society of India (OSI), India, a life fellow of the International Society of Research and Development (ISRD), UK, a fellow of the Institution of Engineering and Technology (IET), UK, a fellow of Institute of Electronics and Telecommunication Engineers (IETE), India and a fellow of Institution of Engineers (IEI), India. He is also a senior member of the Institute of Electrical and Electronics Engineers (IEEE), USA, the International Institute of Engineering and Technology (IETI), Hong Kong, and the Association for Computing Machinery (ACM), USA.

He is a life member of the Cryptology Research Society of India (CRSI), Computer Society of India (CSI), Indian Society for Technical Education (ISTE), Indian Unit for Pattern Recognition and Artificial Intelligence

(IUPRAI), Center for Education Growth and Research (CEGR), Integrated Chambers of Commerce and Industry (ICCI), and Association of Leaders and Industries (ALI). He is a member of the Institution of Engineering and Technology (IET), UK, International Rough Set Society, International Association for Engineers (IAENG), Hong Kong, Computer Science Teachers Association (CSTA), USA, International Association of Academicians, Scholars, Scientists and Engineers (IAASSE), USA, Institute of Doctors Engineers and Scientists (IDES), India, The International Society of Service Innovation Professionals (ISSIP) and The Society of Digital Information and Wireless Communications (SDIWC). He is also a certified Chartered Engineer of the Institution of Engineers (IEI), India. He is on the Board of Directors of the International Institute of Engineering and Technology (IETI), Hong Kong.

Contributors

Alexander E. Arkhipov
Southwest State University
Russia

Hema Banati
Dyal Singh College, University of Delhi
New Delhi, India

Rajib Banerjee
Department of Electronics and Communication Engineering
DR. B. C. Roy Engineering College Durgapur India

Sharmila Banu K.
School of Computer Science and Engineering
VIT Vellore
India

Abhishek Basu
Department of Electronics and Communication Engineering
RCC Institute of Information Technology
Kolkata, India

Saikat Basu
Department of Computer Science & Engineering
Maulana Abul Kalam Azad University of Technology
Kolkata, India

Andrey Batranin
School of Non-Destructive Testing
Tomsk Polytechnic University
Tomsk, Russia

Rupal Bhargava
Learning Experience
upGrad Education Pvt. Ltd.
Mumbai, India

Siddhartha Bhattacharyya
Rajnagar Mahavidyalaya
Birbhum, India

K. Bhima
Information Technology
B. V. Raju Institute of Technology
Narsapur, India

Arindam Biswas
Centre of IoT and AI Integration with Education-Industry-Agriculture and School of Mines and Metallurgy
Kazi Nazrul University Asansol
Asansol, India

Maxim Bobyr
Southwest State University
Russia

Jaiyesh Chahar
Solution Engineering
Siemens Technology Services Pvt. Ltd.
India

Sergei Chakhlov
School of Non-Destructive Testing
Tomsk Polytechnic University
Tomsk, Russia

Debashish Chakravarty
Department of Mining Engineering
Indian Institute of Technology
Kharagpur, India

Anindya Chatterjee
Solution Engineering
Siemens Technology Services Pvt. Ltd.
India

Moumita Chatterjee
Department of Computer Science & Engineering
Aliah University
Kolkata, India

Avik Chattopadhyay
Institute of Radio Physics and Electronics
University of Calcutta
Kolkata, India

Fangfang Chen
Department of Mathematics and Information Engineering
Chongqing University of Education
Chongqing, China

Abhijit Das
Department of IT
RCC Institute of Information Technology
Kolkata, India

Poulami Das
K. C. College of Engineering and Management Studies and Research
Thane, India

Rik Das
Solution Engineering
Program of Information Technology
Siemens Technology Services Pvt. Ltd.
India
Xavier Institute of Social Service
Ranchi, India

Susmita Das
Electronics and Instrumentation Engineering
Narula Institute of Technology
Kolkata, India

Arghya Dasgupta
Department of Computer Science Engineering
DR. B. C. Roy Engineering College Durgapur
India

Debashis De
Department of Computer Science & Engineering
Maulana Abul Kalam Azad University of Technology
Kolkata, India

Sourav De
Department of Computer Science and Engineering
Cooch Behar Govt. Engineering College
India

Abhirup Deb
Department of Computer Science and Engineering
Cooch Behar Government Engineering College
Cooch Behar, India

Soumyaratna Debnath
Department of Computer Science and Engineering
Cooch Behar Government Engineering College
Cooch Behar, India

Sandip Dey
Department of Computer Science
Sukanta Mahavidyalaya
Dhupguri, India

Sayantani Ghosh
Department of IT
University of Calcutta, Technology Campus
Kolkata, India

Sergey Gorbachev
Faculty of Innovative Technologies, National Research Tomsk State University
Tomsk State University
Tomsk, Russia

Ayush Gupta
Department of computer Science and Engineering
Cooch Behar Govt. Engineering College
India

Lakshmi Shankar Iyer
School of Business and Management
Christ University
Bangalore, India

A. Jagan
Computer Science and Engineering
B. V. Raju Institute of Technology
Narsapur, India

Rui Jiang
Department of Mathematics and Information Engineering
Chongqing University of Education
Chongqing, China

Shuvam Kabiraj
Department of Electronics and Communication Engineering
DR. B. C. Roy Engineering College Durgapur
India

Indrajit Kar
Solution Engineering
Siemens Technology Services Pvt. Ltd.
India

Pravar Kulbhushan
Solution Engineering
Siemens Technology Services Pvt. Ltd.
India

Piyush Kumar
Amity Institute of Information Technology
Amity University Kolkata
India

Long Li
School of Mathematics and Big Data, Chongqing University of Education
Chongqing, China

Xiaoqing Li
Department of Mathematics and Information Engineering
Chongqing University of Education
Chongqing, China

Yan Ma
Department of Mathematics and Information Engineering
Chongqing University of Education
Chongqing, China

Tripti Mahara
School of Business and Management
Christ University
Bangalore, India

Koushik Majumder
Department of Computer Science & Engineering
Maulana Abul Kalam Azad University of Technology
Kolkata, India

Ashish Mani
Amity Innovation & Design Centre
Amity University Noida
India

Jitin Jain Mathew
School of Business and Management
Christ University
Bangalore, India

Victor Medennikov
Federal Research Center, Computer Science and Control of the Russian Academy of Sciences
Russia

Natalia Milostnaya
Southwest State University
Russia

Anwesha Mukherjee
Department of Computer Science
Mahishadal Raj College
Mahishadal, India

Sunetra Mukherjee
Department of Electronics and Communication Engineering
DR. B. C. Roy Engineering College Durgapur
India

Sudipta Mukhopadhyay
Solution Engineering
Siemens Technology Services Pvt. Ltd.
India

Ludmila Muratova
VIAPI n.a. A.A. Nikonov - branch of the FSBSIFRC AESDRA VNIIESH
Russia

Anchal N G
School of Business and Management
Christ University
Bangalore, India

Dalia Nandi
Electronics and Communication Engineering
Indian Institute of Information Technology Kalyani
Kalyani, India

Biswarup Neogi
Electronics and Communication Engineering
JIS College of Engineering
Kalyani, India

Nivedita
School of Computer Science and Engineering
VIT Vellore
India

Cristian Nolivos
Southwest State University
Russia

Sergey Osipov
School of Non-Destructive Testing
Tomsk Polytechnic University
Tomsk, Russia

M. S. Prasad
Amity Institute of Space Science & Technology
Amity University Noida
India

K. Dasaradh Ramaiah
Information Technology
B. V. Raju Institute of Technology
Narsapur, India

Avishek Ray
K. C. College of Engineering and Management Studies and Research
Thane, India

Pulakesh Roy
Department of Mining Engineering
Kazi Nazrul University Asansol
Asansol, India

Soumit Roy
Analytics Presales and Solution Practice Lead, Jade Global
Chicago, USA

Souvik Roy
Department of Electronics and Communication Engineering
DR. B. C. Roy Engineering College Durgapur
India

Subhrajit Sinha Roy
Department of Electronics and Communication Engineering
RCC Institute of Information Technology
Kolkata, India

Sudipta Saha
Department of Computer Science & Engineering
Maulana Abul Kalam Azad University of Technology
Kolkata, India

Rahul Sai R. S.
School of Computer Science and Engineering
VIT Vellore
India

Ravi Saini
ASET
Amity University Noida
India

Sergey Salnikov
VIAPI n.a. A.A. Nikonov - branch of the FSBSIFRC AESDRA VNIIESH
Russia

Poulomi Samanta
Amity Institute of Information Technology
Amity University Kolkata
India

Dhrubasish Sarkar
Amity Institute of Information Technology
Amity University Kolkata
India

Anirbit Sengupta
Department of ECE
Dr. Sudhir Chandra Sur Institute of Technology and Sports Complex
Kolkata, India

Chengxiang Shi
Department of Mathematics and Information Engineering
Chongqing University of Education
Chongqing, China

Sudershan Sridhar
School of computer Science and Engineering
VIT Vellore
India

Akshay Sriram
School of Business and Management
Christ University
Bangalore, India

B. K. Tripathy
School of Information technology and Engineering
VIT Vellore
India

Asha Yadav
University of Delhi
New Delhi, India

Alexander Zatsarinny
Federal Research Center, Computer Science and Control of the Russian Academy of Sciences
Russia

Yanxia Zhang
School of Mathematics and Big Data, Chongqing University of Education
Chongqing, China

Chapter 1

Creating ratings of agricultural universities based on their digital footprint

Victor Medennikov
Federal Research Center, Computer Science and Control of the Russian Academy of Sciences

Ludmila Muratova and Sergey Salnikov
VIAPI n.a. A.A. Nikonov - branch of the FSBSIFRC AESDRA VNIIESH

Alexander Zatsarinny
Federal Research Center, Computer Science and Control of the Russian Academy of Sciences

CONTENTS

1.1 INTRODUCTION

The active development of digital technologies, as well as a number of serious restrictions associated with the COVID-19 pandemic, has become a driver for rethinking almost all aspects of both global and Russian educational activities. At the same time, the problem of the competitiveness of universities has become especially acute. To increase competitiveness, it was necessary to make significant investments in digital technologies for distance education, which was an obstacle to declining revenues due to lower

DOI: 10.1201/9781003381167-1

incomes of consumers of university education as well as a decrease in the number of applicants from other countries. In the transition to self-isolation, the image of universities has become most influenced by their reasonable, effective representation in the Internet space, and not just the quality of educational services, which has become more difficult to control during the pandemic. The representation of universities in the Internet space is called the digital footprint.

As a rule, the concept of a digital footprint is understood as having two types: active and passive.

The active type includes information that is published in the Internet space directly by its owner or by a person authorized by him for this activity. A passive trace reflects information that is collected from his site without his participation and knowledge. In practice, this division is rather arbitrary, since the activities of universities can be displayed from many other sources, particularly those created by automated means. Such a digital trace will be called semi-active.

In our work, under the active digital footprint in relation to agricultural universities in Russia, we mean the amount of data that is posted on the website of each of them in the form of developments, publications, consultations, legal and regulatory materials, distance learning, application software, as well as databases and information about electronic employment agencies (EEAs) and digital marketplaces (DMPs) [1].

As the indicated digital footprints were not taken into consideration, the decree of Russian President was not implemented to achieve the following indicator of improving the activities of Russian universities by 2020: five should be included in the ranking of the 100 best universities in the world. The Accounts Chamber of the country issued a disappointing opinion on the unattainability of the target indicators of the decree, summing up at the end of its implementation period, having analyzed the institutional ratings QS [2] and THE [3] developed in the UK, as well as ARWU [4] used in China, since not a single Russian university participating in the program to implement the presidential decree could get into the first group of the hundred ratings considered above, although large sums were spent on the implementation of the program – over 80 billion rubles [5]. Note that such a result was evident long before the deadline for the completion of the specified program of state support for Russian universities in accordance with the presidential decree [6].

The analysis shows that the main reason for not implementing the decree is that the adopted program, focused on the implementation of the presidential decree, is based on the requirements for the information content of the educational environment in Russia in the field of university education. These requirements are enshrined in the relevant order No. 662 dated August 5, 2013, by the Ministry of Science and Higher Education of the Russian Federation "On Monitoring the Education System". In accordance with the order, these requirements are focused on accounting only for educational activities. In developed countries, scientific research is essential. Such a digital divide could be eliminated by the presence of some integrated scientific and educational environment, the top of which is the formed single digital platform in the form of an information Internet space of scientific and educational resources [7]. In addition, in the context of active digitalization, the image of the university, which influences its ranking, is increasingly dependent on its comprehensive display in the Internet space. That is, the image in these conditions is most determined by the passive digital footprint.

The purpose of this study is to form and analyze the ratings of Russian universities using agricultural universities as an example, on the basis of their digital traces with the calculation of the integral rating. The resulting integral rating is an assessment of a complex indicator of the effectiveness of the use of information, scientific, and educational resources by agricultural universities. The methodology for forming these rankings based on calculated digital traces is the basis for harmonizing Russian and Western approaches to the development of university rankings in the context of the inevitability of the transition to a single digital platform of scientific and educational resources as the digital economy develops, smoothing out the shortcomings of the requirements of the Ministry of Science and Higher Education of the Russian Federation to the assessment of educational organizations.

1.2 METHODOLOGY FOR THE FORMATION OF UNIVERSITY RANKINGS BASED ON THE PASSIVE DIGITAL FOOTPRINT

For this technique, we use the so-called method of sitemetric assessment of universities, based on the analytical capabilities of the SiteAuditor site analysis and audit service program [8]. This program makes it possible to obtain the characteristics of the sites under study that are of interest to us (Table 1.1). The weighting factors in the last column are based on the results of expert judgments [9–11].

The values of indicators included in the "indexation" group were formed according to the metrics of search services included in Bing, Google, Yandex, and Seznam. The values of indicators included in the "directories" group are the result of a simple sum of indices (1-yes, 0-no) of the presence of university sites in the DMOZ, Mail.ru, Rambler TOP, and Yandex catalogs. The values of indicators included in the "problems" group are the result of a simple sum of indices (1-yes, 0-no) from the corresponding characteristics on the Spamhaus (IP) and Yandex.AGS sites. The values of indicators included in the "rating" group were formed on the basis of data from

Table 1.1 Group indicators of the sitemetric method for calculating ratings

No.	*Group indicators*	*Number in the group*	*Weight in %*
1	Indexation	4	8,1
2	Directories	4	7,9
3	Problems	2	5,1
4	Alexa rating (global)	1	3,9
5	Alexa rating (local)	1	3,9
6	Google rating PR	1	8,1
7	Yandex rating	1	8,1
8	Social services	3	4,9
9	Links on the website	4	39,9
10	Links from the site	2	10,1
	Total	23	100,0

Alexa sites (both global and local), Google PR, and Yandex. The values of the "social services" group's indicators were the result of summing up the corresponding characteristics in social networks such as Facebook, Google Plus, and "My World" (Mail.ru). Site link group metrics were generated from relevant sections of Alexa, Google, Linkpad, and Majestic. The values of the indicators of the group of links from the site were determined as the average expression of the total number of corresponding indicators from the sites such as Bung and Linkpad. More detailed studies with detailed calculations can be found in [6,7]. The remaining indicators that can be obtained using the site-auditor program (another 17 indicators from various groups of indicators) were excluded from consideration in this chapter due to their low relevance and/or zero value for all the sites under study.

For the mathematical description of the methodology for evaluating universities on the basis of the considered passive digital footprint, we introduce the following expressions.

P_2^m – the criterion for evaluating the m-th university, calculated on the basis of a passive digital footprint;
d_{rm}^2 – the size of the r-th indicator of site evaluation, calculated on the basis of a passive digital traces of the m-th university;
q_{rm}^2 – the size of the r-th indicator of the site assessment, calculated on the basis of a passive digital traces of the m-th university;
ω_r^2 – weight of the size of the r-th indicator of site evaluation, calculated on the basis of a passive digital traces;

$$q_{rm}^2 = d_{rm}^2 / \max_m d_{rm}^2;$$

Then we get

$$P_2^m = \sum_k \omega_k^2 q_{km}^2 \qquad (1.1)$$

In accordance with the abovementioned mathematical description of the methodology for evaluating universities, ten private ratings were formed based on sitemetry data. The obtained private ratings were further summed up with the weights given in Table 1.1. The results of calculations of assessments and ratings of universities obtained on the basis of their passive digital footprint are summarized in Table 1.10.

1.3 METHODOLOGY FOR GENERATING RATING BASED ON A SEMI-ACTIVE DIGITAL FOOTPRINT

The information required to generate ratings based on a semi-active digital footprint has been taken from self-examination reports according to [1]. Due to their large number, Table 1.2 shows only eight indicators as an example. The weights of all indicators in the final rating calculations have been determined by using mathematical statistics methods based on correlation analysis, the Kendall concordance factor, the probabilistic estimation model, and calculating the competency matrix. The calculation results show a high consistency of all ratings, which provide ample opportunities for using any of the abovementioned methods in further research, as well as their combination, such as average grades and ratings.

Table 1.2 Groups of indicators for a semi-active digital footprint

No.	*Indicators*	*Weight*
1	Specific weight of the number of RAS with an academic degree of candidate of sciences in the total RAS number at the university	4.27
2	Specific weight of the number of RAS with an academic degree of doctor of sciences in the total RAS number at the university	4.59
3	Total area of classrooms for training per student	4.70
4	Number of computers per student	4.57
5	University income for all types of activities per RAS member	4.67
6	University income generated by commercial activities per RAS member	4.35
7	Share of the RAS average income at a university, taking into account all types of activities, to the average salary in the region	4.44
8	Number of students studying under the federal budget funding program	4.43

In Table 1.2, RAS stands for research academic staff. The mathematical description of the methodology for evaluating universities based on semi-active digital footprints is as follows:

P_5^m – the evaluation criterion of the mth university, calculated on the basis of a semi-active digital footprints;
d_{hm}^5 – the size of the hth indicator, calculated on the basis of a semi-active digital footprints of the mth university;
q_{hm}^5 – the size of the hth indicator, calculated on the basis of a semi-active digital footprints of the mth university;
ω_{hm}^5 – the size weight of the hth indicator, calculated on the basis of a semi-active digital footprints of the mth university;

$$q_{hm}^5 = d_{hm}^5 / \max_m d_{hm}^5$$

Then, we get:

$$P_5^m = \sum_h \omega_h^5 q_{hm}^5 \qquad (1.2)$$

The resulting evaluation, with the corresponding ratings of universities based on their semi-active digital footprint, is comparatively given in Table 1.10.

1.4 METHODOLOGY FOR GENERATING RATINGS BASED ON AN ACTIVE DIGITAL FOOTPRINT

As already stated in the Introduction section, in addition to the traditional image role and the role of the university's business card, the role of the abovementioned digital footprint gains significantly greater importance in modern conditions because, with competent, scientifically grounded representation of universities on the Internet – in addition to providing faculty members with qualitatively new opportunities for an enhanced sharing of ideas with each other and their digital interaction with scientific and educational resources – it provides an effective system for introducing this knowledge into the economy and contributes to an increase in the intellectual level of society through the improvement of the education system [6]. Such a triunique role of scientifically grounded representation of universities on the Internet is most fully implemented when creating the above space of information scientific and educational resources (ISERs) on the Internet, integrating the following types of them posted on the websites of universities, scientific institutions, and various consulting centers of the agro-industrial complex: publications, development, consultations, legal and statutory materials, training in a remote format, custom software, and databases. It is these scientific and educational resources that are most in demand in the agricultural economy [6,7,12].

In recent years, in addition to ISERs, these organizations on their websites have started developing, in one way or another, EEAs and DMPs, which reflect modern trends in rendering Internet services in the form of such digital services. Thus, we can consider seven types of ISERs, as well as EEAs and DMPs, as manifestations of an active digital footprint.

In view of the abovementioned reasons, we present the parameter values of this methodology:

Therefore, based on the above data, we present a mathematical description of the method:

i – identifier defining the level of ISERs integration, $i \in I$ (Table 1.3);
l – identifier specifying the storage form for ISERs, $l \in L$ (Table 1.4);
n – identifier defining the kind of representation of ISERs, $n \in N$ (Table 1.5);

Table 1.3 Indicators of the IR integration

No.	*Designation*	*Weight (%)*
1	Unordered list	10
2	Ordered electronic presentation	90
Total		100

Table 1.4 Indicators of the IR storage forms

No.	*Designation*	*Weight (%)*
1	Catalog	30
2	Full-format electronic presentation	70
Total		100

Table 1.5 Indicators of the ISER presentation types

No.	*Designation*	*Weight (%)*
1	Developments	30
2	Publications	20
3	Databases	5
4	Application software packages	5
5	Remote learnings	5
6	Consultants	30
7	Regulatory information	5
Total		100

Table 1.6 Particular criteria for evaluating an active digital footprint

No.	*Designation*	*Weight (%)*
1	Evaluation criteria for ISER representation types	70
2	Private criterion for evaluating IRs based on the DMP's status	15
3	Private criterion for evaluating IRs based on the EEA's status	15
	Total	100

Table 1.7 Indicators of the IR evaluation criterion based on the DMP's status

No.	*Designation*	*Weight (%)*
1	Unstructured message board	5
2	Structured message board	10
3	Automated search for a trading partner according to a given indicator	20
4	Automated information processes for all trading operations	25
5	Complete automation of e-commerce	40
	Total	100

Table 1.8 Indicators of the IR evaluation criterion based on the EEA's status

No.	*Designation*	*Weight (%)*
1	Unstructured message board	10
2	Structured message board	20
3	Electronic employment agency (automated search)	60
4	Links to other employment agencies	10
	Total	100

m – university code, $m \in M$;

P_j^m – the private criterion for evaluating, calculated on the basis of a active digital traces of the mth university based on the jth indicator, $j \in J$ (Table 1.6);

P^m – the integral criterion for evaluating, calculated on the basis of a active digital traces of the mth university;

α_i^1 – the value weight of the ith indicator of the scientific and educational IR integration;

α_l^2 – weight of the lth storage form ISERs;

α_n^3 – the value weight of the nth indicator of the scientific and educational IR presentation form;
β_j – weight of the jth private criterion of assessment, calculated on the basis of a active digital traces (Table 1.6);
$v_{i\ln 0}^m$ – the size of ISERs of the ith integration level, the lth storage form, n^{th} presentation kind of the mth university;
$\lambda_{i\ln}^m$ – the normalized value of the estimation of the size of the ISERs of the ith integration level, the lth storage form, n^{th} presentation kind of the mth university;

$$\lambda_{i\ln}^m = v_{i\ln 0}^m \,/\, \max_m v_{i\ln 0}^m;$$

d_{sm}^3 – the size of the sth indicator of the website evaluation criterion based on the DMP's status of the mth university (Table 1.7);
ω_s^3 – the value weight of the sth indicator of the website evaluation criterion based on the DMP's status (Table 1.7);
d_{gm}^4 – the size of the gth indicator of the website evaluation criterion based on the EEA's status of the mth university (Table 1.8);
ω_g^4 – the value weight of the gth indicator of the website evaluation criterion based on the EEA's status (Table 1.8). Then:

$$P^m = \sum_j \beta_j \, P_j^m, \tag{1.3}$$

where $P_1^m = \sum_{i,l,n} \lambda_{i\ln}^m \alpha_i^1 \alpha_l^2 \alpha_n^3$, $P_3^m = \sum_s \omega_s^3 d_{gm}^3$,

$$P_4^m = \sum_g \omega_g^4 d_{gm}^4 \tag{1.4}$$

To demonstrate the underestimation of the active and passive digital footprints, Table 1.9 provides information about the low information content of ISERs on the universities' websites, where the column numbers mean the following indicators: C1 – percentage of websites having this type of ISER, C2 – unordered list, C3 – digital catalog, C4 – unordered full-format presentation, C5 – ordered full-format digital presentation, ASPs – application software packages, RL – remote learning, and RI – regulatory information.

Table 1.9 Quality and quantity of ISERs on universities' websites

Types of ISERs	*C1*	*C2*	*C3*	*C4*	*C5*
Developments	85	3,684	391	337	248
Publications	89	18,649	408	344	0
Databases	11	530	45	0	0
ASPs	2	828	2	25	0
RL	12	1,195	0	0	3
Consultants	25	216	43	9	0
RI	89	18,649	408	344	0

Table 1.10 Comparison of university ratings based on passive, semi-active, and active digital footprints

University	*N1/N2/N3/N4*	*University*	*N1/N2/N3/N4*
Russian State Agrarian University – MTAA	1/1/16/4	Belgorod State Agrarian University	6/43/10/7
Orel State Agrarian University	2/20/14/5	Krasnoyarsk State Agrarian University	7/22/32/2
Novosibirsk State Agrarian University	3/4/18/28	Saint Petersburg State Agrarian University	8/3/19/34
Bryansk State Agrarian University	4/21/2/20	Saratov State Agrarian University	9/6/11/9
Kazan State Agrarian University	5/28/15/8	Bashkir State Agrarian University	10/19/7/32

1.5 RATING CALCULATION RESULTS

Table 1.10 presents a comparison of the rankings of the first ten universities of the integral ranking (N1) based on the passive (N2), semi-active (N3), and active (N4) digital footprint.

Our analysis of the comparison results shows that the majority of universities virtually do not pay attention to their image. Thus, not a single university from the top-10 of the N1 rating has retained its leadership in the other three ratings. Only three institutions have been included in the leaders of three ratings out of four: Russian State Agrarian University – MTAA, Belgorod State Agrarian University, and Saratov State Agrarian University.

1.6 CONCLUSION

Our research has shown that universities are currently forced by the regulation to develop their websites, focusing only on those resources that are required by regulatory authorities and leaving a largely semi-active digital footprint on the Internet. Therewith, they pay little attention to tools that enhance their active and passive digital footprints. Underestimation of these tools was the reason for not fulfilling the Russian President's instruction to achieve the goal; five Russian universities should be included in the ranking of the 100 best world universities by 2020. However, focusing only on indicators included in the calculations of the most famous university rankings, QS Rankings [2] and Times Higher Education (THE) [3], poses a great threat to the Russian economy. The indicators of these ratings are reflected in the self-survey report [1]. They are taken into account in the form of semi-active digital traces in this article. If in developed countries, thanks to market relations, the transfer of innovations to the economy is well established, in our country it was destroyed during the years of perestroika. Business in Russia is not interested in the indicators from the QS and THE ratings; they need the indicators included in the calculation of the active digital traces [6,13,14].

REFERENCES

1. Order of the Ministry of Education and Science of the Russian Federation of June 14, 2013 N 462 "On approval of the procedure for conducting self-examination by an educational organization", https://base.garant.ru/70405358/, last accessed 2021/09/1.
2. QS World University & Business School Rankings, https://www.qs.com/rankings/, last accessed 2021/08/20.
3. World University Rankings, https://www.timeshighereducation.com/world-university-rankings, last accessed 2021/08/20.
4. Academic Ranking of World Universities, www.shanghairanking.com, last accessed 2021/08/20.
5. Universities from the project "5–100" did not enter the top 100 international rankings, https://www.rbc.ru/society/18/02/2021/602cbdff9a7947765cbb58e5?from=from_main_10, last accessed 2021/08/20.
6. Medennikov, V., Muratova, L., Salnikov, S. *Methodology for Assessing the Effectiveness of the Use of Information Scientific and Educational Resources.* Moscow, Analitik (2017).
7. Zatsarinny, A., Ereshko, F., Medennikov, V. Scientific and methodological approaches to the generation of the internet information space of scientific and educational resources. In: *IEEE Xplore Digital Library. Proceedings of the 1st International Conference of Information Systems and Design*, Moscow, Russia, December 5 (2019).
8. Site-Auditor, https://freesoft.ru/windows/siteauditor, last accessed 2021/08/20.
9. Sirotkin, G. System analysis of factors of quality of education in higher education institutions. *Caspian Journal: Management and High Technologies* 2(22), 109–118 (2013).
10. Guzaeva, M. The use of information resources of science and education to improve the effectiveness of the implementation of new forms of education, http://pedsovet.su/publ/164-1-0-1048/, last accessed 2021/08/20.

11. Antyukhov, A., Fomin, N. Development of the fund of evaluation funds in the context of the Federal State Educational Standard for Higher Education, https://cyberleninka.ru/article/n/razrabotka-fonda-otsenochnyh-sredstv-v-kontekste-fgos-vpo/viewer, last accessed 2021/08/20.
12. Medennikov, V., Raikov, A. Creating the requirements to the national platform "Digital Agriculture". In: *Proceedings of the 8th International Scientific Conference on Computing in Physics and Technology*, Moscow region, Russia, November 9–13, 2020, pp. 13–18. CEUR Workshop Proceedings, 2763 (2020).
13. Valero, A., Reenen, J. V. The economic impact of universities: Evidence from across the globe. *Economic of Education Review* 68, 53–67 (2019).
14. Hamdan, A., Sarea, A., Khamis, R., Anasweh, M. A causality analysis of the link between higher education and economic development: empirical evidence. *Heliyon* 6(6), Article 04046 (2020).

Chapter 2

Mechatronic complex's fuzzy system for fixating moving objects

Maxim Bobyr
Department of Computer Science, Kursk, RF

Cristian Nolivos
Southwest State University, Postgraduate of Computer Engineering Department, Kursk, RF

Natalia Milostnaya
Southwest State University, Leading Researcher of Computer Engineering Department, Kursk, RF

Sergey Gorbachev
National Research Tomsk State University, Faculty of Innovative Technologies, Senior Researcher, Tomsk, RF

Alexander E. Arkhipov
Senior Researcher of Computer Engineering Department, Kursk, RF

CONTENTS

DOI: 10.1201/9781003381167-2

2.1 INTRODUCTION

This chapter presents a hardware and software model for a mechatronic complex (MC) designed to automate technological processes and performing the following technological operations (TO): actuating mechanisms (AM), detection, fixation, and storage of moving objects. In an earlier published scientific study [1], the team of authors considered the issues of synergetic integration of MC components into a single system and the process of its operation in manual control mode.

The manual control mode of the MC implies that the human operator controls the above TO using commands on the interactive touch panel (ITP) screen. The automatic control mode differs; in that mode, the human operator starts the MC using real and/or virtual buttons, and the rest of the actions occur due to the program embedded in the programmable logic controller (PLC). When performing a working cycle in this mode of operation of the MC, the most difficult technological operation is fixing a moving object. The successful implementation of this operation depends on the prediction and control of the trajectory of movement of the pneumatic AM with nonlinear characteristics [2]. If, as a result of the movements, the central axes of the movable object and the pneumatic gripper are not aligned, then the object will, with a high degree of probability, not be fixed in the gripper jaws. In this case, the further course of the working cycle is disrupted, which affects the operability of the MC [3–6].

In manual mode, the human operator, using visual control, determines the moment of alignment of the axes of the pneumatic gripper relative to the moving object and then executes the command to grip the moving object. In the automatic control mode of the MC, an optical sensor is used to fix the moving object. It only senses the presence of a moving object in the detection zone, and it cannot determine the moment when the axes of the controlled objects are aligned. Thus, for the automatic control mode of the MC, a problematic situation arises associated with finding a balance between the minimum time required for aligning the axes of the moving object and the pneumatic gripper, which is controlled by system timers. The results analysis of which that are presented in this work was devoted to solve the following problems: development of program code for controlling the MC in automatic mode [7]; creation of a human–machine interface model and verification of its performance; analysis of the kinematic model; comparison of the obtained experimental results for two control modes; assessment of forecasting the moment in time when the axes of the pneumatic gripper and the moving object are aligned.

2.2 COMPOSITION AND PRINCIPLE OF OPERATION OF THE MC

The main components of the MC include: PLC (Siemens S7-1200), pneumatic actuators; magnetic position sensors (MPS); optical sensor (OS);

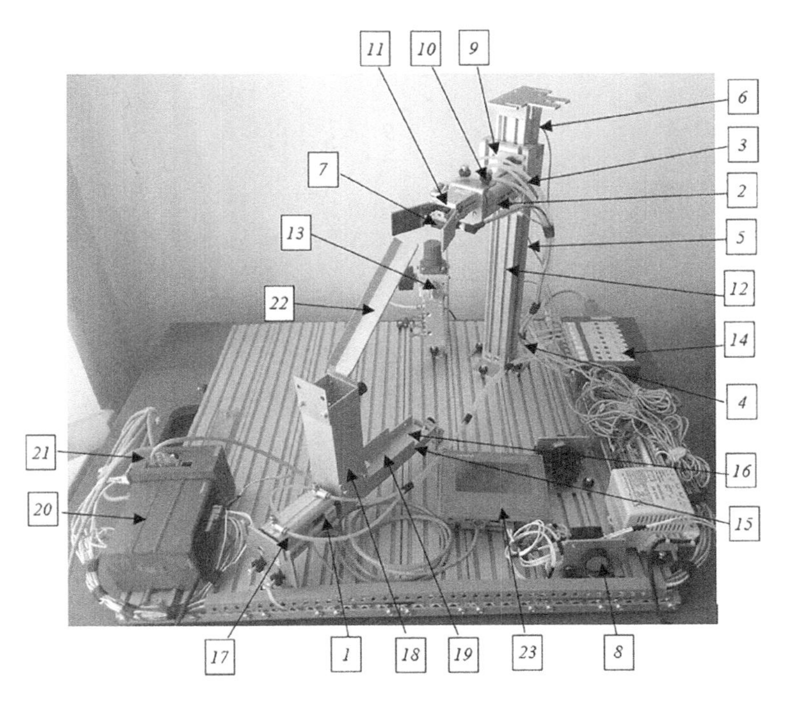

Figure 2.1 Appearance and composition of MK: 1–6: magnetic position sensors; 7: optical sensor; 8: "Start" button; 9: rodless cylinder with carriage; 10: double-acting cylinder; 11: pneumatic gripper; 12: guide rodless cylinder; 13: filter regulator; 14: valve terminal; 15: throttle-regulator; 16: nozzle; 17: single-acting cylinder; 18: storage tray; 19: guide; and 20: programmable logic controller Siemens S7-1200; 21: Ethernet switch Scalance XB005; 22: gutter; 23: interactive touch panel.

interactive touch panel (Simatic HMI KTP 400); Ethernet switch (Scalance XB005); personal computer (PC); valve terminal, filter pressure regulator, pressure gauges, and compressor. The hardware and software model of the MC is shown in Figure 2.1 (PC and compressor are not tagged) [8–10].

To carry out the technological operation of fixation in the automatic mode of operation of the MC, it is necessary to move the actuators from the manual position. After hitting a moving object with a single-acting cylinder rod, the ball moves from the storage tray along the guide to the nozzle and begins to float in the compressed air stream (from the third to the fifth stroke). The carriage of a rodless cylinder (RC) with a double-acting cylinder (DAC) and a pneumatic gripper (PG) are moved until a floating (moving) object is detected by an OS. When the OS has detected the object and the centers of coordinates of the object and the PG coincide as shown in Figure 2.2, the condition for the alignment of the axes is met. The required time delay is formed in the program code loaded into the PLC from the PC via the Ethernet switch using the system timers. Then, the object is secured with guaranteed reliability using a PG. After that, the air supply to the

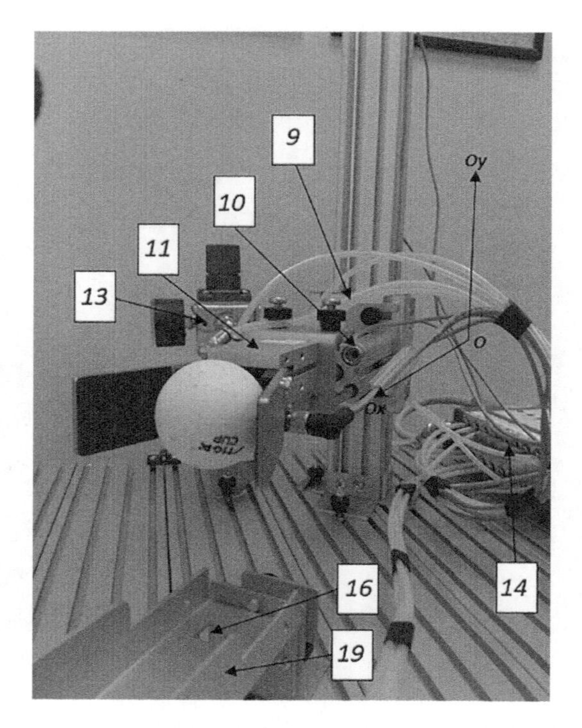

Figure 2.2 Alignment of the central axes of the moving object and the AM.

nozzle is stopped, and the fixed object is moved for storage in the storage tray. The actuators return to the starting position, as shown in Figure 2.1, before starting a new work cycle.

When designing the MC control system, the team of authors developed a model of a human–machine interface (HMI) in the Simatic WinCC specialized software package, which is part of the Totally Integrated Automation (TIA) Portal V14.1 integrated environment for technological process automation systems. Visualization of the process of movement of actuators, detection, fixation, and storage of a moving object is implemented on the screen of an interactive touch panel. Figure 2.4 shows a view in the operator's working window – control buttons and the actual location of the MC actuators.

The human operator controls the process of moving the carriage of a RC and a DAC with a PG. On the screen, rectangles depict MPS and OS with a color indication function, indicating the current location of the AM relative to the moving object. If the MPS is in the value of a logical 1, that is, it is triggered, and the AM is in the contact area of this sensor, then the rectangle corresponding to the sensor lights up in green. In the case when the AM is absent in the contact area and the sensor contact is open, meaning, is in the value of logical 0, the indication changes to red [11–13].

2.3 KINEMATIC MODEL OF THE MC

The kinematic model was developed to describe the movements of the MK actuators. It allows you to predict the time and coordinates for aligning the axes of the PG and the moving object. The acceleration when moving the PG is calculated by the formula

$$a_{cur} = \left(A_0 + \sin\left(\upsilon\pi/\upsilon_{max}\right)\right)a_{max} \tag{2.1}$$

where a_{max} is the maximum value of the acceleration of the MI, m/s^2; υ is the current speed of the MI movement, m/s; υ_{max} is the maximum speed of the MI movement, m/s, A_0 is the adjustment factor.

The speed of the MI movement in a given period of time is described by the equation

$$\upsilon_{cur} = \upsilon_{last} + a_{cur}/t \tag{2.2}$$

where υ_{last} is the speed at the previous step of integration, m/s; t is the travel time of the actuators, s. The current position of the PG is calculated for the coordinates S_x and S_y, it is equal to the distance of movement of the MI along the axes O_x, O_y and is presented in general form as

$$S = S_{last} + \upsilon_{cur}/t \tag{2.3}$$

where S_{last} is the previous position of the aircraft, m. The following two formulas make it possible to determine the coordinates of the moving object at the current time:

$$x = x_0 + \sin\left(x_{last} + \upsilon_x/t\right) \times l_x \tag{2.4}$$

$$y = y_0 + \sin\left(y_{last} + \upsilon_y/t\right) \times l_y \tag{2.5}$$

where x_0, y_0 is the position of the moving object when the axes are aligned, m; x_{last}, y_{last} - previous positions at the moment of oscillation, m; l_x, l_y - swing range, m. Based on formulas (Equations 1.3–1.5), the PG fixes the moving object when the following condition is met

$$\left(\left|S_x - x_0\right| < a_x\right) \& \left(\left|S_y - y_0\right| < a_y\right) \tag{2.6}$$

where a_x, a_y is the distance, m, along the axes O_x, O_y between the center of the PS and the center of the moving object.

To automate the processes of MC functioning, the team of authors has developed a program code for MC control, which provides synchronization of the TO performed in accordance with the kinematic model.

For this, in the Simatic Step 7 environment (TIA Portal v14), process identifiers (Table 2.1) have been created, which are associated with memory addresses and PLC I/O channels. In Figure 2.3 in the Simatic Step 7 inspector window, the device tab shows an example of a PLC configuration with assigned identifier labels. Input variables I refer to the channels for inputting information into the PLC and are responsible for processing data from the OS and position sensors. The output variable Q is associated with the channels for outputting information from the PLC and is used to transfer control action to the MC actuators. Local variable M is used to link the functional buttons on the operator's ITP screen with the PLC as shown in Figure 2.3. Table 2.1 shows the identifier marks required to stabilize the axes of the PG and the movable object when performing the technological operation of fixing the movable object. The columns contain a symbolic address; designation (name); data type; and a note in which you can add comments to the identifier tag. For example, if an OS is connected to the fifth input of the PLC as demonstrated in Figure 2.4, then it will be indicated by a variable of the form% I0.5, where the symbol I indicates that this is an input signal, 0.5- that the sensor is connected to the fifth input of the PLC. The % symbol indicates that the variable I0.5 is an absolute operand to be assigned a name, for example, laz_datchik.

When writing the program code, the Ladder Diagram (LD) language was used. The MC control program code consists of segments that are connected into logical chains using graphic elements of the language (contacts, coils, timers, counters, etc.). In this case, the series-connected contacts perform the logical "AND" operation, and the parallel-connected contacts perform the logical "OR" operation. Figure 2.5 shows the implementation of the technological operation of fixing the moving object. To do this, it is necessary

Table 2.1 Process ID labels

Variables	*Symbol address*	*Designation*	*Type*	*Notes*
Input	%I0.2	arm_outside	bool	Pneumatic gripper extended
	%I0.3	arm_inside	bool	Pneumatic gripper retracted
	%I0.5	laz_datchik	bool	Optical sensor
	%I0.6	gerkon_vpered	bool	Magnetic position sensors (reed switch)
	%I0.7	gerkon_up	bool	
	%I1.0	gerkon_niz	bool	
Output	%Q0.1	shvat	bool	Pneumatic gripper
	%Q0.0	arm	bool	Advancement of PG
	%Q0.5	down	bool	Moving PG down
Local	%M0.1	arm_down	bool	Moving (IT)
	%M10.2	trig2	bool	Cancel fixation
	%M10.4	trig4	bool	Reset movement
	%M5.2	c4_1	bool	Movement with PG advancement
	%DB2	timer_2	real	System delay timer
	%DB6	timer_7	real	

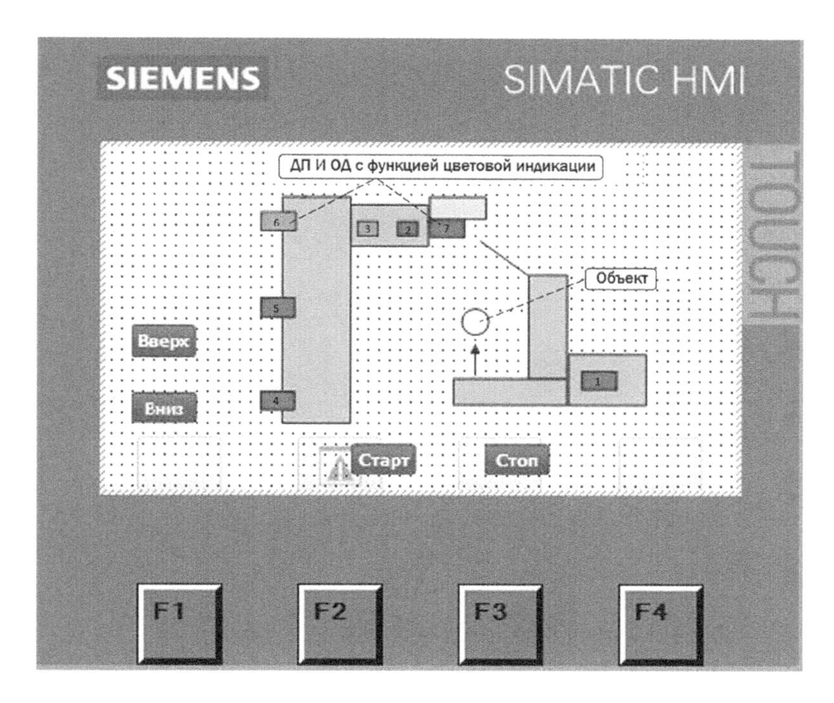

Figure 2.3 View in the operator's working window.

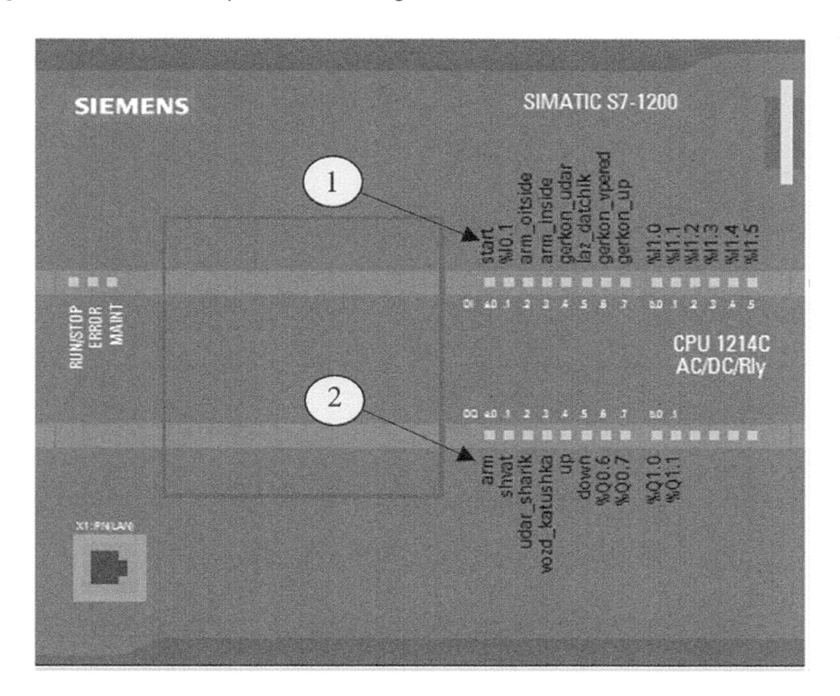

Figure 2.4 An example of a PLC configuration with assigned identifier labels: 1: input variables; 2: output variables.

to control the response time of the PG that equals 1.35 s. This time is predicted using a kinematic model. In the code, this value is set by adding the delays on the% DB6 "timer_7" and% DB2 "timer_2" system timers.

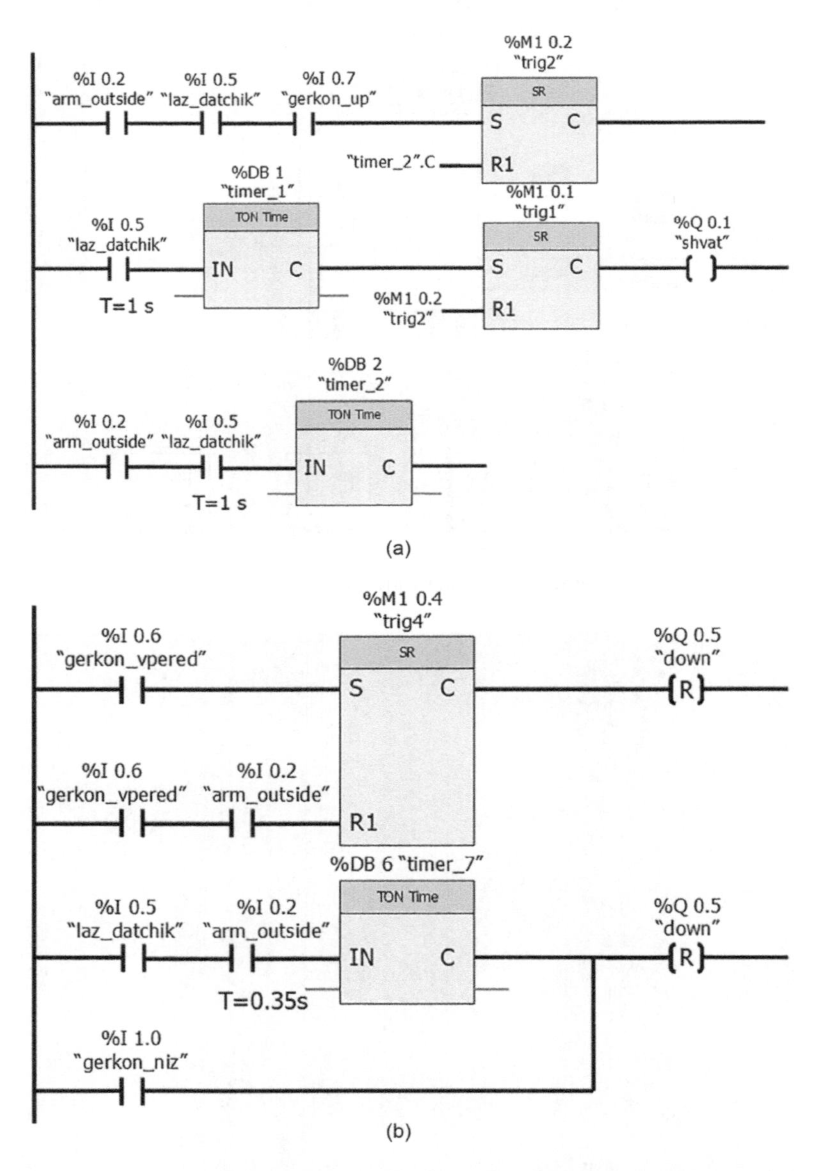

Figure 2.5 Segments of the program code for the implementation of the technological operation of fixation: (a) Stabilization of the IM axes. (b) Capture of a moving object.

The total delay time is 1+0.35=1.35 s (see Figure 2.5). In this case, the timer% DB6 "timer_7" after 0.35 s sets the reset coil% Q0.5 to the active state. The carriage of a RC with a DAC and a PG stops moving; the axes of the PG and the moving object are centered relative to each other, as shown in Figure 2.2. As a result, the ball is fixed by the PG.

2.4 FUZZY SISO-SYSTEM

To increase the efficiency of determining the moment of alignment of the axes of the PG and the moving object, a fuzzy system was developed. Due to the high degree of uncertainty of the moving object's position, the delay of the timer elements (see Table 2.2) was obtained while developing the kinematic model, and the values obtained from the kinematic model assume the maximum time of delay possible for the PG and the moving object to align. Using fuzzy set theory helps the MC to deal with the high degrees of uncertainty while stablishing the time necessary to trigger the TO of the PG to fix the moving object in position.

2.4.1 Fuzzification

The proposed fuzzy logic inference system considers one input; the input variable is set as the amplitude of oscillation of the moving object when dropped to be caught by the PG. The output value is set as the delay for the PG to activate. For the fuzzification process, the amplitude values are translated into linguistic terms presented in the form of If...Then rules, presented in Table 2.2.

2.4.2 Input and output membership functions

The input variable includes five terms: h1, h2, h3, h4, and h5 in the range from 10 to 20 cm, which are defined by the set of equations

Table 2.2 Linguistic rules for the fuzzy inference system

	Definition	
Fuzzy rule	*If*	*Then*
FR_1	amplitude low	delay min
FR_2	amplitude low-mid	delay min-average
FR_3	amplitude mid	delay average
FR_4	amplitude mid-high	delay average-max
FR_5	amplitude high	delay max

$$\mu(h_1) = \left\{ \begin{array}{c} 1 \text{ if } h < h_1, \\ \dfrac{h_2 - h}{h_2 - h_1} \text{ if } h_1 \leq h \leq h_2, \\ 0 \text{ if } h > h_2 \end{array} \right\},$$

$$\mu(h_2) = \left\{ \begin{array}{c} \dfrac{h - h_1}{h_2 - h_1} \text{ if } h_1 \leq h \leq h_2, \\ \dfrac{h_3 - h}{h_3 - h_2} \text{ if } h_2 \leq h \leq h_3, \\ 0 \text{ if } h > h_3 \end{array} \right\},$$

$$\mu(h_3) = \left\{ \begin{array}{c} \dfrac{h - h_2}{h_3 - h_2} \text{ if } h_2 \leq h \leq h_3, \\ \dfrac{h_4 - h}{h_4 - h_3} \text{ if } h_3 \leq h \leq h_4, \\ 0 \text{ if } h > h_4 \end{array} \right\}, \tag{2.7}$$

$$\mu(h_4) = \left\{ \begin{array}{c} \dfrac{h - h_3}{h_4 - h_3} \text{ if } h_3 \leq h \leq h_4, \\ \dfrac{h_5 - h}{h_5 - h_4} \text{ if } h_4 \leq h \leq h_5, \\ 0 \text{ if } h > h_5 \end{array} \right\},$$

$$\mu(h_5) = \left\{ \begin{array}{c} \dfrac{h - h_4}{h_5 - h_4} \text{ if } h_4 \leq h \leq h_5, \\ 1 \text{ if } h > h_5, \end{array} \right\}$$

where h is the amplitude of the moving object's oscillation shown in Figure 2.6.

The output variable includes five terms: t1, t2, t3, t4, and t5 in the range from 20 to 50 ms, which are defined by the set of equations

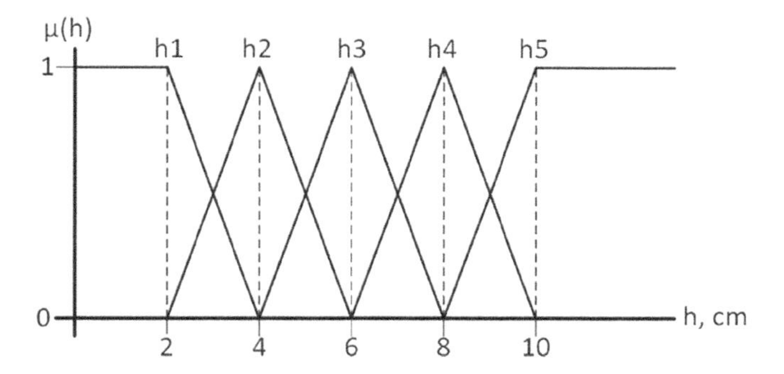

Figure 2.6 Input variable.

$$\mu(h_1)=\left\{\begin{array}{c} 1 \text{ if } h<h_1, \\ \dfrac{h_2-h}{h_2-h_1} \text{ if } h_1\le h\le h_2, \\ 0 \text{ if } h>h_2 \end{array}\right\},$$

$$\mu(h_2)=\left\{\begin{array}{c} \dfrac{h-h_1}{h_2-h_1} \text{ if } h_1\le h\le h_2, \\ \dfrac{h_3-h}{h_3-h_2} \text{ if } h_2\le h\le h_3, \\ 0 \text{ if } h>h_3 \end{array}\right\},$$

$$\mu(h_3)=\left\{\begin{array}{c} \dfrac{h-h_2}{h_3-h_2} \text{ if } h_2\le h\le h_3, \\ \dfrac{h_4-h}{h_4-h_3} \text{ if } h_3\le h\le h_4, \\ 0 \text{ if } h>h_4 \end{array}\right\},$$

$$\mu(h_4)=\left\{\begin{array}{c} \dfrac{h-h_3}{h_4-h_3} \text{ if } h_3\le h\le h_4, \\ \dfrac{h_5-h}{h_5-h_4} \text{ if } h_4\le h\le h_5, \\ 0 \text{ if } h>h_5 \end{array}\right\},$$

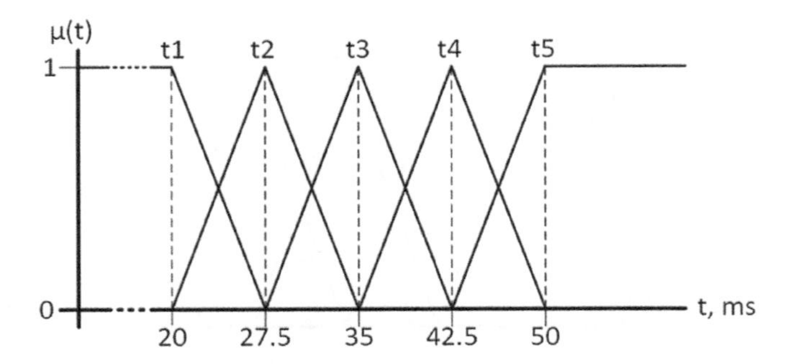

Figure 2.7 Output variable.

$$\mu(h_5) = \left\{ \begin{array}{c} \dfrac{h - h_4}{h_5 - h_4} \text{ if } h_4 \le h \le h_5, \\ 1 \text{ if } h > h_5, \end{array} \right\} \tag{2.8}$$

where t is the delay to trigger the PG as shown in Figure 2.7.

2.4.3 Defuzzification

For the defuzzification process, the output values are defined by the equation

$$t_{defuzzyy} = \frac{\sum_{j=20}^{50} \beta_j t_j}{\sum_{j=20}^{50} \beta_j} \tag{2.9}$$

where composition of the fuzzy inference for the system is defined by the equations

$$\alpha_j = \max_{i=20}^{50}(\mu(t_i), h_i) \tag{2.10}$$

$$\beta_j = \max_{i=1,j=20}^{5.50}(a_{ij}) \tag{2.11}$$

2.5 EXPERIMENTAL RESEARCH

The MC kinematic model was developed using the Simulink environment included in MATLAB. In the course of its implementation, two experiments were carried out. During the first experiment, using expressions

(Equations 2.1–2.6), three-dimensional and two-dimensional models were built to analyze the moment of alignment of the axes of the PS and the moving object as can be seen in Figure 2.8. This experiment makes it possible to determine the location of the AM where the moving object is fixed, as well

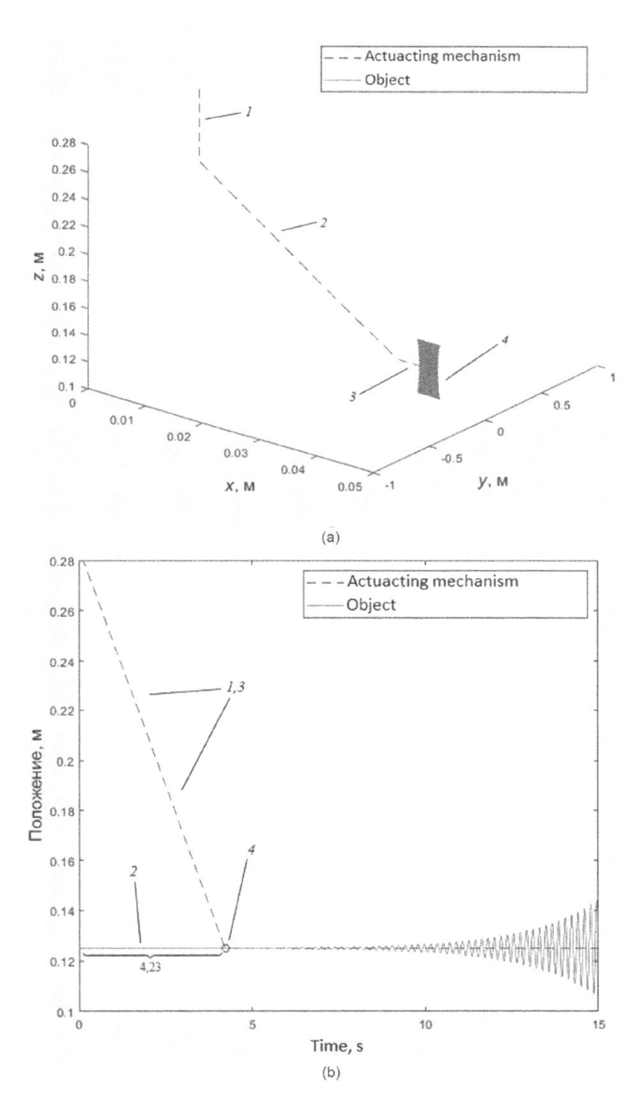

Figure 2.8 Kinematic model of the alignment of the axes of the pneumatic gripper and the moving object: (a) 3D model. (b) 2D model; 1, 3: movement of the rodless cylinder carriage along the Oy axis; 2: movement of the carriage of the rodless cylinder along the Ox axis; 4: the moment of alignment of AM with the moving object.

as to predict the response time of the PG, the kinematic model for this process is presented in Figure 2.8b. The gripper is set in the program code using the system timers shown in Figure 2.5. The last value is taken into account in the system timers% DB6 "timer_7" and% DB2 "tim-er_2. The accuracy of predicting the fixation time of a moving object using the developed kinematic model was estimated based on the regression method and the coefficients MSE, MAPE, and R 2 [14]. For this, linear and power regression equations were synthesized and are presented in Figure 2.9 [11,14], showing the relationship between time and movement of actuators $t=f(S)$. In Figure 2.9, a is presented an analysis of the movement in time of the carriage of a RC along the Y-axis referenced in Figure 2.8 Section 2.1. The travel time of the AM shown in Figure 2.9a is 2.04. In Figure 2.9b is presented an analysis of the movement in time of a DAC along the X-axis.

The travel time of the AM shown in Figure 2.9b is 0.84 s. In Figure 2.9c is shown the analysis of the movement of the carriage of a RC along the Y-axis shown in Figure 2.8, until the axes of the moving object and the PG are aligned. The travel time of the MI is 1.35 s. In Figure 2.8b, the abscissa shows the time of movement of the actuators for the implementation of the technological operation of fixing the moving object. The specified time interval was 4.23 s, it was predicted using the developed kinematic model. Table 2.3 shows the analysis of indicators for predicting the moment in time when the axes of the PG and the moving object are aligned, as referenced in Figure 2.2. Based on the data in Table 2.3, it can be concluded that the proposed kinematic model in relation to regression models makes it possible to more accurately predict the time of alignment of the axes for fixing a moving object in a PG. Another positive feature of the kinematic model is its continuity [15–19]. When using regression models, at the points of change in the axes of movement of the actuators of the MC, a break occurs in the points of the trajectory. As a consequence, if we build regression models for the entire process of displacements of actuators, then the statistical coefficients will have lower values, and the execution time of the TO is about 45% less if the MC works in automatic mode, as shown in the data in Table 2.4.

As a result, the inference system determines the delay necessary for the PG to fixate the moving object in a more efficient manner compared to the calculations of the kinematic model.

2.6 CONCLUSIONS

The hardware and software elements of the MC that perform TO in two control modes – manual and automatic – are considered. A model of the human–machine interface and its software implementation have been created, which provides the operator with information about the current location of the actuators, signal states of the sensors using color indication,

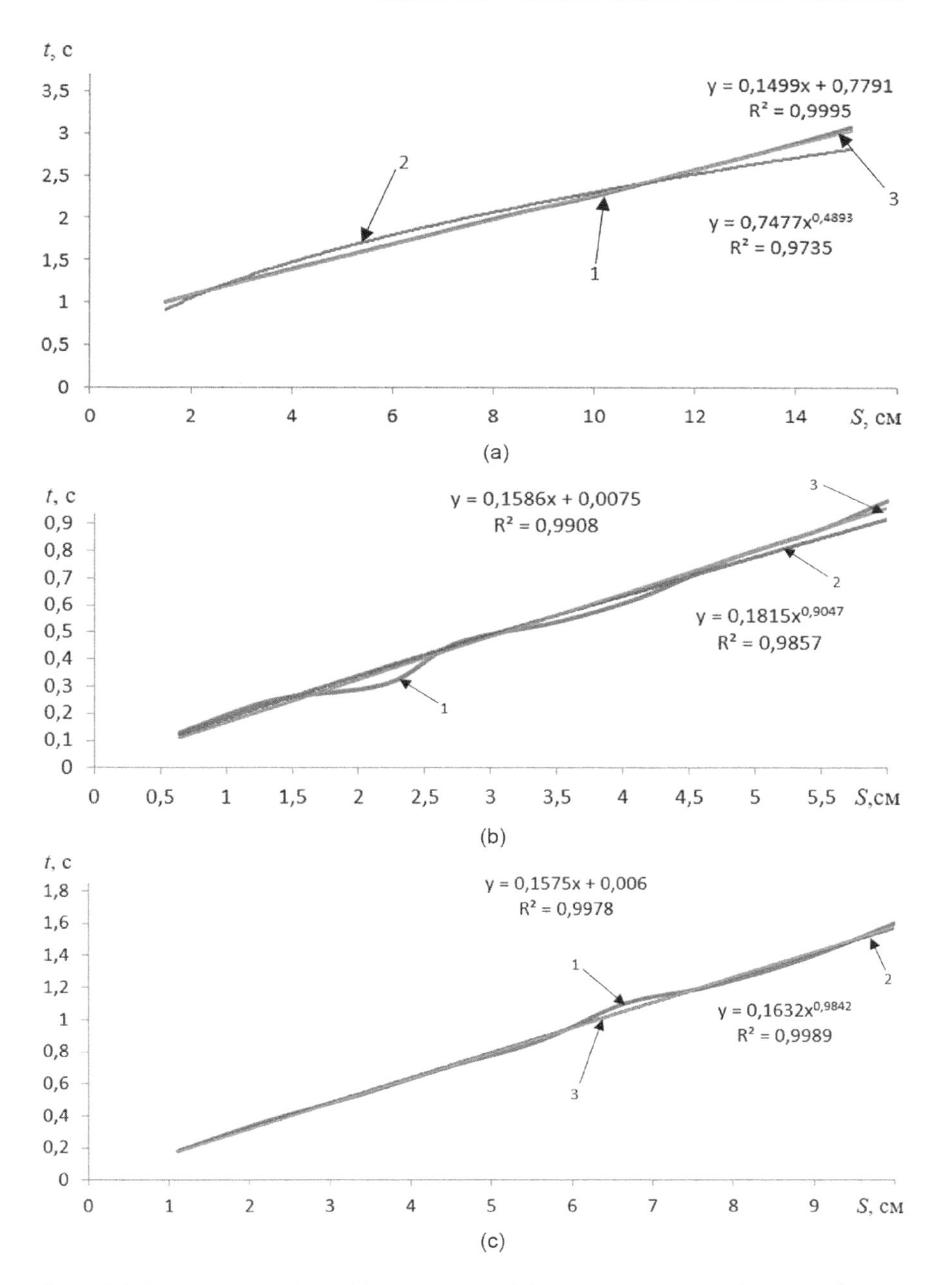

Figure 2.9 Regression analysis: (a) Movement of the rodless carriage along the O_y axis from the starting position. (b) Movement of the pneumatic gripper along the Ox axis. (c) Movement of the rodless carriage along the O_y axis until the alignment of the axes of the AM and the moving object.

Table 2.3 Regression analysis

Analysis method	*AM movement*	*MSE, s*	*MAPE*	R^2	*Accuracy, %*
Linear regression	Rodless carriage (Figure 2.9a)	0,0002	0,56	0,9995	99,44
	Pneumatic gripper (Figure 2.9b)	0,0005	1,92	0,9978	98,08
	Rodless carriage (Figure 2.9c)	0,0007	6,55	0,9908	93,45
Power regression	Rodless carriage (Figure 2.9a)	0,039	5,11	0,974	94,89
	Pneumatic gripper (Figure 2.9b)	0,003	1,97	0,9989	98,03
	Rodless carriage (Figure 2.9c)	0,013	6,03	0,9857	93,97
Kinematic model	Rodless carriage (Figure 2.9a)	0,0002	0,37	0,9996	99,63
	Pneumatic gripper (Figure 2.9b)	0,0004	1,99	0,999	98,01
	BC carriage (Figure 2.9c)	0,0002	5,63	0,991	96,37

Table 2.4 TO execution time while controlling the MC

	Operation time, s	
	Management mode	
Cycle number	*Manual*	*Automatic*
1	22,00	14,00
2	22,04	13,96
3	21,97	14,04
4	22,00	13,92
5	21,95	14,02
6	21,83	14,00
7	21,86	13,92
8	21,90	13,94
9	21,89	13,90
10	21,97	14,02
Average arithmetic value, s	21,94	13,97

and the ability to control the MC on the ITP screen. A program code has been developed in the language of relay-contact circuits, which ensures the execution of movements of the actuators in order to fix the moving object.

Using a kinematic model and a fuzzy inference system, the problem of the alignment of the axes of the PG and the moving object has been solved. The results of the analysis of the accuracy of the actuator response time, which is estimated by the regression method, are presented. An experimental study was carried out with fixing the time of the working cycle. The average execution time was 13.97 seconds, which increased the operating speed of the MC by 7.97 seconds in comparison with the manual mode. The reliability of the MC has been checked; the results are presented based on the performance of 100 cycles.

The work was prepared as part of the implementation of the RSF project No. 23-21-00071. The authors are grateful to the Foundation for their support.

REFERENCES

1. Bobyr, M.V., Kulabukhov, S.A. Simulation of control of temperature mode in cutting area on the basis of fuzzy logic. *Journal of Machinery Manufacture and Reliability*, 2017, 46(3), 288–295.
2. Beater, P. *Pneumatic Drives: System Design, Modelling and Control*, Berlin Heidelberg, Springer-Verlag, 2007, p. 324.
3. Pfeffer, A., Glück, T., Schausberger, F., Kugi, A. Control and estimation strategies for pneumatic drives with partial position information. *Mechatronics*, 2018, 50, 259–270. DOI: 10.1016/j.mechatronics.2017.09.01.01.
4. Wang, Y., Ying, Y., Xie, S., Wang, H., Feng, X. Mechatronics education at CDHAW of Tongji University: Laboratory guidelines, framework implementations aCnd improve-ments. *Mechatronics*, 2009, 19, 1346–1352. DOI: 10.1016/j.mechatronics.2009.09.001.
5. Saravanakumar, D., Mohan, B., Muthuramalingam, T., Sakthivel, G. A review on recent research trends in servo pneumatic positioning systems. *Precision Engineering*, 2017, 49, 481–492.
6. Yuhimets, D.A., Yudinkov, E.E. Development of an application software interface for controlling a Mitsubishi RV-2FB robotic arm. *Programmnaya Ingeneria*, 2018, 9(6), 253–261. DOI: 10.17587/ prin.9.253-261 (in Russian).
7. Backe, W. The application of servo-pneumatic drives for flexible mechanical handling techniques. *Robotics*, 1986, 2(1), 45–56.
8. Östring, M., Gunnarson, S., Norrlöf, M. Closed-loop identification of an industrial robot containing flexibilities. *Control Engineering Practice*, 2003, 11(3), 291–300. DOI: 10.1016/S0967-0661(02)00114-4.
9. Arreguin, J.M.R., Ortega, J.C.P., Gorrostieta, E., Troncoso, R.J.R. Artificial intelli-gence applied into pneumatic flexible manipulator. *Proceedings of 7th Mexican Inter-national Conference on Artificial Intelligence 2008*, 2008, pp. 339–345. DOI: 10.1109/MICAI.2008.76.
10. Elgeneidy, K., Loshe, N., Jackson, M. Bending angle prediction of control soft pneumatic actuators with embedded flex sensor – A data driven approach. *Mechatronics*, 2018, 50, 234–237.
11. Salamandra, B.L. Analysis of label position stabilization methods on automatic packaging machines. *Journal of Machinery Manufacture and Reliability*, 2017, 46(2), 181–186.
12. Shelekhov, V.I., Tumurov, E.G. Applying automata-based software engineering for the lift control program. *Programmnaya Ingeneria*, 2017, 8(2), 99–111. DOI: 10.17587/prin.8.99-111 (in Russian).
13. Bobyr, M.V., Yakushev, A.S., Dorodnykh, A.A. Fuzzy devices for cooling the cutting tool of the CNC machine implemented on FPGA. *Measurement*, 2020. DOI: 10.1016/j.measurement.2019.107378.
14. Kruglova, T.N. Intellectual method of diagnosing and forecasting of a technical condition of the mechatronic complexes for extreme conditions. *Mehatronika, avtomati-zatsiya, upravlenie*, 2011, 3, 47–51 (in Russian).

15. Chunikhin, A.Yu., Glazunov, V.A. Developing the mechanisms of parallel structure with five degrees of freedom designed for technological robots. *Journal of Machinery Manufacture and Reliability*, 2017, 46(4), 313–321.
16. Bobyr, M.V., Emelyanov, S.G. A nonlinear method of learning neuro-fuzzy models for dynamic control systems. *Applied Soft Computing*, 2020, 88, 106030.
17. Loshitskiy, P.A., Shehovcova, E.E. Calculation and simulation of an industrial manipulator on the power of the shell elements. *Mekhatronika, avtomatizatsiya, upravlenie*, 2015, 16(7), 470–475. DOI: 10.17587/mau.16.470-475 (in Russian).
18. Skvortsov, A.A. The use of finite state machines in the development of software for machines with complex structure. *Programmnaya Ingeneria*, 2018, 9(7), 332–336. DOI: 10.17587/prin.9.332-336 (in Russian).
19. Bobyr, M., Titov, V., Belyaev, A. Fuzzy system of distribution of braking forces on the engines of a mobile robot. *MATEC Web of Conferences*, 2016, 79, 01052.

Chapter 3

Quad sensor-based soil-moisture prediction using machine learning

Anirbit Sengupta
Dr. Sudhir Chandra Sur Institute of Technology and Sports Complex

Sayantani Ghosh
University of Calcutta, Technology Campus

Anwesha Mukherjee
Mahishadal Raj College

Abhijit Das
RCC Institute of Information Technology

Debashis De
Maulana Abul Kalam Azad University of Technology

CONTENTS

DOI: 10.1201/9781003381167-3

3.1 INTRODUCTION

Soil is a rudimentary integrant in gratifying the increasing stipulation for food and further necessities [1]. The health of the soil is correlated with agricultural sustainability, and thus the common people in India are reliant on their sustenance. For growing micronutrients, soils of India are less productive. Thus, the yielding of crops can be enhanced by preserving the fertility of the topsoil. For the cultivation of crops, agriculture requires a decision support system. The water level is lowering every day. Optimization of water usage can be done by soil moisture monitoring. Even soil moisture is one of the most significant factors of climate change. Water and energy are circulated by soil moisture between the atmosphere and the land. It has a distinct correlation with climate change. Along with the climatic factors, heterogeneity in precipitation and temperature directly affects the diversification of soil moisture. Since soil moisture is favorable for crop yield, the procedures involved for their growth can be improved if the content of the soil moisture of any area or location is predicted successfully [2]. This enlightens a farmer to know more about whether adequate watering has been done to the yields for its proper growth or not, the most appropriate time of dispersing and harvesting the yields, soil percolation is suitable or not, etc. Remote sensing technology gathers information regarding ground surface and a basic spatial resolution is provided. Other than the temperature, pH, and turbidity, there are too many factors such as atmospheric association, chemical processes of soil, the crudeness of soil, vegetation, etc.

Soil moisture monitoring is highly significant from the agricultural perspective. The efficient use of water plays an important role in cultivation. Proper water usage can be possible by soil moisture prediction. For better soil moisture content estimation, various parameters have to be considered such as soil moisture, turbidity, pH, temperature, and humidity. The objective of this work is to monitor the soil moisture using these parameters. To accomplish the objective, four different sensors are used in our work, and a comparative analysis between a few machine learning methods is performed for detecting the moisture content in the soil.

3.2 RELATED WORKS

There are various soil moisture estimation models such as the SPAW (Soil Plant Atmosphere Water) model [3], USDAHL (U.S. Department of Agriculture Hydrograph Laboratory) model [4], SAC-SMA (Sacramento Soil Moisture Accounting) model [5], etc. Recently, remote sensing methods have also been used for soil moisture estimation, such as microwave remote sensing measurements [6]. Data-driven forecasting tools such as artificial neural networks (ANNs) can also participate in soil moisture estimation [7]. Error Propagation Learning Back Propagation (EPLBP) neural network can be used for soil moistness retrieval [8]. A data-driven model – Support Vector Machine (SVM) [9] can be used for soil moisture prediction using meteorological variables. A data-driven scheme for predicting soil moisture was discussed in [10].

The use of IoT in soil moisture monitoring was explored in [11]. An IoT model was developed in [12] for monitoring soil moisture using the Watermark 200SS sensors. The authors in [13] proposed a system to use an inexpensive wireless communication scheme for monitoring soil humidity using Decagon EC-5 sensors. For effective crop prediction, different regression models such as linear, multilinear, and SVM are tested. It facilitated the farmers in planting the crops appropriate for their farmland. It helps to improve the production of yields. In [14], the authors' objective was to forecast a crop whose cultivation can offer maximum yield after harvesting in the proper soil under appropriate climatic conditions. In [15], the authors gave an outline of soil monitoring schemes. The sensor data is first transmitted to the MCP3204 A/D converter. From the A/D converter, the data is sent to the cloud using Raspberry pi. The information helps select the suitable crop to be cultivated with a given soil parameter.

3.3 PROPOSED SYSTEM

In this work, we have used four different types of sensors for temperature, pH, turbidity, and moisture data collection, and the data analysis is performed using machine learning. We have used different machine learning models to perform a comparative analysis between them to select the one providing a better result. We have highlighted the use of edge-cloud architecture as well. To store and analyze a huge volume of data, cloud is used. However, edge computing provides the facility to bring the resources to the network edge, which helps to reduce the latency [16].

To reduce the latency, the networking philosophy of edge computing is to bring computing as close as possible to the data source. Edge computing usually refers to the execution of fewer processes in the cloud and the migration of those processes to local devices, like an edge server, an IoT device, a user's computing gadget, etc. Bringing the computation to the edge reduces the probability of long-distance client-server communication. This helps reduce the latency. Edge computing facilitates data storage and computation nearer to the devices where it is being aggregated, instead of at a central location. The benefit of using edge computing is the capability to process and store data faster, enabling the effective real-time applications that we are using. The exponential growth of IoT devices leads to the development of edge computing since the IoT devices link to the internet to have two-way communication with the cloud. Various IoT devices produce a massive amount of data in runtime. Here, the edge devices include various things, such as an IoT device with a quad sensor, a computer, and a smartphone.

3.3.1 Sensors used in the proposed framework

The temperature, moisture, pH, and turbidity sensors play a significant role in crop production data collection. In [17], we have designed a device FarmFox. In the present work, we have used temperature, moisture, pH, and turbidity sensors in the IoT device. These sensors are described as follows.

Temperature sensor: To keep track of the humidity and temperature of the environment, we have used a temperature sensor where crop plantation is intended. The sensor operates in the voltage of 12 V, 1 A current supply, 00–500 Centigrade temperature, and 20%–90% humidity, allied with Arduino Nano R3.

Moisture sensor: Water content in the soil is measured using a moisture sensor. The operating voltage of the moisture sensor is 12 V, and it works on less than 1 A current supply. It uses capacitance to measure dielectric permittivity.

pH sensor: To measure the amount of alkalinity and acidity in water, a pH sensor is used. The operating voltage of the pH sensor is 12 V. This sensor measures the soil volumetric water content based on dielectric constant. By using dielectric permittivity of the soil using capacitance, it estimates the water volume existing in the soil.

Turbidity sensor: The turbidity sensors are on the land in such a way that they can measure the opacity or haziness of the water which is used for irrigation.

3.3.2 Hardware architecture

The temperature, pH, turbidity, and soil moisture sensors are connected to the Arduino Nano R3 board. With this board, the ThingSpeak Cloud with inbuilt Wi-Fi and the ThingView Application are connected, allowing agrarians to access all data stored in the ThingSpeak Cloud. The data extracted from the sensors and collected from ThingSpeak Cloud are used in the analysis of soil moisture prediction.

3.3.3 Dataset

The dataset generation and processing are discussed as follows.

Dataset generation: In this research work, we have used four analog sensors, measuring temperature, pH, turbidity, and soil moisture. The data has been collected in real-time directly from the fields of different districts of West Bengal and stored in the ThingSpeak Cloud for future use. The data is recorded every 30 minutes and there are 48

Table 3.1 Collected and merged data from different districts of West Bengal

	ID	*pH*	*Temperature*	*Turbidity*	*Moisture*
0	151	6.23	27	1,554	96
1	152	6.23	27	1,554	98
2	153	6.23	27	1,554	98
3	154	6.20	27	1,554	97
4	155	6.20	27	1,554	97

records collected from each sensor daily. Table 3.1 shows the integrated data from different districts of West Bengal.

Data pre-processing: The four considered parameters (temperature, pH, turbidity, and soil moisture) will be responsible for monitoring whether the soil is suitable for the growth of a particular crop or not. Derivation of sensor data and uploading the data to ThingSpeak generates four different CSV files. These files are merged into one. xlsx file, based on their ID numbers. The correlation between the data is checked. Among several measures of the correlation, Pearson's Product-Moment coefficient is used. It produces a value within the range of –1 to +1 so that the dependence between two variables can be quantified. When the correlation is positive, variables change in the same direction, whereas when the correlation is positive, variables change in opposite directions. But when the correlation is 0 or neutral, it means that there is no correspondence between the variables and the variables are unrelated. The data is checked for null value or cell. To decrease the variability in the data, outliers are removed to make the prediction statistically more significant. As a result, the quality of the prediction is improved. All the pre-processed data is time windowed, i.e., it takes into account the values in a time frame into a set of contiguous values and predicts the next values. Then in the learning phase, the algorithm tries to find a set of parameters so that it can model the training dataset. In the running phase, the algorithm applies the same model on each input set.

3.3.4 Data analysis

We have used the following algorithms for the data analysis.

Linear regression (LR): LR [18] is a statistical way of measuring the association among variables, a dependent variable, and single or multiple explanatory variables, and finding the future values. With this relationship, it aims to fit best the total observed data points with a straight line which eventually reduces the remaining error. Dependent variables, which we want to explain or focus on, are denoted as y. Independent or explanatory variables are the variables that explain other variables, denoted as x. The relation between y and x can be explained as:

$$y = mx + c \tag{3.1}$$

where m is the slope and c is the intercept. These are the parameters required for the estimation.

Support vector regression (SVR): SVR [19] is a regression procedure that we can use for continuous values. It examines the limits of the dataset and considers a decision boundary known as hyperplane, near the limits in the datasets. It performs linear regression in higher-dimensional space. It uses the perception of vectors and its dependence on the data points has made it lesser susceptible to outliners. SVM considers the points which are present within the decision boundary line at a distance e from the original hyperplane. Hyperplane that contains a maximum number of points is the best-fit line considered here.

Equation of the hyperplane:

$$y = wx + b \tag{3.2}$$

Two equations of the boundary lines are:

$$wx + b = +e \tag{3.3}$$

$$wx + b = -b \tag{3.4}$$

The equation that satisfies SVR is:

$$-e < y - wx + b < +e \tag{3.5}$$

where e is the epsilon, weight factor is w, and b stands for the bias vector.

Decision tree regression (DTR): DTR [20] is a nonlinear regression technique. It is used to forecast the continuous-valued target variable. It uses faster divide and conquer greedy algorithms. It recurrently separates the data into smaller subsets, and an associated decision tree is developed. The final tree consists of the decision nodes and leaf nodes.

Random forest (RF): RF [21] is another type of supervised machine learning technique that is very effective when other types of machine learning algorithms seem to be computationally expensive. It is an ensemble learning technique that makes accurate forecasts by taking forecasts from more than one machine learning algorithms or taking forecasts from one single algorithm several times. Forecasting from one solo-specific model may not produce more accuracy in the results, and hence we use ensemble learning. This solves the problem of over-fitting the training sets. RF Regression constructs multiple decision trees in the training time and combines the result of forecasts from individual decision tree to determine the ultimate outcome. It can be written as:

$$g(x) = f0(x) + f1(x) + f2(x) + \ldots \tag{3.6}$$

where *fi* are the simple base models and *g* is the final model.

3.3.5 Performance evaluation measures

We have selected the subsequent evaluation measures to indicate the performance of different machine learning models in the data analysis.

Mean squared error (MSE): MSE processes the average squared errors of machine learning predictions. For each of the observations, it calculates the sum of the square of the variance between actual and predicted values, i.e., the errors, and then it averages out the values. To remove negative signs, squaring is necessary here. It also gives more weight to larger differences. The smaller the value of the MSE, the improved accurateness of the forecast model. The MSE is calculated as:

$$\frac{1}{m}\sum_{i=1}^{m}\left(Z_i - \hat{Z}_i\right)^2 \tag{3.7}$$

where m is the total number of predictions, Z_i is the vector of actual values, and $\hat{Z}_i$ is a vector of predicted values.

Root mean squared error (RMSE): RMSE is the arithmetic square root of MSE. It specifies how the data concentration round the line of best fit is. To bring errors on the same scale as the target value scale, we are using square root on the MSE. It helps in penalizing large errors more so that it can be more appropriate in some cases.

$$\sqrt{\frac{1}{m}\sum_{i=1}^{m}\left(Z_i - \hat{Z}_i\right)^2} \tag{3.8}$$

Mean absolute error (MAE): Here, we calculate the error as an average of the absolute variance between the actual value and the predicted values. MAE is a linear score and it signifies that all the individual variances are weighted similarly in the average.

$$\frac{1}{m}\sum_{i=1}^{m}\left|\left(Z_i - \hat{Z}_i\right)\right| \tag{3.9}$$

3.4 RESULTS AND DISCUSSION

In this section, we have measured soil moisture based on the collected data from soil temperature, pH, moisture, and turbidity sensors. We have used four machine learning models LR, SVR, RF, and DTR, for data analysis purpose, and the compared the results in terms of accuracy, MAE, MSE, RMSE, and time consumption.

We have determined the correlation between the collected data. Thus, by blending all the inputs, the prediction has been improved. The regression techniques used for predicting water content in the soil contain four different sensors (temperature, pH, moisture, and turbidity). To help the farmers in detecting the soil quality for the growth of a variety of crops, it is essential that we look into all the benchmark algorithms to check which works better and by how much. In this chapter, the efficiency of the models is considered in terms of their efficacy and the least amount of error they produce.

Therefore, the same algorithm has been tested on the production dataset, selected randomly from the available data. The expected moisture means the moisture to be predicted, known from the historical data. The predicted moisture represents the value predicted by the algorithm. The aforementioned errors represent the errors between the expected and predicted moisture. Considering dual sensors, the efficiency of the models is obtained in Tables 3.2–3.4.

From Tables 3.2–3.4, we observe that at a time two sensors are considered. However, for better prediction, the consideration of only two parameters is not sufficient. In Table 3.5, we have presented the MAE, MSE, and RMSE of the predicted soil moisture values using different machine learning models. The average values considering these models are also presented

Table 3.2 MAE, MSE, and RMSE of LR, SVR, DTR, and RF considering pH and temperature

	LR	*SVR*	*DTR*	*RF*
MAE	0.47	0.25	0.075	0.088
MSE	0.36	0.45	0.0375	0.037
RMSE	0.60	0.67	0.194	0.191

Table 3.3 MAE, MSE, and RMSE of LR, SVR, DTR, and RF considering pH and turbidity

	LR	*SVR*	*DTR*	*RF*
MAE	0.541	0.25	0.05	0.363
MSE	0.696	0.45	0.025	0.329
RMSE	0.834	0.67	0.158	0.164

Table 3.4 MAE, MSE, and RMSE of LR, SVR, DTR, and RF considering temperature and turbidity

	LR	*SVR*	*DTR*	*RF*
MAE	0.54	0.25	0.3625	0.36
MSE	0.025	0.45	0.3281	0.32
RMSE	0.626	0.67	0.5728	0.57

Table 3.5 MAE, MSE, and RMSE of LR, SVR, DTR, and RF using quad sensor

	LR	*SVR*	*DTR*	*RF*
MAE	0.46	0.25	0.05	0.20
MSE	0.34	0.45	0.025	0.211
RMSE	0.59	0.67	0.15	0.395

in Table 3.5. As we have considered four different sensors to collect temperature, humidity, moisture, pH, and turbidity, it results in a comparatively better prediction of the soil moisture.

Table 3.6 presents a comparative study among the current work and existing state-of-the-art models for predicting soil moisture based on the MAE, MSE, and RMSE.

As we observe from Table 3.6, the MAE, MSE, and RMSE are much less in our case than the existing works. This is because, the consideration of all the parameters such as temperature, humidity, moisture, pH, and turbidity, provides better soil moisture prediction than the existing works. We have used k-fold cross-validation to verify the accuracy of the models and the results are presented in Table 3.7. We observe that the average accuracy is above 90%. We also observe that the average accuracy in RF is better than the other methods.

Table 3.6 A comparison chart of the present work and existing models based on MAE, MSE, and RMSE

Model	*MAE*	*MSE*	*RMSE*
LR [18]	6.89	–	–
SVM [22]	3.65	16.4025	4.05
AGNN [23]	1.26	–	–
Multiple linear regression (average) [24]	–	1.59(7 days ahead)	1.260 (7 days ahead)
SVR [19]	–	1.991 (7 days ahead)	1.411 (7 days ahead)
Recurrent neural network [25]	–	2.033 (7 days ahead)	1.425 (7 days ahead)
Relevance vector machine [10]	3.93	0.29	0.538
Present work (average)	0.20	0.211	0.395

Table 3.7 A comparison chart of the present work and existing models based on accuracy observed using k-fold cross validation

	Models	*Accuracy*
$k=1$	LR	91
	SVR	93
	RF	96
	DTR	94
$k=2$	LR	94
	SVR	93
	RF	95
	DTR	95
$k=3$	LR	92
	SVR	92
	RF	94
	DTR	95
$k=4$	LR	95
	SVR	93
	RF	95
	DTR	95
$k=5$	LR	93
	SVR	95
	RF	96
	DTR	96
Average	LR	93
	SVR	93.2
	RF	95.2
	DTR	95

3.5 CONCLUSION

In this chapter, we have measured the soil moisture by analyzing the data collected using temperature, pH, moisture, and turbidity sensors. The temperature sensor collects both temperature and humidity data. The data storage takes place inside the cloud due to the high volume. On the collected temperature and humidity, pH, moisture, and turbidity data, we have used machine learning-based methods for analysis. Based on the outcome, we have compared the machine learning-based methods to select the best-suited one.

Data Availability: All the data used in the analysis of this research work can be obtained from the corresponding author on request.

REFERENCES

1. Hemageetha, N. (2016). A survey on application of data mining techniques to analyze the soil for agricultural purpose. In *2016 3rd International Conference on Computing for Sustainable Global Development (INDIACom)* (pp. 3112–3117). IEEE.
2. Prakash, S., Sharma, A., & Sahu, S. S. (2018). Soil moisture prediction using machine learning. In *2018 Second International Conference on Inventive Communication and Computational Technologies (ICICCT)* (pp. 1–6). IEEE.
3. Ouyang, W., Wu, Y., Hao, Z., Zhang, Q., Bu, Q., & Gao, X. (2018). Combined impacts of land use and soil property changes on soil erosion in a mollisol area under long-term agricultural development. *Science of the Total Environment*, *613*, 798–809.
4. Mishra, S. K., Singh, V. P., & Singh, P. K. (2018). Revisiting the soil conservation service curve number method. In V. P. Singh, S. Yadav, & R. N. Yadava (Eds.), *Hydrologic Modeling* (pp. 667–693). Springer, Singapore.
5. Katsanou, K., & Lambrakis, N. (2017). Modeling the Hellenic karst catchments with the sacramento soil moisture accounting model. *Hydrogeology Journal*, *25*(3), 757–769.
6. Zhang, K., Chao, L. J., Wang, Q. Q., Huang, Y. C., Liu, R. H., Hong, Y., Tu, Y., Qu, W., & Ye, J. Y. (2019). Using multi-satellite microwave remote sensing observations for retrieval of daily surface soil moisture across China. *Water Science and Engineering*, *12*(2), 85–97.
7. Han, H., Choi, C., Kim, J., Morrison, R. R., Jung, J., & Kim, H. S. (2021). Multiple-depth soil moisture estimates using artificial neural network and long short-term memory models. *Water*, *13*(18), 2584.
8. Lu, Z., Chai, L., Liu, S., Cui, H., Zhang, Y., Jiang, L., Jin, R., & Xu, Z. (2017). Estimating time series soil moisture by applying recurrent nonlinear autoregressive neural networks to passive microwave data over the Heihe River Basin, China. *Remote Sensing*, *9*(6), 574.
9. Zhu, Q., Wang, Y., & Luo, Y. (2021). Improvement of multi-layer soil moisture prediction using support vector machines and ensemble Kalman filter coupled with remote sensing soil moisture datasets over an agriculture dominant basin in China. *Hydrological Processes*, *35*(4), e14154.

10. Hong, Z. (2015). *A Data-Driven Approach to Soil Moisture Collection and Prediction Using a Wireless Sensor Network and Machine Learning Techniques*. Dissertation/thesis of Z. Hong, Electrical & Computer Engineering, University of Illinois at Urbana-Champaign.
11. Kodali, R. K., & Sahu, A. (2016). An IoT based soil moisture monitoring on Losant platform. In *2016 2nd International Conference on Contemporary Computing and Informatics (IC3I)* (pp. 764–768). IEEE.
12. Payero, J. O., Mirzakhani-Nafchi, A., Khalilian, A., Qiao, X., & Davis, R. (2017). Development of a low-cost Internet-of-Things (IoT) system for monitoring soil water potential using Watermark 200SS sensors. *Advances in Internet of Things*, *7*(3), 71–86.
13. Payero, J. O., Nafchi, A. M., Davis, R., & Khalilian, A. (2017). An arduino-based wireless sensor network for soil moisture monitoring using Decagon EC-5 sensors. *Open Journal of Soil Science*, *7*(10), 288.
14. Mhaiskar, S., Patil, C., Wadhai, P., Patil, A., & Deshmukh, V. (2017). A survey on predicting suitable crops for cultivation using IoT. *International Journal of Innovative Research in Computer and Communication Engineering*, *5*, 318–323.
15. Badhe, A., Kharadkar, S., Ware, R., Kamble, P., & Chavan, S. (2018). IOT based smart agriculture and soil nutrient detection system. *International Journal on Future Revolution in Computer Science & Communication Engineering*, *4*(4), 774–777.
16. Mukherjee, A., De, D., Ghosh, S. K., & Buyya, R. (2021). *Mobile Edge Computing*.Springer Nature, Switzerland AG.
17. Sengupta, A., Debnath, B., Das, A., & De, D. (2021). FarmFox: A quad-sensor based IoT box for precision agriculture. *IEEE Consumer Electronics Magazine*, *10*(4), 63–68.
18. Seber, G. A., & Lee, A. J. (2012). *Linear Regression Analysis* (Vol. 329). John Wiley & Sons, Hoboken, NJ.
19. Awad, M., & Khanna, R. (2015). Support vector regression. In M. Awad & R. Khanna (Eds.), *Efficient Learning Machines* (pp. 67–80). Apress, Berkeley, CA.
20. Xu, M., Watanachaturaporn, P., Varshney, P. K., & Arora, M. K. (2005). Decision tree regression for soft classification of remote sensing data. *Remote Sensing of Environment*, *97*(3), 322–336.
21. Pal, M. (2005). Random forest classifier for remote sensing classification. *International journal of remote sensing*, *26*(1), 217–222.
22. Meyer, D., Leisch, F., & Hornik, K. (2003). The support vector machine under test. *Neurocomputing*, *55*(1–2), 169–186.
23. Selvi, C., & Sivasankar, E. (2019). A novel Adaptive Genetic Neural Network (AGNN) model for recommender systems using modified k-means clustering approach. *Multimedia Tools and Applications*, *78*(11), 14303–14330.
24. Uyanık, G. K., & Güler, N. (2013). A study on multiple linear regression analysis. *Procedia-Social and Behavioral Sciences*, *106*, 234–240.
25. Medsker, L. R., & Jain, L. C. (2001). Recurrent neural networks. *Design and Applications*, *5*, 64–67.

Chapter 4

Stability analysis for a diffusive ratio-dependent predator-prey model involving two delays

Yanxia Zhang and Long Li
Chongqing University of Education

CONTENTS

4.1 INTRODUCTION

In nature, any biological population has a certain relationship with other populations. Except in the laboratory, there is no single population. It is of great significance to study the change law of population and the factors affecting the change of population quantity for the healthy development of population. Biologists and mathematicians have been using mathematical theories and methods to reveal the change law of individual number and structure of the population, predict the future development trend of the population, and strive to reflect the reality more truly through continuous revision and improvement, which not only directly promotes the development of ecology, but also has an important impact on other fields of biomathematics. People usually use differential equations to describe the evolution process of some characteristics of the actual object with time, analyze its change law, predict its future behavior, and study its control means. Of course, the relationship between biological populations can be described by differential equations, especially the predator-prey system, which plays a very important role in the dynamic system.

As a result of the universality of individual attack and attacked predator-prey phenomenon in nature, predator-prey system has always been one of the research topics in the field of population dynamics. Predation theory originates from the classical Lotka-Volterra model, which is generally expressed by differential equations [1] to present the rules followed by

DOI: 10.1201/9781003381167-4

predator and prey in a stock of fish when there is no human fishing activity. The early mathematical model is simple on account of its absence of the constraints of population density in the establishment process. With more and more precise requirements for the system, it is found that the evolution process of biological system is not only related to the current state of the sys-tembut also related to the state at a certain time or several times in the past. That is, there is a certain time lag in the influence of the state of the system in the past on the current state of the system. Therefore, it is natural to consider the time lag factor. It suggeststhat a time-delay dynamic system is more complex than the one described by ordinary differential equations, and its properties are more abundant. Even so, its dynamic properties may change qualitatively because of the influence of time delay. Therefore, in order to better reflect the change law of biological dynamic system, some predator-prey systems with delays have received extensive attention [2–7].

Later, considering that the predator population is affected by the feeding rate, many researchers [8–12] proposed the idea that bringing a ratio-dependent factor into the predator-prey system is more in line with the natural law, so the following system iswildly considered.

$$\begin{cases} \dfrac{dN}{dt} = r_1 N(t) - \varepsilon N(t)P(t), \\ \dfrac{dP}{dt} = P(t)\left(r_2 - \theta \dfrac{P(t-\tau_2)}{N(t-\tau_1)} \right), \end{cases} \tag{4.1}$$

where these constants r_1, r_2, ε and θ are positive and $\tau_1, \tau_2 \geq 0$ refer to the delays for the prey and predator, respectively. The densities of predator and prey populations at time t are symbolized as $P(t)$ and $N(t)$, respectively. $N(t-\tau_1)$ refers to the larvae of prey born at time τ_1, $P(t-\tau_2)$ represents the larvae of predators born at time τ_2, and both of them survive at time t. It means that they need to mature for a period of time to have enough ability to capture or be captured. That is to say, the predators need a certain maturity time τ_2 to obtain the capture ability, and they catch only the prey with a certain maturity time τ_1. In this system, the predator-prey relationship between two populations is described by a model of ordinary differential equations. On the one hand, the prey population has an aptness of infinite exponential growth, $r_1 N(t)$, which is restricted to predator: the influence of predators on prey is reflected by the response function $\varepsilon P(t)N(t)$. On the other hand, in addition to the natural growth of the predator population, the carrying capacity is in proportion to the number of preysthat are matureand is also affected by the number ofmature prey of each predator.

At present, some scholars have studied the stability of the solution of system (Equation 4.1) for different cases of time delays. Zhou et al. [10] showed that the equilibrium is asymptotically stable under $\tau_1 = \tau_2 = 0$. Also they took Allee effects into account to mainly analyze the influence of these effects on the stable state of the system. Celik [11,12] discussed the impact of a single delay on the stability of system when $\tau_1 = 0, \tau_2 \neq 0$ and $\tau_2 = 0, \tau_1 \neq 0$, as well as the direction and stability of Hopf bifurcation. Later, the case $\tau_1 = \tau_2 = \tau$ for model (Equation 4.1) was studied in detail by Karaoglu and Merdan [8] who regarded τ as a parameter to make stability and bifurcation analysis. However, they ignored the fact that the differences of geographical location and living environment in the nature world may lead to some spatial movements. As the distribution of populations in their environment is uneven, so are their immigration and emigration and the resources allocated to each individual. Therefore, in a population system, in addition to the evolution of time direction, we should also consider the changes in spatial direction, which can be described by the diffusion system. It is found that diffusion can be used to simulate the migration of population from a crowded region to a few and scattered region and Laplace operation Δ is usually used to characterize this diffusion state in the ordinary differential equation system [13–15]. Motivated by this, a ratio-dependent delayed predator-prey model with reaction-diffusion under Neumann boundary condition is mainly considered:

$$\begin{cases} N_t = d_1 \Delta N(t,x) + r_1 N(t,x) - \varepsilon P(t,x) N(t,x), \ x \in \Omega, t > 0, \\ P_t = d_2 \Delta P(t,x) + P(t,x)\left(r_2 - \theta \dfrac{P(t-\tau_2,x)}{N(t-\tau_1,x)} \right), \ x \in \Omega, t > 0, \\ \dfrac{\partial N(t,x)}{\partial \upsilon} = \dfrac{\partial P(t,x)}{\partial \upsilon} = 0, \ x \in \partial\Omega, t > 0, \\ N(t,x) = \Phi(t,x), P(t,x) = \Psi(t,x), \ x \in \overline{\Omega}, t \in \left[-\max\{\tau_1,\tau_2\}, 0\right], \end{cases} \tag{4.2}$$

where $P(t,x)$ and $N(t,x)$ are the densities of predators and prey at time t and position x, respectively. The two positive constants, d_1 and d_2, denote diffusion coefficients of the two populations, respectively. The domain Ω is bounded in $R^N (N \geq 1)$, andυrefers to an outward unit normal vector on smooth boundary $\partial\Omega . \Phi(t,x)$ and $\Psi(t,x)$ are nonnegative and Holder continuous initial conditions. For convenience, we assume $\Omega = (0,\pi)$ and regard $\tau_1 = \tau_2 = \tau$ as bifurcation parameter.

The formation of this work is as follows. The stability conditions are considered for system (Equation 4.2) in Section 4.2 and some theorems about the stability and the existence of Hopf bifurcation are also presented. In Section 4.3, some particular systems and simulations are displayed to support the theoretical results. Finally, a conclusion and prospect are given in Section 4.4 to end this work.

4.2 STABILITY ANALYSIS

In the current section, we concentrate on the stability analysis and the existence of Hopf bifurcation of system (Equation 4.2). First, we calculate the coexisting positive equilibrium, and then derive the linear approximate equation of system (Equation 4.2). Next, we analyze the distribution of roots of the characteristic equation to the linearized model. Finally, we obtain some conditions to ensure the stability and Hopf bifurcating existence of the system near the equilibrium.

According to equations

$$\begin{cases} r_1 N - \varepsilon PN = 0, \\ P\left(r_2 - \theta \dfrac{P}{N}\right) = 0, \end{cases} \tag{4.3}$$

the unique coexisting steady-state point $E(N^*, P^*) = \left(\dfrac{r_1\theta}{r_2\varepsilon}, \dfrac{r_1}{\varepsilon}\right)$ of system (Equation 4.2) is easily obtained. We denote $f(u,v) = r_1 u - \varepsilon uv$ and $g(u,v,w) = u\left(r_2 - \theta \dfrac{v}{w}\right)$ for system (Equation 4.2). It is clear that $f(N^*, P^*) = 0$, $g(P^*, P^*, N^*) = 0$, and easy to calculate the following partial derivatives at point $E(N^*, P^*)$

$$\left.\frac{\partial f}{\partial u}\right|_E = r_1 - \varepsilon P^* = 0, \left.\frac{\partial f}{\partial v}\right|_E = -\varepsilon N^*,$$

$$\left.\frac{\partial g}{\partial u}\right|_E = r_2 - \theta\frac{P^*}{N^*} = 0, \left.\frac{\partial g}{\partial v}\right|_E = -\theta\frac{P^*}{N^*}, \left.\frac{\partial g}{\partial w}\right|_E = \theta\frac{\left(P^*\right)^2}{\left(N^*\right)^2}.$$

By using Taylor formula for functions f and g at the point E, their linearized functions are expressed as $f \approx -\varepsilon N^* v$ and $g \approx -\theta \dfrac{P^*}{N^*} v + \theta \dfrac{\left(P^*\right)^2}{\left(N^*\right)^2} w$. So, we can get the following linear approximate equation of system (Equation 4.2) at the equilibrium

$$\begin{pmatrix} \dfrac{\partial N}{\partial t} \\ \dfrac{\partial P}{\partial t} \end{pmatrix} = D\Delta\begin{pmatrix} N(t) \\ P(t) \end{pmatrix} + L_1\begin{pmatrix} N(t) \\ P(t) \end{pmatrix} + L_2\begin{pmatrix} N(t-\tau) \\ P(t-\tau) \end{pmatrix}, \tag{4.4}$$

where

$$D=\begin{pmatrix} d_1 & 0 \\ 0 & d_2 \end{pmatrix}, L_1=\begin{pmatrix} 0 & -\varepsilon N^* \\ 0 & 0 \end{pmatrix}, L_2=\begin{pmatrix} 0 & 0 \\ \theta\frac{(P^*)^2}{(N^*)^2} & -\theta\frac{P^*}{N^*} \end{pmatrix}.$$

The characteristic equation of system (Equation 4.4) is

$$\det\left(\lambda I_2 - M_n - L_1 - L_2 e^{-\lambda\tau}\right)=0, \tag{4.5}$$

where

$$I_2=\begin{pmatrix} 1 & 0 \\ 0 & 1 \end{pmatrix}, M_n=-\frac{n^2}{l^2}\begin{pmatrix} d_1 & 0 \\ 0 & d_2 \end{pmatrix}, n\in N_0=\{0,1,2,\ldots\}.$$

Then, the following characteristic equation is hold through deducing (Equation 4.5)

$$\lambda^2 + A_n\lambda + B_n + \left(C_n + r_2\lambda\right)e^{-\lambda\tau}=0, n\in N_0, \tag{4.6}$$

where $A_n=(d_1+d_2)\frac{n^2}{l^2}\geq 0$, $B_n=\frac{n^4}{l^4}d_1d_2\geq 0$, and $C_n=\frac{n^2}{l^2}d_1r_2+\varepsilon r_2P^*>0$.

Note that $r_2=\theta\frac{P^*}{N^*}>0$ by equation (Equation 4.3). The roots of (Equation 4.6) are used to determine the stability of system (Equation 4.2); in other words, the system (Equation 4.2) is stable at equilibrium if all eigenvalues of (Equation 4.6) have negative real parts, and it is unstable if at least one eigenvalue has a positive real part [8,12,13,16].

Obviously, $\lambda=0$ is not the root of (Equation 4.6) for $\forall n\in N_0$. Next, we discuss the stability of $E(N^*,P^*)$ for the two different cases $\tau=0$ and $\tau>0$.

Case 1 $\tau=0$

In this case, equation (Equation 4.6) can be reduced to

$$\lambda^2 + \left(A_n + r_2\right)\lambda + B_n + C_n = 0, n\in N_0. \tag{4.7}$$

Denote $\lambda_{1,2}^{(n)}$ are eigenvalues of equation (Equation 4.7), then $\lambda_1^{(n)}+\lambda_2^{(n)}=-\left(A_n+r_2\right)<0$ and $\lambda_1^{(n)}\lambda_2^{(n)}=B_n+C_n>0$, then all eigenvalues of (Equation 4.7) always have negative real parts, which implies $E(N^*,P^*)$ is locally asymptotically stable. Hence, the following conclusion is obtained.

Theorem 4.1

For system (equation 4.2), the equilibrium $E(N^*, P^*)$ is asymptotically stable when $\tau = 0$.

Case 2 $\tau > 0$

Let $\lambda = iw(w > 0)$ be a root of (Equation 4.6). It can be obtained by substituting $\lambda = iw$ into (Equation 4.6) and performing a simple calculation

$$\begin{cases} C_n \cos \omega\tau + \omega r_2 \sin \omega\tau = \omega^2 - B_n, \\ -C_n \sin \omega\tau + \omega r_2 \cos \omega\tau = -A_n\omega. \end{cases} \tag{4.8}$$

Taking square on both of (Equation 4.8) and adding them up, it leads to the following equation

$$\omega^4 + \left(A_n^2 - 2B_n - r_2^2\right)\omega^2 + B_n^2 - C_n^2 = 0. \tag{4.9}$$

Set $z = \omega^2$, (Equation 4.9) is written as

$$z^2 + \left(A_n^2 - 2B_n - r_2^2\right)z + B_n^2 - C_n^2 = 0. \tag{4.10}$$

Assume $z_n^{\pm}$ are the roots of (Equation 4.10), then

$$z_n^{\pm} = \frac{1}{2}\left[-\left(A_n^2 - 2B_n - r_2^2\right) \pm \sqrt{\Delta_n}\,\right],$$

where

$$\Delta_n = \left(A_n^2 - 2B_n - r_2^2\right)^2 - 4\left(B_n^2 - C_n^2\right).$$

Note that

$$z_n^+ + z_n^- = -\left(A_n^2 - 2B_n - r_2^2\right) = -\left[\left(d_1^2 + d_2^2\right)\frac{n^4}{l^4} - r_2^2\right],$$

$$z_n^+ z_n^- = B_n^2 - C_n^2 = (B_n + C_n)(B_n - C_n) = (B_n + C_n)\left(\frac{n^4}{l^4}d_1d_2 - \frac{n^2}{l^2}d_1r_1 - \varepsilon r_2P^*\right).$$

Denote $\omega_n^{\pm} = \sqrt{z_n^{\pm}}$, and according to (Equation 4.8), we can obtain

$$\tau_n^{0,\pm} = \frac{1}{\omega_n^{\pm}} \arccos \frac{(C_n - A_n r_2)\left(\omega_n^{\pm}\right)^2 - B_n C_n}{C_n^2 + r_2^2 \left(\omega_n^{\pm}\right)^2},$$

$$\tau_n^{j,\pm} = \tau_n^{0,\pm} + \frac{2j\pi}{\omega_n^{\pm}} (j \in N_0).$$

Here, we define the following sets to discuss the roots of (Equation 4.10).

$$S_1 = \{n \mid B_n - C_n < 0, n \in N_0\},$$

$$S_2 = \{n \mid B_n - C_n > 0, A_n^2 - 2B_n - r_2^2 < 0, \Delta_n > 0, n \in N_0\},$$

$$S_3 = \{n \mid B_n - C_n > 0, A_n^2 - 2B_n - r_2^2 > 0, \Delta_n > 0, n \in N_0\},$$

$$S_4 = \{n \mid B_n - C_n > 0, \Delta_n < 0, n \in N_0\}.$$

Since $B_0 - C_0 = -\varepsilon r_2 P^* < 0$, S_1 is a nonempty set. In particular, when $n = 0$, the discussion of the root of (Equation 4.6) is detailed in [8]. More generally, the following conclusion is held.

Theorem 4.2

For the transcendental equation (4.6), the following conclusions are true.

i. If $n \in S_1$, then $\pm i\omega_n^+$ are two pure imaginary roots of (Equation 4.6) at $\tau_n^{j,+} (j \in N_0)$.
ii. If $n \in S_2$, then $\pm i\omega_n^{\pm}$ are two pairs of pure imaginary roots of (Equation 4.6) at $\tau_n^{j,\pm} (j \in N_0)$.
iii. If $n \in S_3 \cup S_4$, then (Equation 4.6) has no pure imaginary root.

Proof:

i. When $n \in S_1$, we have $z_n^+ z_n^- < 0$, which implies that (Equation 4.10) has a positive root z_n^+. So (Equation 4.6) has two pure imaginary roots $\pm i\omega_n^+$ at $\tau = \tau_n^{j,+} (j \in N_0)$, where $\omega_n^+ = \sqrt{z_n^+}$.
ii. When $n \in S_2$, we have $z_n^+ + z_n^- > 0$ and $z_n^+ z_n^- > 0$, which implies that (Equation 4.10) has two positive roots $z_n^{\pm} > 0$. So $\pm i\omega_n^{\pm}$ are two pairs

of pure imaginary roots of (Equation 4.6) at $\tau = \tau_n^{j,\pm} (j \in N_0)$, where $\omega_n^{\pm} = \sqrt{z_n^{\pm}}$.

iii. If $n \in S_3$, then $z_n^+ + z_n^- < 0$ and $z_n^+ z_n^- > 0$, we can know that the two real roots of (Equation 4.10) are $z_n^{\pm} < 0$. So for (Equation 4.6), there is no pure imaginary root. Similarly, if $n \in S_3$, there is no such root for (Equation 4.6).

Theorem 4.3

The following results are true for $n \in S_1 \cup S_2, j \in N_0$.

$$\mathrm{Re}\left(\frac{d\lambda}{d\tau}\right)_{\tau=\tau_n^{j,+}} > 0, \mathrm{Re}\left(\frac{d\lambda}{d\tau}\right)_{\tau=\tau_n^{j,-}} < 0,$$

that is, it is satisfied for transversality condition.

Proof: Differentiating (Equation 4.6) with respect to τ yields to

$$\frac{d\lambda}{d\tau} = \frac{\lambda^2 r_2 e^{-\lambda\tau} + C_n \lambda e^{-\lambda\tau}}{2\lambda + A_n + r_2 e^{-\lambda\tau} - (C_n + \lambda r_2) e^{-\lambda\tau} \tau},$$

thus,

$$\left(\frac{d\lambda}{d\tau}\right)^{-1} = \frac{(2\lambda + A_n) e^{\lambda\tau} + r_2}{r_2 \lambda^2 + C_n \lambda} - \frac{\tau}{\lambda}.$$

From Equations (4.8) and (4.10), and further calculating, we obtain

$$\mathrm{Re}\left[\left(\frac{d\lambda}{d\tau}\right)^{-1}\right]_{\tau=\tau_n^{j,\pm}} = \mathrm{Re}\left[\frac{(2\lambda + A_n) e^{\lambda\tau} + r_2}{r_2 \lambda^2 + C_n \lambda} - \frac{\tau}{\lambda}\right]_{\substack{\lambda = i\omega, \\ \omega = \omega_n^{\pm}, \tau = \tau_n^{j,\pm}}}$$

$$= \left[\frac{2\omega^2 + A_n^2 - 2B_n - r_2^2}{r_2^2 \omega^2 + C_n^2}\right]_{\omega=\omega_n^{\pm}, \tau=\tau_n^{j,\pm}}$$

$$= \pm\left[\frac{1}{r_2^2 \omega^2 + C_n^2} \sqrt{\Delta_n}\right]_{\omega=\omega_n^{\pm}, \tau=\tau_n^{j,\pm}}$$

Therefore, $\mathrm{Re}\left(\frac{d\lambda}{d\tau}\right)_{\tau=\tau_n^{j,+}} > 0$ and $\mathrm{Re}\left(\frac{d\lambda}{d\tau}\right)_{\tau=\tau_n^{j,-}} < 0$ hold.

Denote $\tau^* = \min\left\{\tau_n^{0,\pm} \; or \; \tau_n^{0,+} \mid n \in S_1 \cup S_2\right\}$. Based on the above analysis, the following conclusions are obtained.

Theorem 4.4

For system (Equation 4.2), the following results are true.

i. *If $S_1 \cup S_2 = \varnothing$, then $E(N^*, P^*)$ is locally asymptotically for all $\tau \geq 0$.*
ii. *If $S_1 \cup S_2 \neq \varnothing$, then $E(N^*, P^*)$ is locally asymptotically for $\tau \in [0, \tau^*)$.*
iii. *It appears a Hopf bifurcation at steady-state point $E(N^*, P^*)$ when $\tau = \tau_n^{j,+}(\tau_n^{j,-})$, $j \in N_0$, $n \in S_1 \cup S_2$.*

4.3 EXAMPLES AND SIMULATIONS

Now, some particular systems and simulations are shown to support the validity of the above-mentioned theorems.

Let $d_1 = d_2 = 0.1$, $r_1 = 0.15$, $r_2 = 1$, $\varepsilon = 0.003$, $\theta = 0.005$, then the system (Equation 4.2) with the following initial conditions is given by

$$\begin{cases} N_t = 0.1\Delta N(t,x) + 0.15N(t,x) - 0.003P(t,x)N(t,x), \; x \in (0,\pi), \; t > 0, \\ P_t = 0.1\Delta P(t,x) + P(t,x)\left(1 - 0.005\dfrac{P(t-\tau,x)}{N(t-\tau,x)}\right), \; x \in (0,\pi), \; t > 0, \\ \dfrac{\partial N(t,x)}{\partial v} = \dfrac{\partial P(t,x)}{\partial v} = 0, \; x = 0,\pi, t > 0, \\ \Phi(t,x) = 0.25 \times (1 + 0.3\sin(1.4x - 0.6) + 0.2\sin(3.7x)), x \in (0,\pi), t \in [-\tau, 0], \\ \Psi(t,x) = 50 \times (1 + 0.3\sin(0.74x + 0.5) + 0.2\sin(2.7x)), x \in (0,\pi), t \in [-\tau, 0]. \end{cases} \tag{4.11}$$

The unique coexisting steady-state point of system (Equation 4.11) is $E(N^*, P^*) = (0.25, 50)$. We fix $l = 2$ and $n = 0$ in simulations.

When $\tau = 0$, the two roots of (Equation 4.7) are $\lambda_1 = \lambda_2 = -0.5$. By Theorem 4.1, the equilibrium E is asymptotically stable, asillustrated in Figures 4.1 and 4.2, the solutions of system (Equation 4.11) with $\tau = 0$ are always converging to the steady-state point $E(0.25, 50)$. When $\tau > 0$, one can obtain the parameter $\omega = 1.0109$ and the critical value $\tau^* = 1.4081$

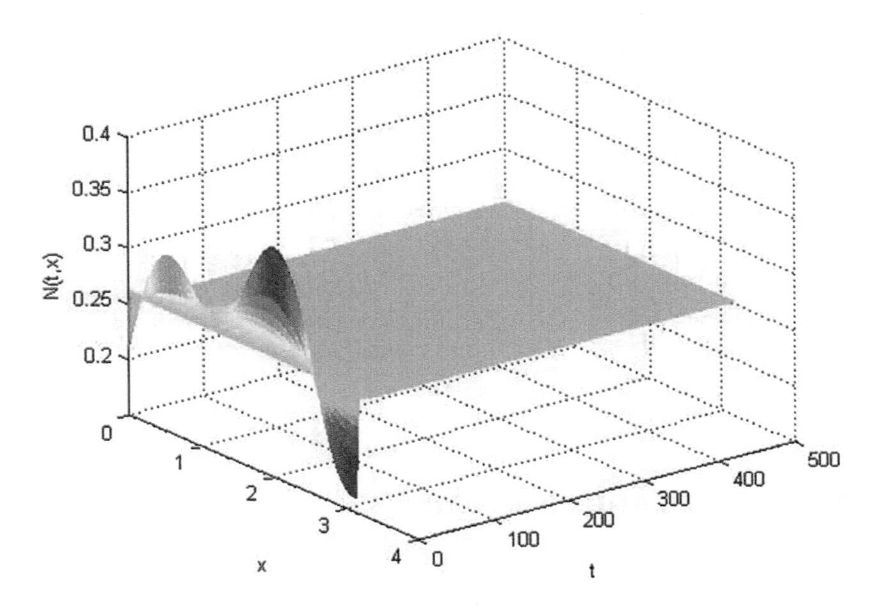

Figure 4.1 The solution $N(t, x)$ of system (Equation 4.11) with $\tau = 0$ is always converging to the steady-state point $N^* = 0.25$.

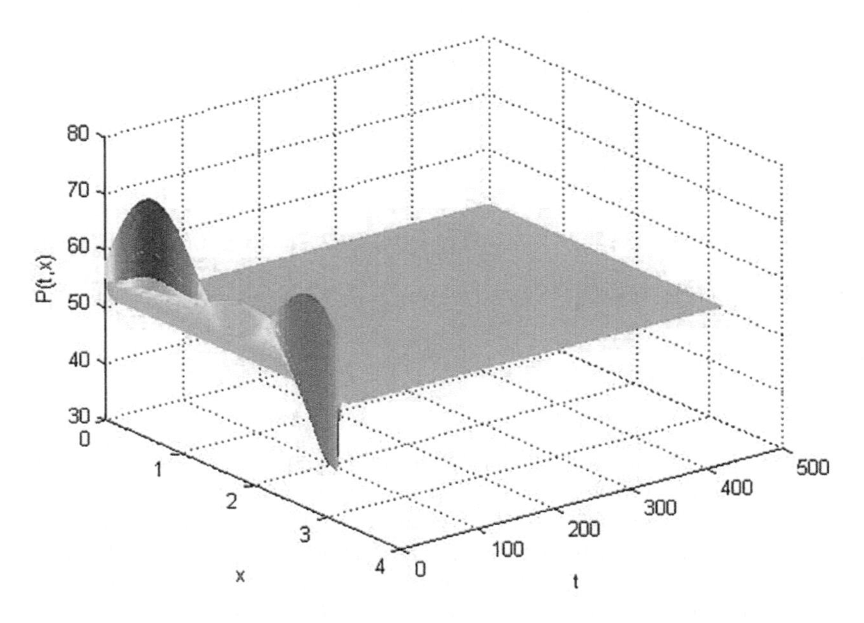

Figure 4.2 The solution $P(t, x)$ of system (Equation 4.11) with $\tau = 0$ is always converging to the steady-state point $P^* = 50$.

through some numerical calculation, that is, (Equation 4.6) has two pure imaginary roots $\pm i\omega$ by Theorem 4.2. According to Theorem 4.3, the transversality condition $\mathrm{Re}\left[\left(\mathrm{d}\lambda / \mathrm{d}\tau\right)^{-1}\right]_{\tau=\tau^*} = 0.9995 > 0$ is satisfied, and $n \in S_1 \cup S_2 \neq \varnothing$. We change the value of τ for (Equation 4.11) and set $\tau = 1.40 < \tau^*$ and $\tau = 1.41 > \tau^*$, respectively. When $\tau = 1.40 < \tau^*$, E is asymptotically stable, as depicted in Figures 4.3 and 4.4, the solutions of system (Equation 4.11) are converging to the stable state $E(0.25, 50)$. When τ exceeds its critical value τ^*, E losses its stability and system (Equation 4.11) occurs Hopf bifurcation, which are depicted in Figures 4.5–4.8. We find that there are a series of periodic solutions branching from the coexisting equilibrium E. These phenomena show that the change of time delay has an important impact on the stability of a system.

To demonstrate the effect of diffusion on the spatial distributions of predator and prey population, we fix $d_1 = 0.1$, set $d_2 = 0.001$, and other parameters remain unchanged, including l and n. The system (Equation 4.11) becomes

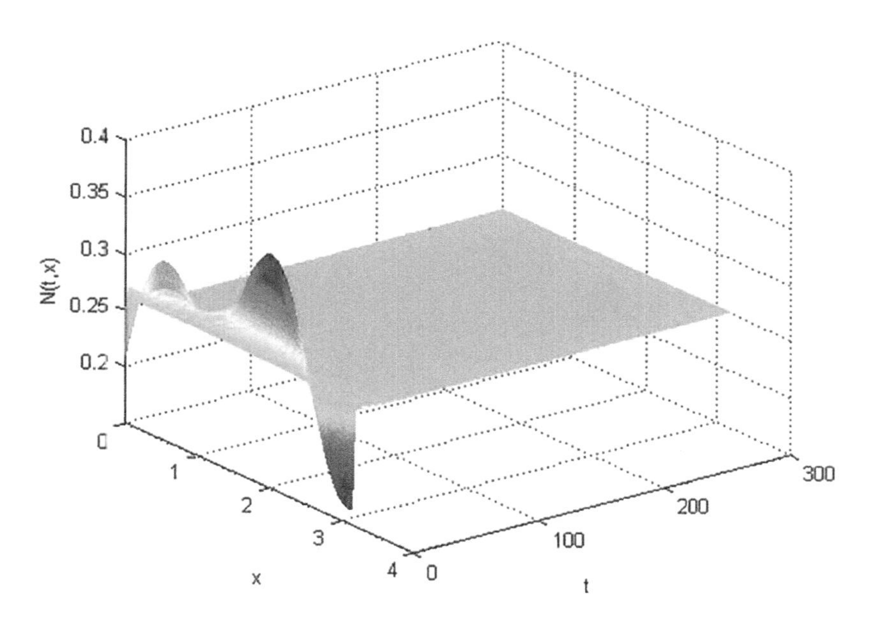

Figure 4.3 The solution $N(t, x)$ of system (Equation 4.11) is converging to the stable state $N^* = 0.25$ with $\tau = 1.40 < \tau^* = 1.4081$, where $d_1 = d_2 = 0.1$.

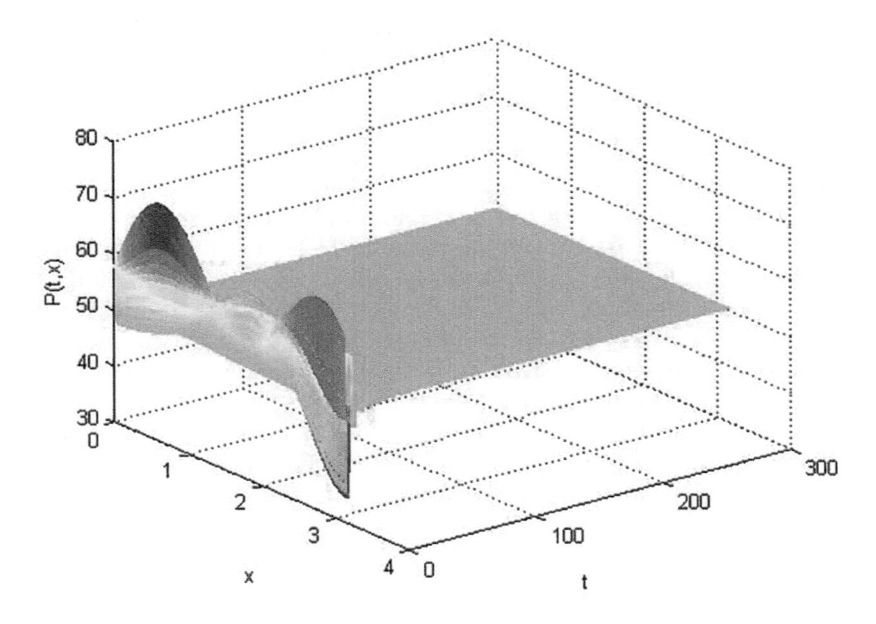

Figure 4.4 The solution $P(t, x)$ of system (Equation 4.11) is converging to the stable state $P^* = 50$ with $\tau = 1.40 < \tau^* = 1.4081$, where $d_1 = d_2 = 0.1$.

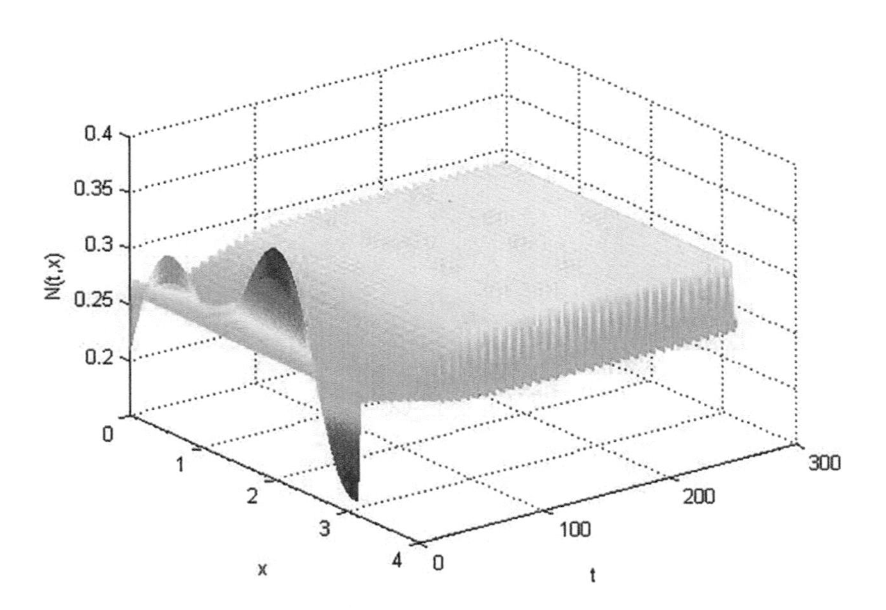

Figure 4.5 The solution $N(t, x)$ of system (Equation 4.11) with $\tau = 1.41 > \tau^* = 1.4081$. The equilibrium $N^* = 0.25$ loses stability and periodic solution occurs, where $d_1 = d_2 = 0.1$.

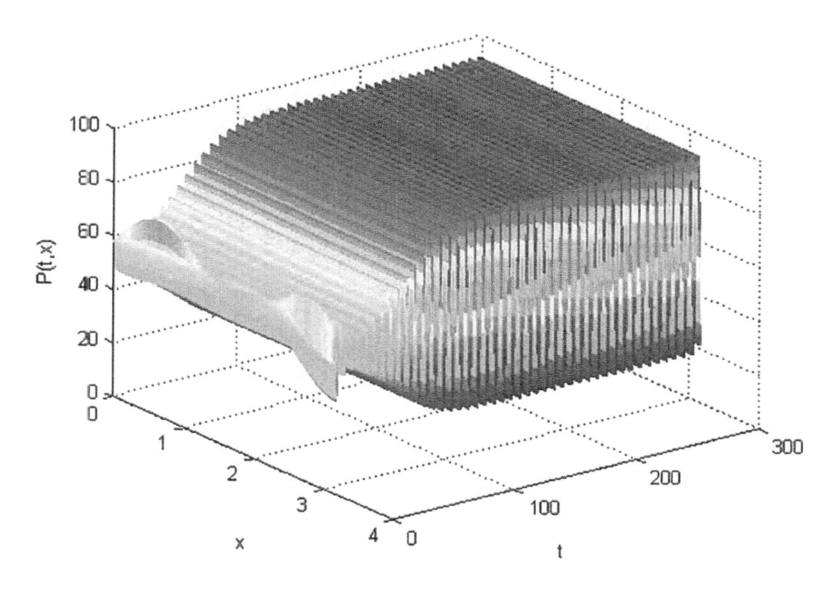

Figure 4.6 The solution $P(t,x)$ of system (Equation 4.11) with $\tau = 1.41 > \tau^* = 1.4081$. The equilibrium $P^* = 50$ loses stability and periodic solution occurs, where $d_1 = d_2 = 0.1$.

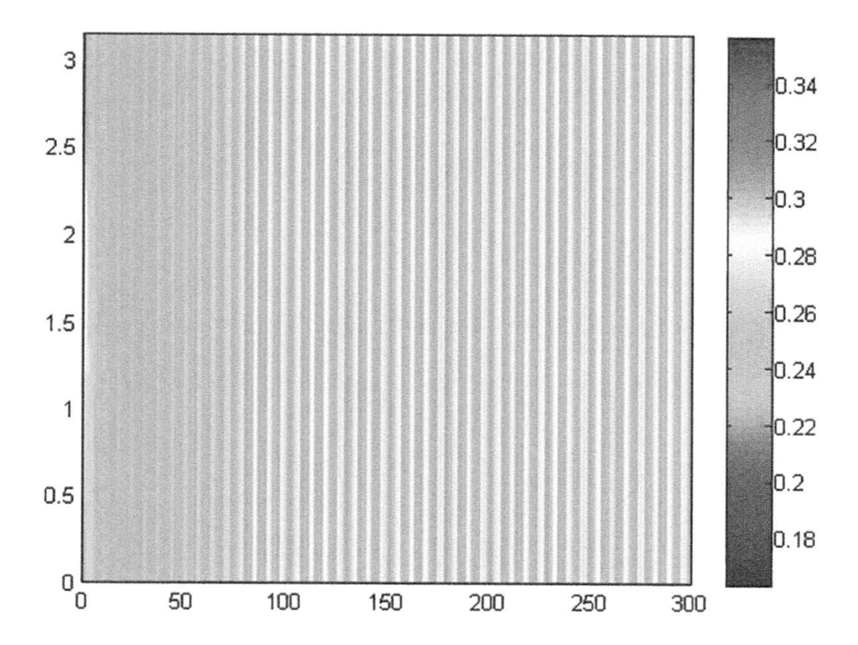

Figure 4.7 The spatial state of prey population of system (Equation 4.11) with $\tau = 1.41 > \tau^* = 1.4081$. Bifurcating periodic solution occurs, where $d_1 = d_2 = 0.1$.

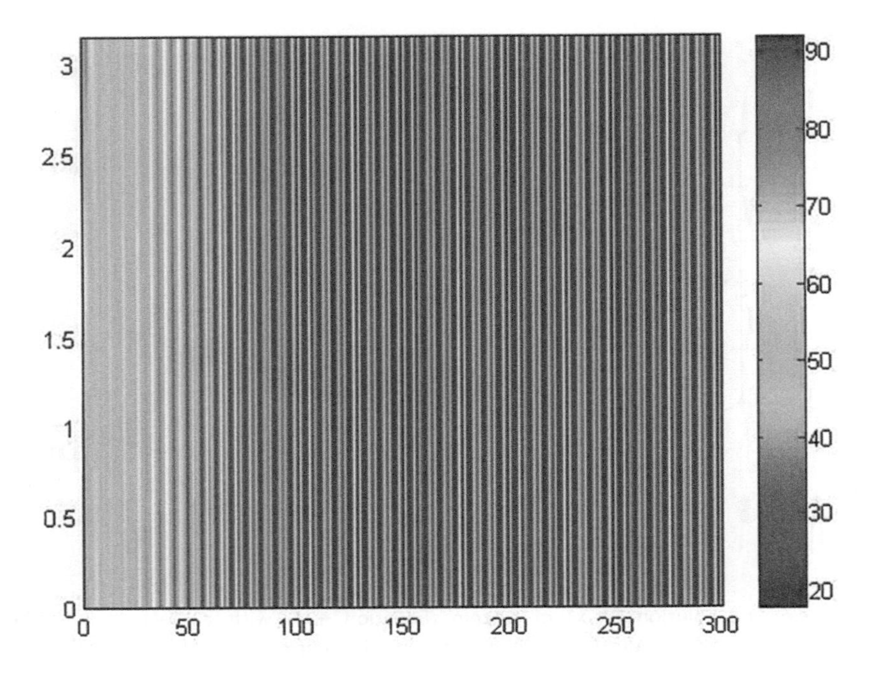

Figure 4.8 The spatial state ofpredator population of system (Equation 4.11) with $\tau = 1.41 > \tau^* = 1.4081$. Bifurcating periodic solution occurs, where $d_1 = d_2 = 0.1$.

$$\begin{cases} N_t = 0.1\Delta N(t,x) + 0.15N(t,x) - 0.003P(t,x)N(t,x),\ x \in (0,\pi),\ t > 0, \\ P_t = 0.001\Delta P(t,x) + P(t,x)\left(1 - 0.005\dfrac{P(t-\tau,x)}{N(t-\tau,x)}\right),\ x \in (0,\pi),\ t > 0, \\ \dfrac{\partial N(t,x)}{\partial v} = \dfrac{\partial P(t,x)}{\partial v} = 0,\ x = 0,\pi, t > 0, \\ \Phi(t,x) = 0.25 \times (1 + 0.3\sin(1.4x - 0.6) + 0.2\sin(3.7x)),\ x \in (0,\pi),\ t \in [-\tau, 0], \\ \Psi(t,x) = 50 \times (1 + 0.3\sin(0.74x + 0.5) + 0.2\sin(2.7x)),\ x \in (0,\pi),\ t \in [-\tau, 0]. \end{cases} \tag{4.12}$$

The steady-state point E is still $(0.25, 50)$. Some corresponding simulations of system (Equation 4.12) are shown in Figures 4.9–4.14. When $\tau = 1.40 < \tau^*$, the solutions of (Equation 4.12) are converging to the stable state, as depicted in Figures 4.9 and 4.10. When $\tau = 1.41 > \tau^*$, the solutions

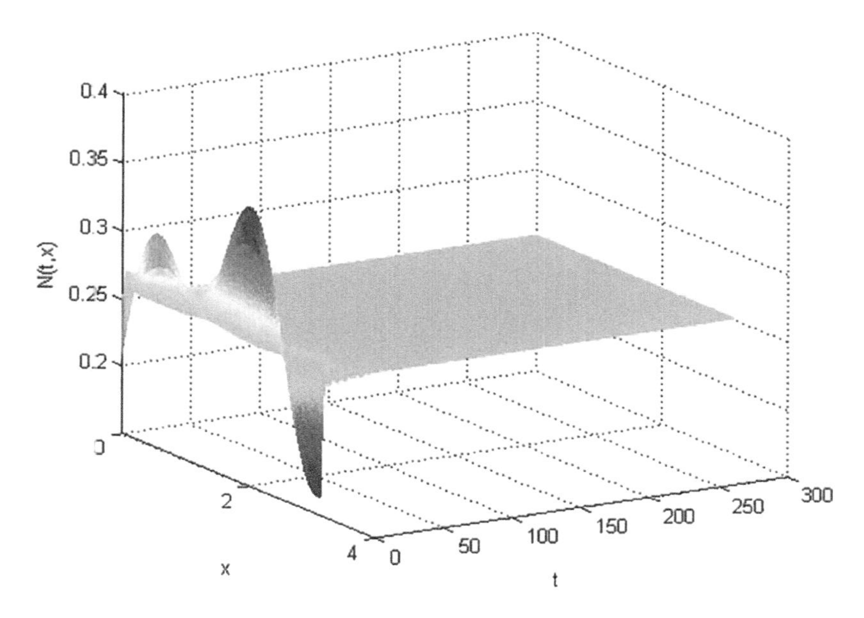

Figure 4.9 The solution $N(t, x)$ of system (Equation 4.12) is converging to the stable state $N^* = 0.25$ with $\tau = 1.40 < \tau^* = 1.4081$, where $d_1 = 0.1$, $d_2 = 0.001$.

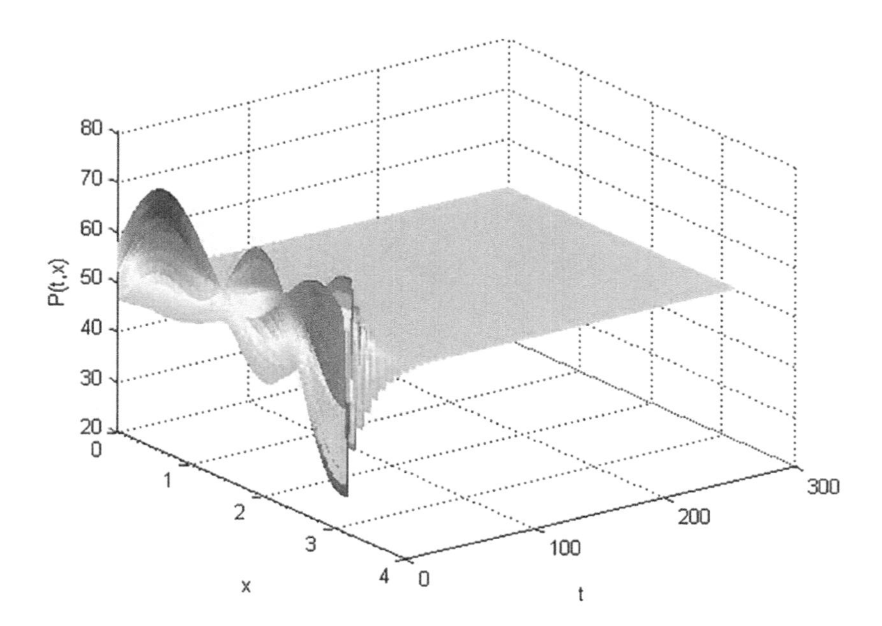

Figure 4.10 The solution $P(t, x)$ of system (Equation 4.12) is converging to the stable state $P^* = 50$ with $\tau = 1.40 < \tau^* = 1.4081$, where $d_1 = 0.1$, $d_2 = 0.001$.

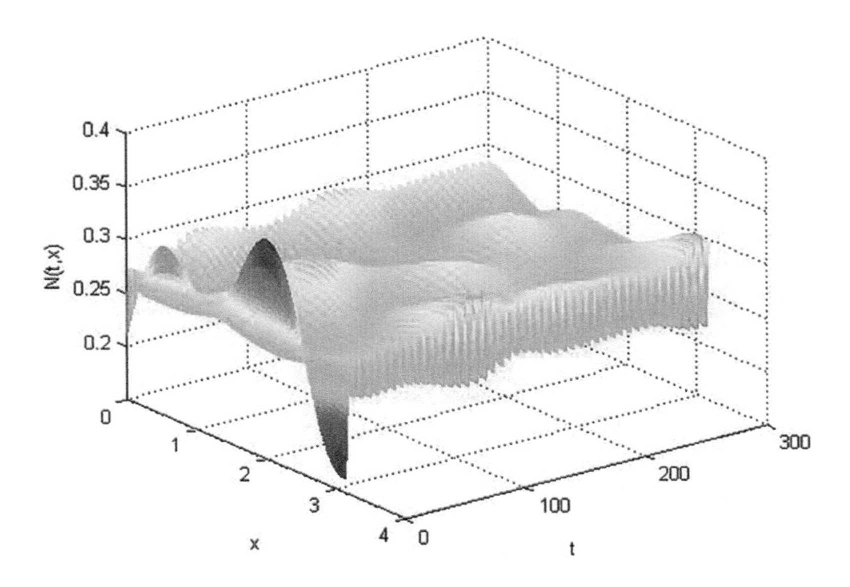

Figure 4.11 The solution $N(t,x)$ of system (Equation 4.12) with $\tau = 1.41 > \tau^* = 1.4081$. The equilibrium $N^* = 0.25$ loses stability and periodic solution occurs, where $d_1 = 0.1$, $d_2 = 0.001$.

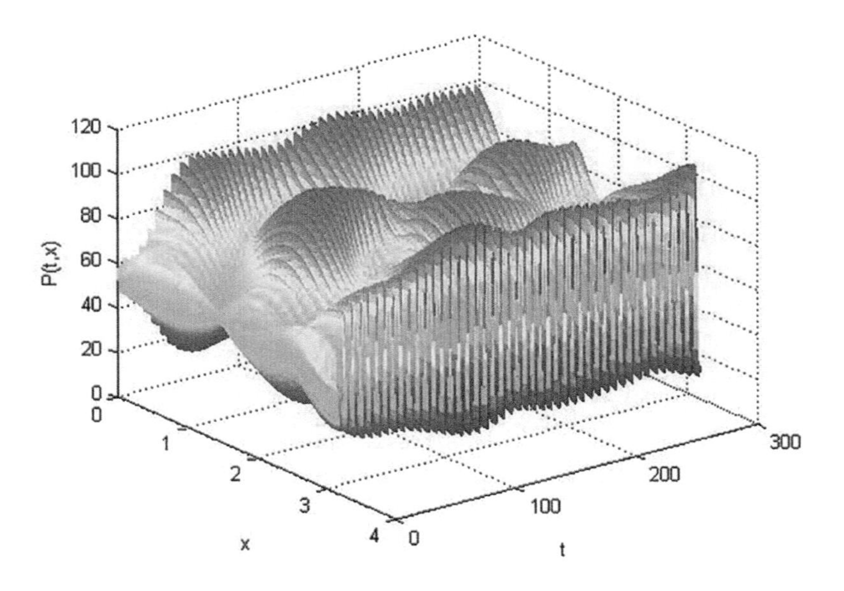

Figure 4.12 The solution $P(t,x)$ of system (Equation 4.12) with $\tau = 1.41 > \tau^* = 1.4081$. The equilibrium $P^* = 50$ loses stability and periodic solution occurs, where $d_1 = 0.1$, $d_2 = 0.001$.

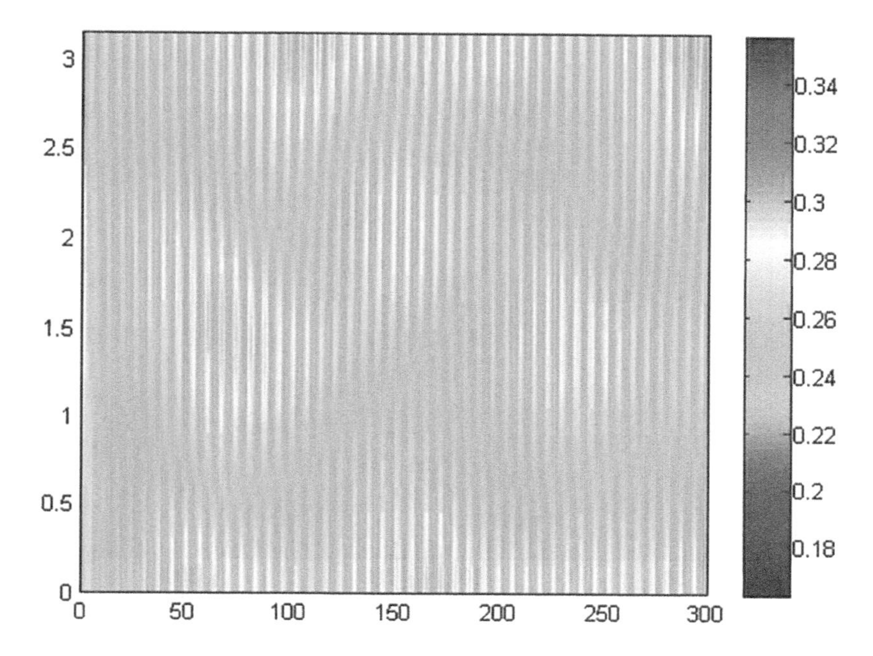

Figure 4.13 The spatial state of prey population of system (Equation 4.12) with $\tau = 1.41 > \tau^* = 1.4081$. Bifurcating periodic solution occurs and the density of this population has richer spatial distribution, where $d_1 = 0.1$, $d_2 = 0.001$.

of (Equation 4.12) are unstable and periodic solutions bifurcating from the equilibrium E arise with more abundant spatial features because the time delay exceeds its critical point, as shown in Figures 4.11–4.14.

4.4 CONCLUSION

Based on the ratio-dependent delayed prey–predator model in the previous article [8], we further consider the factors of diffusion and delays and propose a ratio-dependent prey–predator system with diffusive factors and delays in this work. We regard the two delays as thesame certain maturation time τ, that is to say, the premise that a predator has the ability to capture is that it must be mature enough, and a young prey is protected from capture until it is an adult. The two cases $\tau = 0$ and $\tau > 0$ are discussed byanalyzing the distribution of the eigenvalues. When $\tau = 0$, the equilibrium E is stable, which indicates that the two species maintain an ideal stable without time delay. When $\tau > 0$, some conditions that make the system stable are considered and its stability interval is obtained. On the one hand, when $S_1 \cup S_2 = \varnothing$ hold, the coexisting steady-state point E is

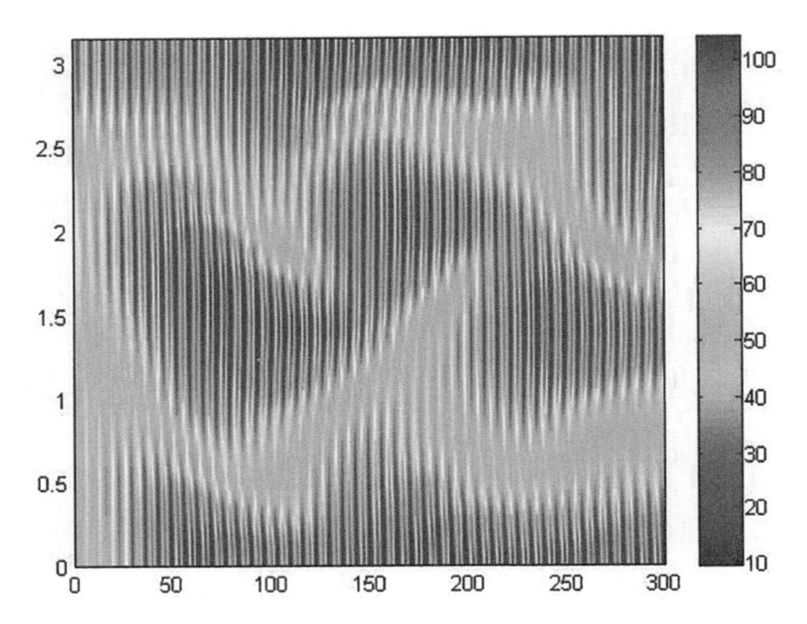

Figure 4.14 The spatial state of predator population of system (Equation 4.12) with $\tau = 1.41 > \tau^* = 1.4081$. Bifurcating periodic solution occurs and the density of this population has richer spatial distribution, where $d_1 = 0.1$, $d_2 = 0.001$.

asymptotically stable for all $\tau \geq 0$. On the other hand, when $S_1 \cup S_2 \neq \emptyset$, the coexisting equilibrium E is asymptotically stable or $0 \leq \tau < \tau^*$ and a Hopf bifurcation takes place at E. Furthermore, for the designated system, we implement corresponding numerical computations and simulations to demonstrate the theoretical results. Besides, the spatial motion behavior of the population is captured. All of these indicate that when we change two diffusion coefficients, the stability of the system remains unchanged because the delay is in its stability interval. Once the delay exceeds its critical value, the bifurcating periodic solution will produce different spatial distribution states with the change of diffusions. One can see that time delay is a great deal to the stability analysis of population dynamic systems and diffusion effect plays a key role in the study of population spatial motion behavior.

Predator–prey model is an important content of mathematical ecology. There are many factors affecting the fluctuation of two populations, of which time delay and diffusion are two important ones. The previous conclusions show that time delay and diffusion play an important role in the properties of system dynamics. In the long-term co-evolution process, the population has gradually formed ecological countermeasures to adapt to the environment. Among them, time delay and diffusion are two forms of species-specific ecological countermeasures. By adjusting the time lag, the population can reach the equilibrium state and make the biological system develop continuously and healthily, which is of practical significance for ecological protection. This work only studies the stability of the modeland does not discuss the branching phenomenon of the system. As a very important nonlinear phenomenon, Hopf bifurcation will lead to the loss of the steady development state of a system, and even cause an inestimable harmful impact on it. Therefore, how to delay or control the Hopf bifurcation for a biological dynamical system will be our further discussion topic in the future.

ACKNOWLEDGMENT

This research is supported by Chongqing Big Data Engineering Laboratory for Children, by Chongqing Electronics Engineering Technology Research Center for Interactive Learning, by Chongqing University Innovation Research Group, by Chongqing key discipline of electronic information, by Chongqing Natural Science Foundation Project (CSTC2021-msxm1993), by the program of Chongqing University of Education (KY202116C) and by the Science and Technology Research Program of Chongqing Municipal Education Commission (KJQN202101614).

REFERENCES

1. Arditi, R., Ginzburg, L. R. Coupling in predator prey dynamics: ratio-dependence. *Journal of Theoretical Biology* 139, 311–326 (1989).
2. He, X. Z. Stability and delays in a predator–prey system. *Journal of Mathematical Analysis & Applications* 198(2), 355–370 (1996).
3. Sun, C., Han, M., Lin, Y., et al. Global qualitative analysis for a predator–prey system with delay. *Chaos, Solitons & Fractals* 32(4), 1582–1596 (2007).
4. Yan, X. P., Chu, Y. D. Stability and bifurcation analysis for a delayed Lotka–Volterra predator–prey system. *Journal of Computational and Applied Mathematics* 196(1), 198–210 (2006).
5. Xu, R. Global stability and Hopf bifurcation of a predator-prey model with stage structure and delayed predator response. *Nonlinear Dynamics* 67(2), 1683–1693 (2012).
6. Kar, T. K., Ghorai, A. Dynamic behaviour of a delayed predator-prey model with harvesting. *Applied Mathematics and Computation* 217(22), 9085–9104 (2011).
7. Deng, L. W., Wang, X. D. et al. Hopf bifurcation analysis for a ratio-dependent predator-prey system with two delays and stage structure for the predator. *Applied Mathematics and Computation* 231(15), 214–230 (2014).
8. Karaoglu, E., Merdan, H. Hopf bifurcation analysis for a ratio-dependent predator-prey system involving two delays. *The ANZIAM journal: the Australian & New Zealand Industrial and Applied Mathematics Journal* 55(3), 214–231 (2014).

9. Leslie, P. H. Some further notes on the use of matrices in population mathematics. *Biometrika* 35(3–4), 213–245 (1948).
10. Zhou, S., Liu, Y., Wang, G. The stability of predator–prey systems subject to the Allee effects. *Theoretical Population Biology* 67, 23–31 (2005).
11. Celik, C. The stability and Hopf bifurcation for a predator–prey system with time delay. *Chaos Solitons and Fractals* 37(1), 87–99 (2008).
12. Celik, C. Hopf bifurcation of a ratio-dependent predator-prey system with time delay. *Chaos Solitons and Fractals* 42(3), 1474–1484 (2009).
13. Liu, F., Yang, R., Tang, L. Hopf bifurcation in a diffusive predator-prey model with competitive interference. *Chaos, Solitons and Fractals* 120, 250–258 (2019).
14. Yang, R. Bifurcation analysis of a diffusive predator-prey system with Crowley-Martin function response and delay. *Chaos, Solitons and Fractals* 95, 131–139 (2017).
15. Rao, F, Castillo, C., Kang, Y. Dynamics of a diffusion reaction prey–predator model with delay in prey: Effects of delay and spatial 95: components. *Journal of Mathematical Analysis and Applications* 461, 1177–1214 (2018).
16. Li, L., Mei, Y., Cao, J. Hopf bifurcation analysis and stability for a ratio-dependent predator-prey diffusive system with time delay. *Chaos, Solitons and Fractals* 30(3), 1–20 (2020).

Chapter 5

Analysis and prediction of physical fitness test data of college students based on grey model

Rui Jiang and Cheng Xiang Shi
Chongqing University of Education

CONTENTS

5.1 INTRODUCTION

Physical health is a worldwide problem, and it is the focus of attention all over the world. As the world pays increasingly more attention to physical health issues, various countries and regions have established corresponding physical health assessment systems. In addition, the new crown epidemic is raging around the world, and the mutant new crown virus is still spreading. It is self-evident that physical health status is important to the development of all human beings. Analyzing physical condition from large amount of physical test data has become a hot topic among scientists, and predictive models have emerged as a demand for time. Over the years, related researchers have looked at different angles, methods, and focuses. Wang Wei [1] analyzed the status of students' physical health in Nanchang Hangkong University in 2014, conducted a factor analysis of its influencing factors, and put forward corresponding countermeasures. Zhao Xinzhi [2], based on the gray GM (1,1) prediction model theory, predicted the national

DOI: 10.1201/9781003381167-5

student physical health survey data and the national student physical health standard data and tested the accuracy of the prediction results. Qi Zhenyu [3] analyzed the degree of mastery of college students' basketball skills and verified the experimental hypothesis that the students' physical fitness level and the learning of sports technology are correlated.

To grasp and monitor the changing trends and characteristics of students' physique, we will thoroughly practice the philosophy of labor education proposed by General Secretary Xi Jinping and build an educational system with full development of moral, intellectual, physical, and artistic aspects, with a focus on the cultivation of students' masculinity. Based on the 4-year physical fitness test data of all students of a certain grade in a university, this chapter establishes a gray GM (1,1) model to analyze the development trend of student physique and provides decision support for the school to formulate and set up a physical test plan. At the same time, students can clarify the shortcomings of their own functions based on the analysis results of their own physical test data and strengthen physical exercise suitable for their own development, which has important practical and social significance.

5.2 GREY MODEL RELATED THEORIES

The information in the gray system is semi-transparent, and only part of the information is known. The gray prediction method is a method of predicting the gray system [4]. By identifying various development tendencies between system factors, performing correlation analysis, and generating raw data showing the law of system changes, we can generate strong regular data sequences. Then we build the corresponding differential equation model in order to predict the tendency of things to develop [5].

5.2.1 Introduction to the grey model

Gray prediction models and BP neural network prediction models are common prediction models. In contrast, the BP neural network requires a great quantity of samples, otherwise it will cause large deviations; the convergence speed of the neural network algorithm is slow. There is no unified and complete theoretical guide for choosing a structure [6]. The gray prediction model can handle less eigenvalue data, does not require a large number of samples, is less computationally expensive, and has a higher prediction accuracy [7]. Even if the data is disordered, it still has practical implications.

5.2.2 Grey GM (1,1) prediction model algorithm

In the gray GM (1,1) prediction model, GM (1,1) refers to a model with one order and one variable [8]. The specific steps for establishing the model are as follows:

The first step is to establish the original time series $X^{(0)}$:

$$X^{(0)} = \{X^{(0)}(1),\ X^{(0)}(2),\ X^{(0)}(3),\ldots,X^{(0)}(n)\} \tag{5.1}$$

$X^{(1)}$ is the 1-AGO (accumulation at one time) sequence of $X^{(0)}$:

$$X^{(1)}(k) = \sum_{i=1}^{k} X^{(0)}(I),\ k = 1,2,L,n. \tag{5.2}$$

The generated sequence is:

$$X^{(1)} = \{X^{(1)}(1),\ X^{(1)}(2),\ X^{(1)}(3),\ldots,X^{(1)}(n)\} \tag{5.3}$$

In the second step, the corresponding differential equation of the GM (1,1) model is:

$$\frac{DX^{(1)}}{dt} + aX^{(1)} = b,\ \mu = \frac{b}{a} \tag{5.4}$$

Below that "a" is the development coefficient, and "b" is the ash effect. The development coefficient reflects the tendency to expand the original sequence and accumulation created by the sequence. The gray effect is a single-sequence data modeling without an external effect sequence, which is equivalent to the effect in the system. The development coefficient "a" must be greater than –2 and less than 2, for this model to be meaningful. As the development coefficient increases, the model error also increases.

The third step is to generate a predictive model ($k=1, 2,\ldots, n-1$):

$$\hat{X}^{(1)}(k+1) = \left(X^{(0)}(1) - \frac{b}{a}\right)e^{-ak} + \frac{b}{a} = \left(X^{(0)}(1) - \mu\right)e^{-ak} + \mu \tag{5.5}$$

The fourth step is to obtain the predicted value expression ($k=1, 2,\ldots, n-1$):

$$\hat{X}^{(0)}(k+1) = X^{(1)}(k+1) - X^{(1)}(k) = \left(X^{(0)}(1) - \mu\right)\left(1 - e^{a}\right)e^{-ak} \tag{5.6}$$

5.2.3 Grey forecasting model accuracy test

Grey prediction model testing generally includes relative residual testing, variance ratio testing, and small error probability testing.

First, relative residual Q test:

$$Q = e^{(0)}(k) = \frac{X^{(0)}(k) - \hat{X}^{(0)}(k)}{X^{(0)}(k)},\ k = 1,2,\ldots,n \tag{5.7}$$

Table 5.1 Reference table of prediction model test standards

Forecast accuracy level	*Variance ratio C*	*Small error probability P*
Level 1 (Excellent)	<0.35	>0.95
Level 2 (good)	<0.5	>0.8
Level 3 (Qualified)	<0.65	>0.7
Level 4 (unqualified)	≥0.65	>0.6

The calculated " $|Q| < 0.1$ " is considered to meet the higher requirements; $|Q| < 0.2$, the general requirements are considered to be met.

In addition, the variance ratio C test and the small error probability P test:

Usually, the variance ratio C test is combined with the small error P test. If the original sequence is "X(0)(*k*)", its sequence standard deviation is "S1", the residual sequence is "*e*(0)(*k*)", its standard deviation is recorded as "S2", and the residual mean is "$\overline{e}$". Then, the variance is denoted as "C", and the probability of small error is denoted as "P".

$$\overline{e} = \frac{1}{n}\sum_{k=1}^{n} e(k) \tag{5.8}$$

$$C = \frac{S_2}{S_1} \tag{5.9}$$

$$P = P\left\{\left|e^{(0)}(k) - \overline{e}\right| < 0.674\right\} \tag{5.10}$$

In theory, the smaller the variance ratio "C", the more accurate, and the larger the probability of small error "P", the more accurate.

The specific reference table of the variance ratio "C" test and the small error "P" test for judging the accuracy of the model is shown in Table 5.1.

If used as a predictive model, the accuracy level should be above level two.

5.3 ANALYSIS AND PREDICTION OF PHYSIQUE BASED ON GREY MODEL

A total of 3,185 students in a certain college and a certain grade who have normally participated in physical examinations and have valid results are selected, of which 720 are boys and 2,465 are girls, as the subject of the research.

5.3.1 Data preprocessing

The physical test items include height and weight, vital capacity, 50-m run, standing long jump, sitting forward bend, endurance running (1,000m run for boys or 800m run for girls), pull-ups for boys or 1-minute sit-ups for girls.

Create a box plot, combine real situations to preprocess data, filter out outliers, and check the rationality of preprocessing. The total number of people in the study was reduced from 3,185 to 2,661. Among them, boys were reduced from 720 to 672, and girls were reduced from 2,465 to 1,989.

5.3.2 Data analysis of student fitness test

Through statistics, the average data table (Table 5.2) of the actual physical test indicators of the group of students in the 4 years of university is obtained.

According to the "National Standards for Student Physical Fitness and Health," the total score on the test is the sum of the weights that match the ratings for each item. The total score of the physical fitness test is expressed as:

$$
\begin{aligned}
&\text{Total score} = \text{BMI score} \times 15\% \\
&+ \text{vital capacity score} \times 15\% \\
&+ 50\text{ m running score} \times 20\% \\
&+ \text{standing long jump score} \times 10\% \\
&+ \text{sitting bodybend forward score} \times 10\% \\
&+ \text{endurance runningscore} \times 20\% \\
&+ \text{pull-ups/sit-upsscore} \times 10\%
\end{aligned}
\tag{5.11}
$$

According to the above formula, calculate the total score for each student's fitness test. The scores are divided into four levels, 90 points or more are

Table 5.2 Average table of students' actual physical test data

Project	*Unit*	*Gender*	*1st year*	*2nd year*	*3rd year*	*4th year*
BMI	kg/m^2	Male	20.8	21.0	20.8	20.8
		Female	20.2	20.3	19.7	19.6
Vital capacity	ml	Male	3,561	3,766	3,876	3,940
		Female	2,412	2,563	2,656	2,693
50 m sprint	Second	Male	8.2	8.0	8.4	7.8
		Female	10.0	9.8	9.9	9.6
Standing long jump	cm	Male	222	221	221	218
		Female	167	166	166	166
Sitting body Bend forwarc	cm	Male	16.8	16.5	16.2	19.2
		Female	19.0	19.5	19.0	21.0
1,000 m	minute	Male	4.16	4.31	4.67	5.07
800 m		Female	4.07	4.10	4.34	4.60
Pull-ups	number	Male	7	9	9	10
Sit-ups		Female	32	37	36	39

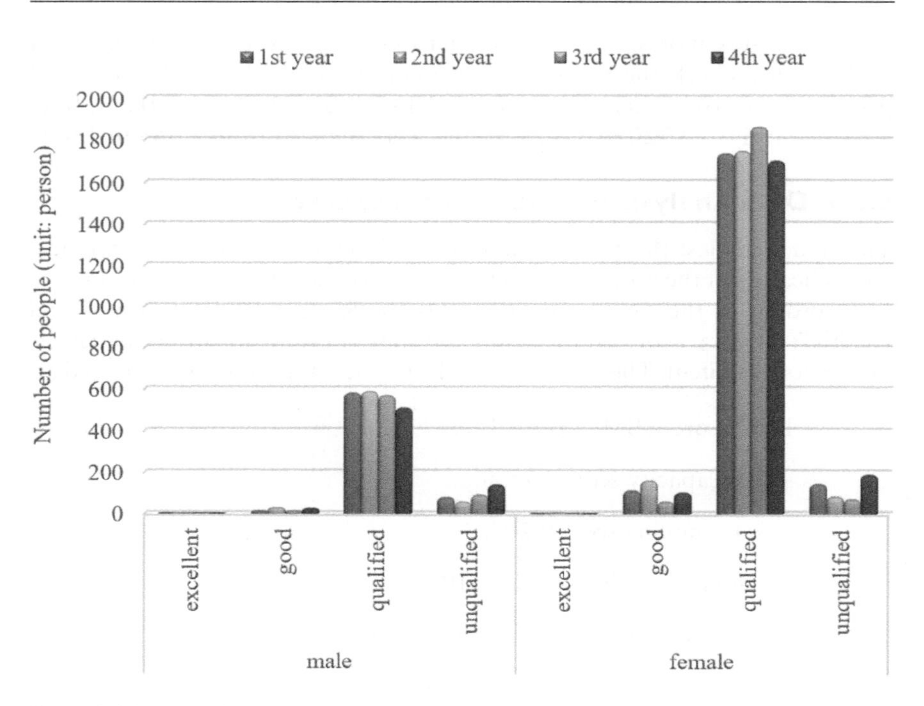

Figure 5.1 Physical fitness test score chart.

excellent, 80 points or more to 90 points are good, 70 points or more to 80 points are passing, and the rest are failed. Finally, get the physique test score chart (see Figure 5.1).

From the image point of view, when comparing the two groups of men and women, there is no significant difference in the ratio of the individual levels. The rate of good grades of girls is slightly higher than that of boys, and the rate of unqualified girls is lower than that of boys. Female students can be more self-disciplined in their daily routines, and most girls also have the habit of fitness plasticity.

From a time perspective, the distribution of the weight of each grade is relatively stable each year, but there are still fluctuations, indicating that the physical health of the same group of students is stable. However, due to changes in diet, work, and rest and increased exercise, physical fitness will also improve.

On the whole, the overall grades of students in the past 4 years have been at the level of passing or above. The number of unqualified students is very low, but there are also very few outstanding students. It can also be associated with the evolution of education today. Students pay more attention to the pursuit of academic studies and neglect the training of sports skills.

Physical fitness primarily includes strength, speed, endurance, agility, and flexibility. Each is independent and complementary to other. The test results of the eight types of fitness test items can reflect the corresponding physical condition. As a modern university student, we should achieve a balanced development of learning and fitness, strengthen the significance of improving physical fitness through regular work and rest, insist on exercise, comprehensively develop various qualities and capabilities, and achieve coordinated physical and mental development.

5.3.3 Prediction of student physical test data based on grey model

Based on the gray GM (1,1) prediction model theory, and according to the average data of the physical test indicators of the 4 years of university students (Table 5.3), the gray prediction model is established by MATLAB.

Obtain the prediction model (according to formula 5), and the results are shown in Table 5.3.

In addition, according to the above Equation (5.6), the grey prediction model of physical test of a certain grade-level student can be obtained. (See Table 5.4.)

The abovementioned tables obtain the prediction model and the prediction value expression and further obtain the corresponding specific prediction value within 8 years (Table 5.5).

Whether the predicted value obtained by the prediction model is accurate needs to be tested for the accuracy of various indicators.

Except for the two items of 50-m running and sitting forward bending, there are three and four levels, and they have reached the first level

Table 5.3 The grey prediction model of physical test of a certain grade student

Project	*Gender*	*Predictive model* $\hat{X}^{(1)}(k+1)$
BMI	Male	$-3,695.260e^{-0.0057k}+3,716.080$
	Female	$-1,150.953\ e^{-0.0177k}+1,171.167$
Vital capacity	Male	$1,66,379.996\ e^{0.0224k}-1,62,819.226$
	Female	$1,03,457.700e^{0.0246k}-1,01,045.450$
50 m sprint	Male	$-532.602e^{-0.0155k}+540.812$
	Female	$-786.509e^{-0.0127k}+796.557$
Standing long jump	Male	$-30,329.115\ e^{-0.0073k}+30,551.585$
	Female	$9,14,526.048e^{0.0002k}-9,14,359.128$
Sitting body Bend forward	Male	$191.905e^{0.0798k}-175.082$
	Female	$495.912e^{0.0378k}-476.873$
1,000 m	Male	$50.405e^{0.0819k}-46.241$
800 m	Female	$70.184e^{0.0568k}-66.111$
Pull-ups	Male	$157.500e^{0.0545k}-150.500$
Sit-ups	Female	$1,313.750e^{0.0273k}-1,281.750$

Table 5.4 The expression of gray predictive value of physical test for a certain grade student

Project	*Gender*	*Predictive value* $\hat{X}^{(0)}(k+1)$
BMI	Male	$-3,695.260\left(1-e^{0.0057}\right)e^{-0.0057k}$
	Female	$-1,150.953\left(1-e^{0.0177}\right)e^{-0.0177k}$
Vital capacity	Male	$1,66,379.996\left(1-e^{-0.0224}\right)e^{0.0224k}$
	Female	$1,03,457.700\left(1-e^{-0.0246}\right)e^{0.0246k}$
50 m sprint	Male	$-532.602\left(1-e^{0.0155}\right)e^{-0.0155k}$
	Female	$-786.509\left(1-e^{0.0127}\right)e^{-0.0127k}$
Standing long jump	Male	$-30,329.115\left(1-e^{0.0073}\right)e^{-0.0073k}$
	Female	$91,4526.048\left(1-e^{-0.0002}\right)e^{0.0002k}$
Sitting body Bend forward	Male	$191.905\left(1-e^{-0.0798}\right)e^{0.0798k}$
	Female	$495.912\left(1-e^{-0.0378}\right)e^{0.0378k}$
1,000 m	Male	$50.405\left(1-e^{-0.0819}\right)e^{0.0819k}$
800 m	Female	$70.184\left(1-e^{-0.0568}\right)e^{0.0568k}$
Pull-ups	Male	$157.500\left(1-e^{-0.0545}\right)e^{0.0545k}$
Sit-ups	Female	$1,313.750\left(1-e^{-0.0273}\right)e^{0.0273k}$

Table 5.5 Gray GM (1,1) model predictive value table of physical test for a certain grade student

Project	*Gender*	*1st year*	*2nd year*	*3rd year*	*4th year*	*5th year*	*6th year*	*7th year*	*8th year*
BMI	Male	20.8	21.0	20.9	20.8	20.7	20.5	20.4	20.3
	Female	20.2	20.2	19.9	19.5	19.2	18.8	18.5	18.2
Vital capacity	Male	3,561	3,774	3,860	3,948	4,037	4,129	4,222	4,318
	Female	2,412	2,572	2,636	2,702	2,769	2,838	2,908	2,981
50 m sprint	Male	8.2	8.2	8.1	8.0	7.8	7.7	7.6	7.5
	Female	10.0	9.9	9.8	9.6	9.5	9.4	9.3	9.2
Standing long jump	Male	222	222	220	219	217	215	214	212
	Female	167	166	166	166	166	166	166	166
Sitting body Bend forward	Male	16.8	15.9	17.3	18.7	20.3	21.9	23.8	25.7
	Female	19.0	19.1	19.8	20.6	21.4	22.2	23.1	23.9
1,000 m	Male	4.16	4.30	4.67	5.07	5.50	5.97	6.48	7.03
800 m	Female	4.07	4.10	4.34	4.60	4.87	5.15	5.45	5.77
Pull-ups	Male	7	9	9	10	10	11	12	12
Sit-ups	Female	32	36	37	38	39	41	42	43

in vital capacity, standing long jump, endurance running, pull-ups, or sit-ups, indicating that the students' physical fitness test. The accuracy of the performance modeling meets the modeling requirements.

5.3.4 Prediction of physique development trend based on grey model

Count the number of "unqualified" for all students of a certain grade in the 4-year physical test of each item (Table 5.6).

Using the same method described above and using MATLAB to establish a gray GM (1,1) predictive model, we obtain a gray predictive model for the number of unqualified physiques (Table 5.7).

Note: BMI values are all at the qualified level, no research is done.

According to the gray prediction model obtained in Table 5.6, the specific prediction results within 8 years are further calculated, and the predicted line chart of the number of people with unqualified physique is drawn, as shown in Figure 5.2.

Intuitively, the number of failures in endurance running has increased dramatically, which directly affects the variation in the total number of ineligible scores, which may be related to the subjective attitude of the students in testing endurance running items. The endurance running test data

Table 5.6 The number of unqualified students in a certain grade

Project	*1st year*	*2nd year*	*3rd year*	*4th year*
Vital capacity	422	196	117	104
50 m sprint	603	351	539	255
Standing long jump	354	398	468	505
Sitting forward bending	78	103	85	57
Endurance running	267	389	830	1243
Pull-ups/Supine	542	422	469	385
Consolidated results	222	140	160	327

Table 5.7 Grey prediction model of the number of unqualified physiques

Project	*Predictive model* $\hat{X}^{(1)}(k+1)$
Vital capacity	$-630.440e^{-0.3553k}+1,052.440$
50 m sprint	$-4,301.021e^{-0.1033k}+4,904.021$
Standing long jump	$-3,305.467e^{0.1155k}-2,951.467$
Sitting forward bending	$-435.674e^{-0.2742k}+513.674$
Endurance running	$675.501e^{0.5060k}-408.501$
Pull-ups/Supine	$-10,939.122e^{-0.0413k}+11,481.122$
Consolidated results	$170.642e^{0.4955k}+51.358$

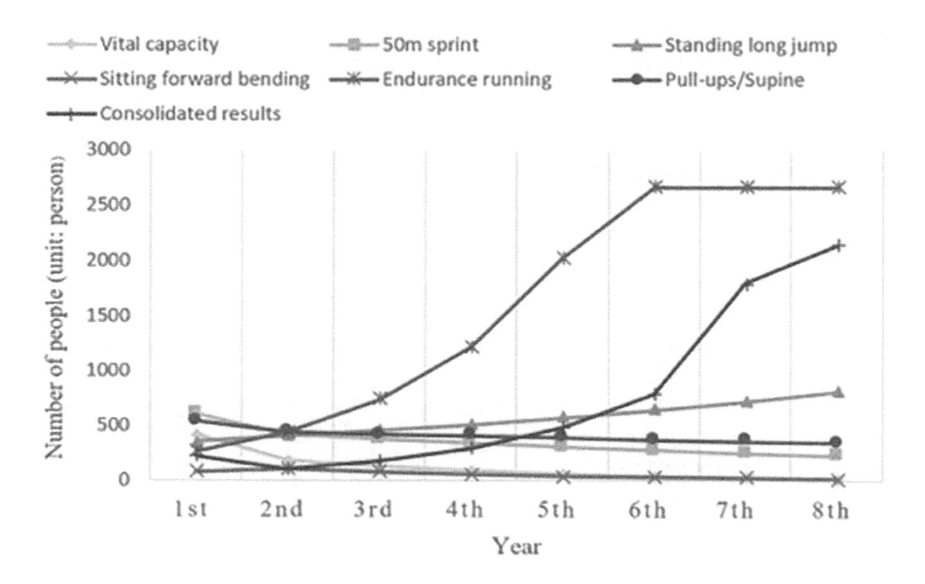

Figure 5.2 Prediction chart for the number of unqualified people.

with objections alone is not enough to reflect changes in overall physical fitness. The number of unqualified standing long jump events is also gradually increasing, reflecting the decline in the explosive power of individual leg muscles with age.

In addition, the number of unqualified people in the other four types of projects has shown a steadily decreasing trend. It shows that the physical fitness of the students has been improved, reflecting the further improvement of the overall level of comprehensive physical fitness.

On the whole, in order to implement the labor education tasks proposed by General Secretary Xi Jinping and strengthen students' physique, it is necessary to strengthen the training of the student endurance running project. Although vital capacity is predicted to show a gradual upward trend, increasing vital capacity alone is not enough to help endurance running to make significant progress. It requires multiple physical fitness to work together. The physical condition reflected in each project is worthy of our study.

The accuracy and credibility of the model need to be further tested. Using the relative residual Q test, variance ratio C test and small error probability P test methods to test the accuracy of the gray prediction model of the number of students with unqualified physique, the results shown in Table 5.8 are obtained. The prediction items except for the 50-m sprint and the cited the body-up or sit-ups are outside the fourth and third levels, and the rest have reached the first level, indicating that the predictive model has a certain degree of credibility. However, the prediction model is also for

Table 5.8 Accuracy test statistics table

Project	*Q*	*C*	*P*	*Grade*
Vital capacity	0.0692	0.0795	1	First level
50 m sprint	0.2105	0.6951	0.75	Fourth level
Standing long jump	0.0141	0.1359	1	First level
Sitting forward bending	0.0348	0.2027	1	First level
Endurance running	0.0675	0.1372	1	First level
Pull-ups/Supine	0.0507	0.4587	0.75	Third level
Consolidated results	0.1098	0.2985	1	First level

reference only. The predicted data are not perfect, and there is room for further improvement.

Through the prediction of the students' physical fitness scores, it is helpful to grasp the changing trend of the students' physical fitness. Remind students that they need to pay more attention to improving physical fitness and strengthening physical exercises in a targeted manner.

In summary, the physical fitness level of college students is showing an upward trend, but there is a lot of room for improvement in some areas. The prediction value obtained by the prediction model of the physical fitness level is based on the data of the original physical fitness measurement value, and the actual measurement value will inevitably have many errors. Therefore, the predictive model can only be used as a reference for making teaching plans, decision-making or physical exercise arrangements.

5.4 CONCLUSION

Student health is a practical matter and is key to the country's prosperity and development. It is necessary for various departments to grasp and monitor the physical health of college students in a timely manner, use student physical test data to analyze the physical health of students, and help college students improve their physical fitness level and enhance their physical health.

Based on the gray GM (1, 1) prediction model theory, this chapter establishes a gray GM (1, 1) prediction model for this batch of student physical fitness test data. Through the prediction results, we will examine in more detail the trend of the development of the student's physical health and find the corresponding physical weaknesses.

Ordinary universities should increase the publicity of students' physical health, enforce health-related laws, and improve the physical exercise system and the corresponding reward system. Provide a comfortable exercise environment, ensure the quality of physical education classroom teaching, create more sports activities, cultivate students, teachers, parents, etc. to attach importance to physical health, and stimulate students' enthusiasm for physical exercise. Ordinary universities should also increase the student's regular diet, work, and rest, organize classes and conferences for healthy living, and encourage students to eat a balanced diet, go to bed early, and get up early. Starting bit by bit in life, forming good habits is an important guarantee for having a good physical fitness.

Students need to clarify the weak aspects of their physical fitness, actively move, actively participate in various sports activities, consciously observe regular hours of work and rest, regular meals and individuals, and implement the goal of improving individual physical fitness into action.

In summary, predictions reflect an increase in the student's overall fitness, but there is still plenty of room for improvement in endurance running and standing long jump. According to the results of the prediction model, it shows that the labor education that General Secretary Xi Jinping attaches importance to and the educational tasks of the comprehensive development of morality, intelligence, physical education, art, and labor have achieved good results.

ACKNOWLEDGMENT

This research was partly financially supported through grants from the Chongqing Science and Technology Bureau Technology Innovation and Application Development Key project (No. cstc2020jscx-dxwtBX0044), the Chongqing Science and Technology Bureau Technology Innovation and Application Development General project (No. cstc2020jscx-msxmX0152), the Chongqing Higher Education Teaching Reform Research Key Project (No. 222166), the Scientific Research Project of Chongqing University of Education (No. KY202107B), the University Student Research Project of Chongqing University of Education (No. KY20200144 and No. KY20210166).

REFERENCES

[1] Wang Wei. *Research on the Status Quo of College Students' Physical Health and its Influencing Factors*. Nanchang Hangkong University, Jiangxi, China, 2016.

[2] Zhao Xinzhi. *Research on the Development Trend and Grey Prediction of the Physical Health of College Students in Yunnan Province*. Yunnan Normal University, Yunnan, China, 2018.

[3] Qi Zhenyu. *Research on the Correlation Between Physical Fitness Level of College Students and Sports Technique Learning*. Shanxi Normal University, Shanxi, China, 2014.

[4] Zhang Shengming, Cai Xiao, Guowei. Analysis of physical fitness indexes of college students' physical fitness test — Taking Guizhou medical university physical examination data as an example. *Stationery and Sports Supplies and Technology* 24: 109–110+141, 2020.

[5] Xie Naiming, Liu Sifeng. Discrete GM (1,1) model and grey prediction model modeling mechanism. *System Engineering Theory and Practice* 1: 93–99, 2005.
[6] Sun Ke. *Forecast of the Labor Supply and Demand Relationship in Qinghai Province based on the GM (1,1) Gray Model: 2019~2030*. Northwest University for Nationalities, Gansu, China, 2019.
[7] Wu Chunguang. *Improvement and Application of GM(1,1) Model and its MATLAB Implementation*. East China Normal University, Shanghai, China, 2010.
[8] Deng Julong. *Basic Methods of Grey System* (Second Edition). Wuhan: Huazhong University of Science and Technology Press, 2005.

Chapter 6

Analysis and research on book borrowing tendency based on Apriori algorithm

Yan Ma, Chengxiang Shi, Rui Jiang, Xiaoqing Li, and Fangfang Chen
Chongqing University of Education

CONTENTS

6.1 INTRODUCTION

Since ancient times, reading can broaden public vision, enhance people's sense of faith, and ability to distinguish right from wrong. College students are in the golden age of reading. The formation of good reading concepts and habits is not only related to the future development of individuals but also related to the prosperity and development of the nation and even the country.

DOI: 10.1201/9781003381167-6

The government report of the two sessions in 2021 emphasizes the significance of "national reading" to social and cultural construction from the perspective of national and national development. At the same time, it expounds the ways and methods of implementing national reading from the perspective of rural library construction and management. With the increasingly prominent position of reading in today's society, the research on reading tendency in academic circles at home and abroad is emerging one after another. Melentieva[1] has found that students' reading behavior decreases with age, switching from paper books to electronic screens and audition materials. Dollinger [2] analyzed that most students prefer reading novels, nature, and science books. Liang Yanbin [3], a domestic scholar, proposed that the library should establish its own book ranking according to the idea of reading guidance in university libraries. Liu Yixiao and Zhang Zhiqiang [4] analyzed book publishing according to library borrowing ranking. In contrast, most domestic research cases are carried out in libraries, the research methods are relatively single, the research perspective is not broad enough, and it is necessary to seek innovation in research means and expand research content.

College students are in the golden age of reading. The formation of good reading concepts and habits is not only related to the future development of individuals but also related to the prosperity and development of the nation and even the country. Therefore, we should pay more attention to the research and in-depth discussion of college students' reading. Based on college students' reading behavior, this chapter discusses the importance and urgency of college library reading promotion based on college students' reading behavior from two aspects of theory and practice. Provide more practical help for stakeholders, promote the progress of education, form a good campus culture and atmosphere, and promote the formation of national cultural atmosphere. The main research object of this chapter is the book borrowing volume of college students from 2015 to 2018.

6.2 RELEVANT THEORIES

Association Rules [5] refer to a large database system that can quickly find the potential and valuable associations between various things and express them in rules.

6.2.1 Association rules

The research and application of association rules is the most active and profound branch of data mining. To make the mined association rules effective, we need to give two most commonly used indicators: minimum support and minimum confidence. Association Rules [6] can be expressed as $X \rightarrow Y$ the implication of shape, where $X$and Y are called the forerunner and successor of association rules, respectively.

Set $I = \{i_1, i_2, ..., i_m\}$ is a collection of items. Remember D as a set of things, things T are a set of items, $T \subseteq I$and. Let A be an itemset of I, if $A \subseteq I$, then it is called Tthing A inclusion. There are several definitions:

1. Association rule [7] is $A \rightarrow B$ an implication of $A \subseteq I, B \subseteq I$, 并且 $A \cap B = \varphi$ shape, here.
2. Rule support: the rule has $A \rightarrow B$ support in D the database, which indicates S the percentage S of D things contained in the AB database at the same time. It $P(AB)$ is the probability, that is: AB Percentage of, which is the $P(AB)$ probability, that is:

$$S(A \rightarrow B) = P(AB) = \frac{|AB|}{|D|} \tag{6.1}$$

where | | it represents the number of things (i.e. frequency).

3. Credibility of rules: rule $A \rightarrow B$ has credibility C, which means that A it contains both the B itemset and the itemset. A Relative to the percentage of the itemset, this $P(B|A)$ is the conditional probability, that is:

$$C(A \rightarrow B) = P(B|A) = \frac{|AB|}{|A|} \tag{6.2}$$

The promotion degree of a rule can be expressed $L(A \rightarrow B)$ as:

$$L(A \rightarrow B) = \frac{C(A \rightarrow B)}{S(A \rightarrow B)} \tag{6.3}$$

4. Critical value: there are two critical values: minimum support and minimum reliability.
5. A collection of items is called an item set, and an k item set containing multiple items is k called an item set. If the itemset meets the minimum support, it is called a frequent itemset.
6. In general, the rules that meet the minimum support and minimum reliability (i.e., two critical values) at the same time are called association rules, that is,

$$S(A \rightarrow B) > \text{min\_sup} \text{ 且 } C(A \rightarrow B) > \text{min\_conf} \tag{6.4}$$

when established, they can also be called strong association rules.

6.2.2 Apriori algorithm

Apriori algorithm [7] is a frequent itemset algorithm for mining association rules. An effective association rule should be satisfied.

$$S_{A\to B} \geq S_{\min}, C_{X\to Y} \geq C_{\min} \quad (6.5)$$

the Apriori algorithm can be divided into three steps:

1. Set the critical values of minimum support and minimum confidence.
2. According to the minimum support, the frequent item sets are calculated, and the strong association rules satisfying the confidence condition are obtained.
3. Set the critical value of the minimum lift. The second step and the minimum confidence are calculated, the strong association rules satisfying the lifting condition are obtained, and the final model results are analyzed.

Apriori algorithm is a process of generating frequent item sets from bottom to top, that is, generating a frequent 1-itemset and a frequent 2-itemset until it cannot be generated, and the candidate set ends. After generating all frequent item sets, the final association rules are generated according to the minimum confidence judgment [8]. For each frequent itemset F, the confidence of all subset combinations is calculated. For example, if F_1 there

$$C_{F_1\to(F\to F_1)} = \frac{S_{F_1\to(F\to F_1)}}{S_{F_1}} \geq C_{min} \quad (6.6)$$

are subsets, the corresponding association rules can $F_1 \to (F \to F_1)$ be generated.

6.3 ANALYSIS OF BOOK BORROWING DATA BASED ON APRIORI ALGORITHM

The data comes from the borrowing data of 22 kinds of books by all students in a university in the past 4 years. Through the Apriori algorithm, according to the association between internal frequent sets found in the student's book borrowing data, explore the potential relationship between data and data, the implicit relationship between major and college and book borrowing, the impact of grade and time on book borrowing and the degree of book borrowing, and use k-means to calculate a method to obtain the distribution of students' borrowing data, and predict the importance of book categories, so as to provide a certain reference basis for library managers to formulate relevant work arrangements [9].

6.3.1 Data preprocessing

The original data information includes student ID, borrowing time, corresponding college and major and book request number of borrowed books.

Table 6.1 Merger and renaming of similar Colleges

Belonging to college	*College*	*Rename*
College of Education	College of Teacher Education College of preschool education	1
School of Management	School of Economics and Business Administration School of Tourism and Service Management	2
Academy of Fine Arts	Academy of Fine Arts	3
Institute of Technology	School of Mathematics and Information Engineering College of Biological and Chemical Engineering	4
College of Literature	College of Foreign Languages and Literature School of Literature and Media	5

To eliminate the "dirty data" in the original data, the borrowing time, student ID, College (weakening the role of major) and book request number are processed. Similar colleges are merged into five colleges (see Table 6.1) and library classification (see Table 6.2), and those that meet the corresponding variables in each column are recorded as T and those that do not meet the variables are recorded as F.

Table 6.2 Corresponding names of library classification

Request number code	*Types of books*	*Request number code*	*Types of books*
A	Marxism Leninism, Mao Zedong thought, Deng Xiaoping theory	N	General theory of natural science
B	Philosophy, religion	O	Mathematical science and chemistry
C	Social sciences	P	Astronomy, earth science
D	Politics and law	Q	Bioscience
E	Military	R	Medicine and health
F	Economics	S	Agricultural science
G	Culture, science, education and sports	T	Industrial technology
H	Language and writing	U	Transportation
I	Literature	V	Aeronautics and astronautics
J	Art	X	Environmental science, safety science
K	History and geography	Z	Comprehensive books

Note: For each column, those satisfying the corresponding variables are recorded as T, and those not satisfying the variables are recorded as F.

6.3.2 Description and analysis of influencing factors of book borrowing volume

Through various analyses and research on the data, it can be seen that the factors affecting the book borrowing volume of students in the university mainly include book type, grade, time, college, and major. The following section analyzes the book borrowing volume based on these five factors.

6.3.2.1 Relationship between book types and book borrowing

It is known that the university has 22 categories of books, and there are certain differences in the borrowing volume of each category of books, which reflect a certain relationship between the borrowing volume of students' books and the types of books. Through analysis, we can further obtain the characteristics of students' book borrowing in the University.

It can be seen from Figure 6.1 above that among the 22 types of books in the sampling data, the borrowing volume of class I – literature books is 7,057 times, while the borrowing times of class A, E, N, P, S, U, V, X, and Z books are less. The average borrowing volume in these four grades is less than ten times, that is, Marxism Leninism, Mao Zedong Thought and Deng Xiaoping theory; Military; General theory of natural science; Astronomy and earth science; Agricultural science; Transportation; Aerospace; Environmental Science, safety science and comprehensive books are borrowed less than 100 times, and the borrowing times are too few. It can be seen that most students in the university are not interested in these nine kinds of books, lack of cognition, are generally more keen on literary books, and recreational reading accounts for a larger proportion.

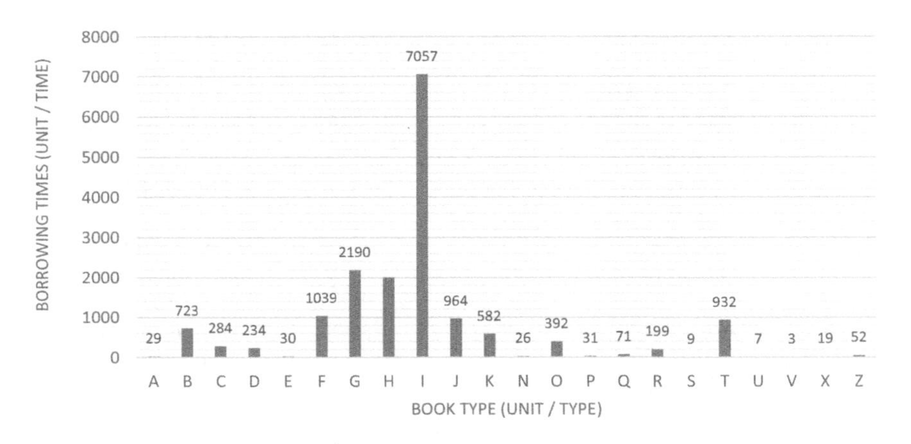

Figure 6.1 Relationship between book types and book borrowing.

6.3.2.2 *Relationship between grade and book borrowing*

As shown in Figure 6.2, among the four grades from 2015 to 2018, 2016 students borrowed the most books (5,896 times), while 2018 students borrowed the least books (2,739 times). Compared with 2016, the number of books borrowed was twice as low. It can be inferred that from 2015 to 2016, students' reading times increased and their reading initiative was improved. The overall reading frequency of 2017 and 2018 students is decreasing year by year, and their reading initiative is declining, indicating that the learning style and quality of 2017 and 2018 students need to be improved.

6.3.2.3 *Relationship between time period and book borrowing*

Due to different curriculum arrangements, there are some differences in students' book borrowing. Therefore, this chapter more systematically describes the university students' book borrowing situation by changing the time point into a time period so as to infer the students' borrowing tendency.

As shown in Figure 6.3, in the four time periods, the students of the University borrow the most in the afternoon, accounting for 47% of the total, followed by the morning and evening, while the books borrowed at noon are the least, accounting for only 9%. Because the time period from 12:00 to 2:00 at noon is lunch and lunch break, and the flow of people in the library is generally the largest in the morning and afternoon, books should also be borrowed more, so this result is reasonable.

6.3.2.4 *Relationship between college and book borrowing*

College is the main factor in the study of the factors affecting the book borrowing of university library. Through the study of the relationship between college and book borrowing, we can better see that the students' study style at those colleges is better and help the school strengthen the students' study

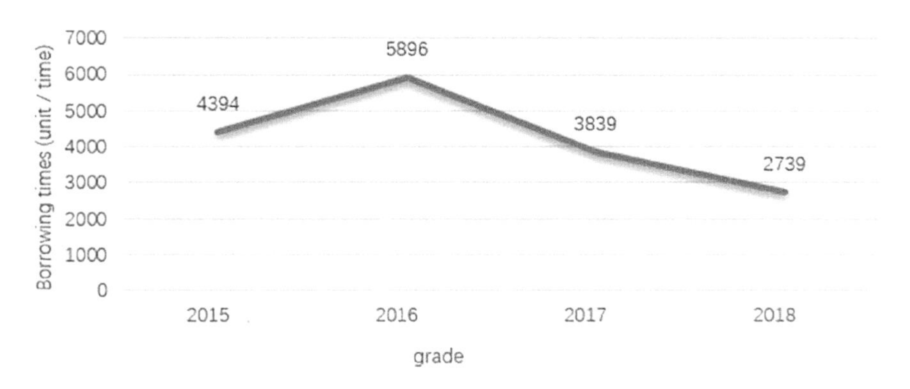

Figure 6.2 Relationship between grade and book borrowing.

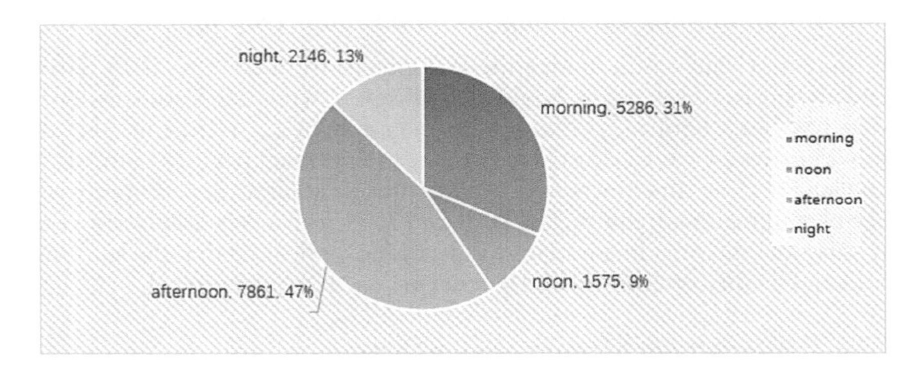

Figure 6.3 Relationship between time period and book borrowing.

style and quality. Colleges are professional, so we can first analyze the borrowing relationship between colleges and book types.

As shown in Figure 6.4, in the borrowing of class I books (literature books), the College of Teacher Education borrows the most times. Among the G-class (culture, science, education, and sports) books, the Academy of Fine Arts has the largest borrowing volume. In the borrowing of T (industrial technology) books, preschool education has the most borrowing times. It can be seen that the students of Teacher Education College prefer literature books; culture, science, education, and sports books are more favored by the Academy of Fine Arts; and the students of preschool education college are more interested in industrial technology books.

In addition, the college is the main factor in the study of the factors affecting the book borrowing of university libraries. By studying the relationship between the college and book borrowing, we can better see that the students of those colleges have a better style of study and help the school strengthen the students' style of study and literacy. Colleges are professional, so we can first analyze the borrowing relationship between colleges and book types.

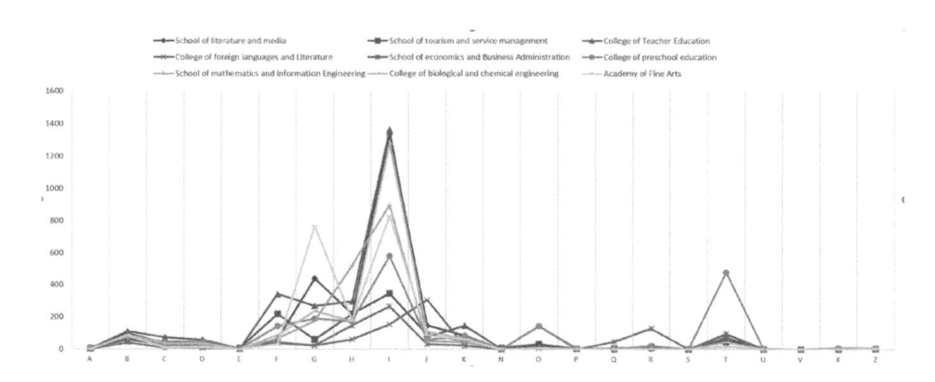

Figure 6.4 Relationship between book types and colleges.

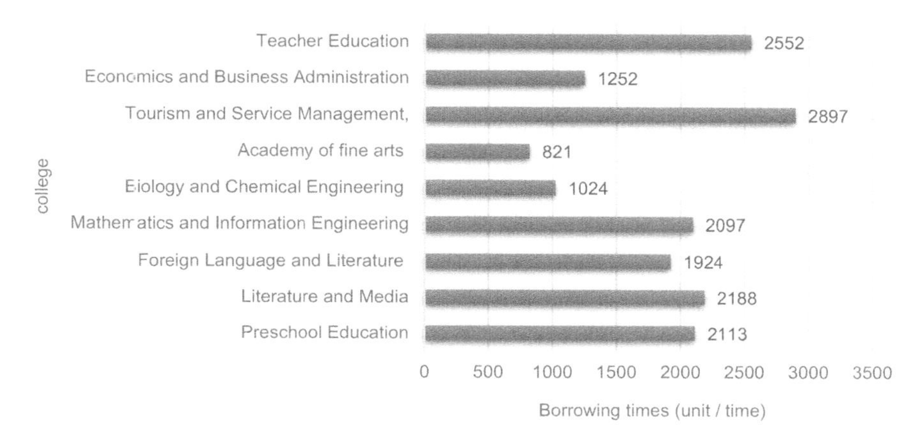

Figure 6.5 Relationship between college and book borrowing.

As shown in Figure 6.5, among the nine colleges, the school of tourism and service management borrowed the most books (2,897 times), and the school of fine arts borrowed the least. Therefore, among the students of grades 2015–2018, the students of the school of tourism and service management are more likely to read and have a good style of study, while the style of study of the school of fine arts needs to be strengthened.

6.3.2.5 Relationship between major and book borrowing

Due to the complexity of majors and numerous directions in the University, in order to facilitate the description of subsequent majors, the majors are treated as follows: primary education (English expertise of primary school general teachers), primary education (Chinese expertise of primary school general teachers), and primary education are collectively referred to as primary education; The Internet of Things Engineering (system design and development direction) and the Internet of things Engineering (ZTE mobile Internet Order class) are collectively referred to as Internet of Things Engineering; English speaking and English (teachers) majors are collectively referred to as English; Computer Science and Technology and Computer Science and Technology (ZTE cloud computing class) are collectively referred to as Computer Science and Technology.

As shown in Figure 6.6 above, among the 30 majors, students majoring in primary education borrowed the most frequently, 3,421 times, followed by preschool education and tourism management. The books on data science and big data technology, economy and finance, clothing and clothing design, e-commerce and information and computing science are borrowed no more than 100 times. The abovementioned situation shows that data science and big data technology, economy and finance, clothing and clothing design College teachers majoring in e-commerce and information and

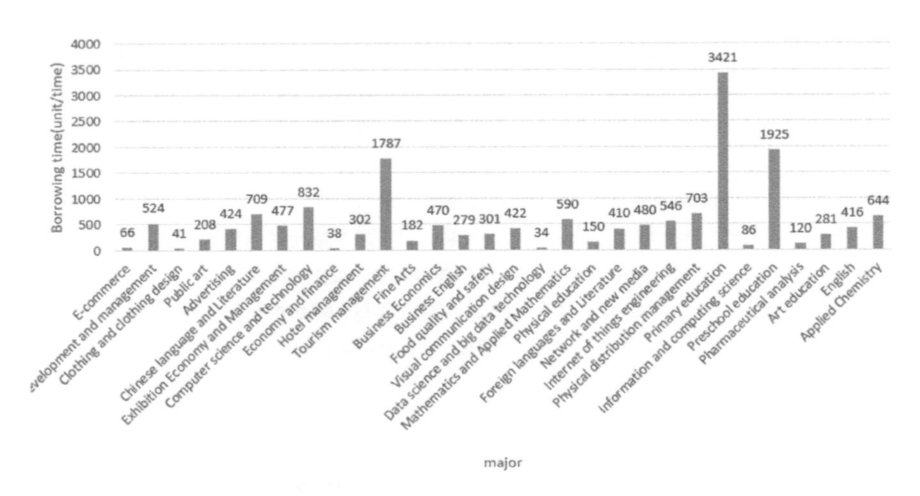

Figure 6.6 Relationship between specialty and book borrowing.

computing science should strengthen the guidance of students' study style, improve students' enthusiasm and enthusiasm for reading. Students majoring in primary education, preschool education, and tourism management have a good study style and love reading.

6.3.2.6 Borrowing preference

From the results of the above analysis on the factors affecting the book borrowing volume, it can be seen that the book borrowing volume of students in this university is unstable. Specifically, in terms of grade, the students of grade 2016 in this university borrow the most books on the whole; in terms of college, the students of the school of tourism and service management borrow the most times; from the professional point of view, students majoring in preschool education borrow the most books; and in terms of borrowing time, students borrow the most in the afternoon. In general, literature books borrow the most. It can be seen that the college students generally love extracurricular reading materials, but lack learning enthusiasm for professional books, and the style of study needs to be strengthened. The relevant teachers should timely urge the improvement of students' style of study and literacy.

6.3.3 Implementation of Apriori algorithm based on Python

According to the book borrowing data, run the Apriori model with Python, and adjust the minimum confidence to 0.5 and the minimum support to 0.1 to obtain the results shown in Table 6.3.

Lift refers to the lifting degree, which is used to avoid the deviation of some unbalanced data labels. In general, it can be used Improve the degree

Table 6.3 Running results based on Apriori algorithm

Rule	*Confidence*	*Support*	*Lifting degree*
{5, I}	0.529	0.129	4.101
{2016, xw}	0.501	0.106	5.010

to illustrate the quality and relevance of f data. On the premise that the promotion degree is greater than 1, the greater the promotion degree, the better the data quality; The smaller the promotion degree, the more unbalanced the data and the weaker the correlation.

The results of Apriori algorithm show that the confidence of the two rules is greater than the minimum confidence of 0.5.

$$\text{Lift}\{5 \rightarrow I\} = 4.101 > 1 \ \text{Lift}\{2016 \rightarrow \text{xw}\} = 5.010 > 1 \quad (6.7)$$

The confidence of the two rules is greater than the minimum confidence of 0.5, under the condition of confidence of 0.5, the corresponding variable 2016 students are strongly correlated with the borrowing time in the afternoon, and the students of the College of Literature are strongly correlated with (class I Books) literature books. It shows that 2016 students prefer to borrow books in the afternoon, while students in the College of Literature (College of foreign languages and College of Literature and media) borrow books more professionally and oriented, which may be related to the arrangement of professional courses.

6.3.4 Visualization of Apriori model

Import all the scores of professional compulsory courses into SPSS modeler and establish the Apriori algorithm model. To avoid the loss of valuable rule information due to too many rules, complex analysis and too few rules, the minimum conditional support is 37% and the minimum rule confidence is 91%. Finally, nine strong association rules are obtained, as shown in Table 6.4. The results obtained by running the Apriori algorithm mining model are shown in Figure 6.7.

Utilize SPSS of Web The complex network analyzes the relevant factors, sets the threshold as absolute and strong link thick, and indicates the strength of the link between grade, college, time period and borrowing type through thick line, thin line and dotted line.

It can be seen that the correlation degree between different influencing factors: The more connections, the closer the connection with other borrowing factors; the thicker the connection, the greater the impact on other borrowing factors. Lending factors closely related to each other should be paid special attention to and given priority, such as 2016 grade and T-class books. For borrowing factors with high mutual influence, it is important

Table 6.4 Apriori mining model

Consequent	*Preamble*	*Support percentage*	*Confidence percentage*
2016	T	38.754	90.194
S	T	38.754	90.194
A	T	38.754	90.194
P	T	38.754	90.194
E	T	38.754	90.194
Q	T	38.754	90.194
R	T	38.754	90.194
D	T	38.754	90.194
O	T	38.754	90.194

Note: 1 ~ 5 refer to the college; 2015 ~ 2018 represent four grades; A ~ Z indicates the type of books; SW (morning), ZW (noon), XW (afternoon) and WS (evening) represent borrowing periods.

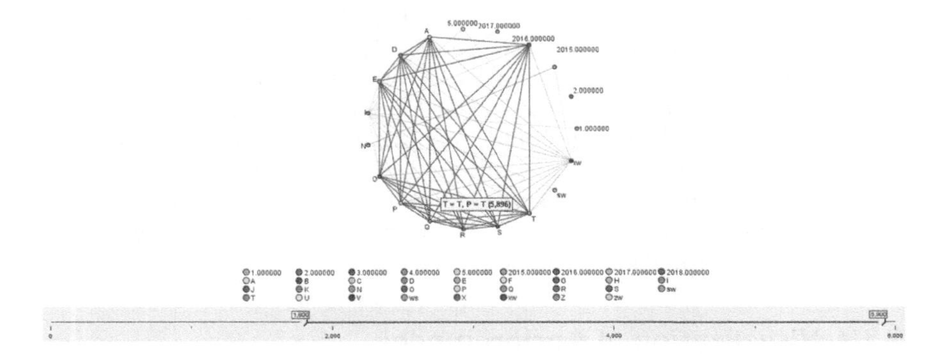

Figure 6.7 Network diagram of borrowing factors.

to consider which borrowing factor is the leading factor and which is the subsequent borrowing factor, such as industrial technology books and 2016 grade, industrial technology books and agricultural science books.

The above results are different from the results of Python code operation. Within the range of the threshold value 1,800–5,900, that is, there are the largest number of coarse students and the largest probability of co-occurrence between 2016 students and book types, and the largest influence on each other. It can be concluded that:

1. The interaction between 2016 students and most books shows that various activities carried out by the library are conducive to promoting the learning of 2016 students as a whole;
2. Grade and borrowing period can reflect the professional orientation of students in the school to a certain extent, but the impact is not great;

3. There is no association between other colleges and book borrowing in the model, which may be due to data integration and cleaning, data conversion errors, or they do not reach the minimum conditional support and minimum rule confidence, or the correlation between the research objects is really not strong;
4. For example, students who borrow industrial technology books are likely to borrow agricultural science books at the same time.

To sum up, it can explain the relationship between book types and students and the importance of book types to a certain extent. The data research results can provide decision support for the setting of library work, the supervision and revision of relevant regulations.

6.3.5 Evaluation and application of borrowing data tendency

In order to more accurately understand the importance of the types of borrowed books and the preference distribution of students' borrowing data, we continue to cluster the data of students' borrowing types through k-Means algorithm. The primary goal of data analysis is to study the logical relationship between borrowed books, obtain the distribution of students' borrowing data, and predict the importance of book categories, in order to guide and suggest students' borrowing direction and learning so as to improve the reading effect; and further assist library managers in carrying out library management. To further understand the relationship between book types and students' borrowing, compare 22 types of books. Next, cluster analysis shall be conducted for book types.

Before running the model, set the number of clusters to 2 ~ 5 for a total of four values in order to find the turning point of clustering quality and select the appropriate number of clusters. After each run, observe the model and record the clustering quality of the model (see Table 6.5).

It can be seen from the above table that the second time, when $k = 3$, is the turning point of clustering quality. It is reasonable to select this turning point to analyze students' borrowing data. The number of clusters is 3. Finally, run the model to get the following results (see Table 6.6).

As shown in Table 6.6, the chart shows that each book type in each category is sorted according to the most frequent proportion of satisfaction T and dissatisfaction F.

Table 6.5 Clustering quality of different *K* values

Frequency	*1*	*2*	*3*	*4*
K value (number of clusters)	2	3	4	5
Clustering quality	0.5	0.4	0.4	0.4

Table 6.6 Mean value of each cluster (part)

Cluster	*Cluster-1*	*Cluster-2*	*Cluster-3*
Label			
Size	50.2% (8,475)	35.0% (5,895)	14.8% (2,497)
Input	I T(55.3%) G F(87.8%) H F(91.0%)	I F(59.8) G F(86.3%) H F(88.8%)	I F(100%) G F(85.9%) H F(77.25)

As the results shown in Table 6.7 (only satisfaction, T is marked in Table 6.7, and the rest are non-satisfaction F.)

The first category has the largest proportion, accounting for 50.2% of the total. This category of students has a narrow reading range and prefers literature books. Teachers should widely cultivate students' interest in reading and better help students develop in many aspects.

The second category of students accounts for 35% of the total. This category of students has the most extensive reading range and strong interest in reading, but there are still some books with less reading volume, such as those on transportation, astronomy, and science and technology. Teachers should continue to support and encourage this category of students and stimulate their comprehensive interest in reading.

The third category of students accounts for the smallest proportion, 14.8%. Relatively speaking, they love natural science books the most. Similar to the first category of students, they have a single interest in reading. Teachers should also strengthen and supervise the reading habits of this category of students.

Table 6.7 Clustering of book types

	Cluster 1 (%)	*Cluster 2 (%)*	*Cluster 3 (%)*		*Cluster 1 (%)*	*Cluster 2 (%)*	*Cluster 3 (%)*
A	99.9	(T)100.0	99.8	H	91.0	88.8	77.2
D	99.1	(T)100.0	97.7	F	96.1	93.1	88.0
E	99.9	(T)100.0	99.9	J	95.8	93.7	90.4
I	(T)55.3	59.8	100.0	K	97.4	96.6	93.7
N	85.5	99.7	(T)100.0	B	96.6	95.7	92.7
O	98.1	(T)100.0	95.7	Z	99.8	99.8	99.2
P	99.9	(T)100.0	99.9	C	98.6	98.4	97.2
Q	99.6	(T)100.0	99.8	G	87.8	86.3	85.9
R	99.2	(T)100.0	98.6	X	100.0	99.8	99.8
S	100.0	(T)100.0	100.0	U	100.0	100.0	99.9
T	96.8	(T)100.0	85.3	V	100.0	100.0	100.0

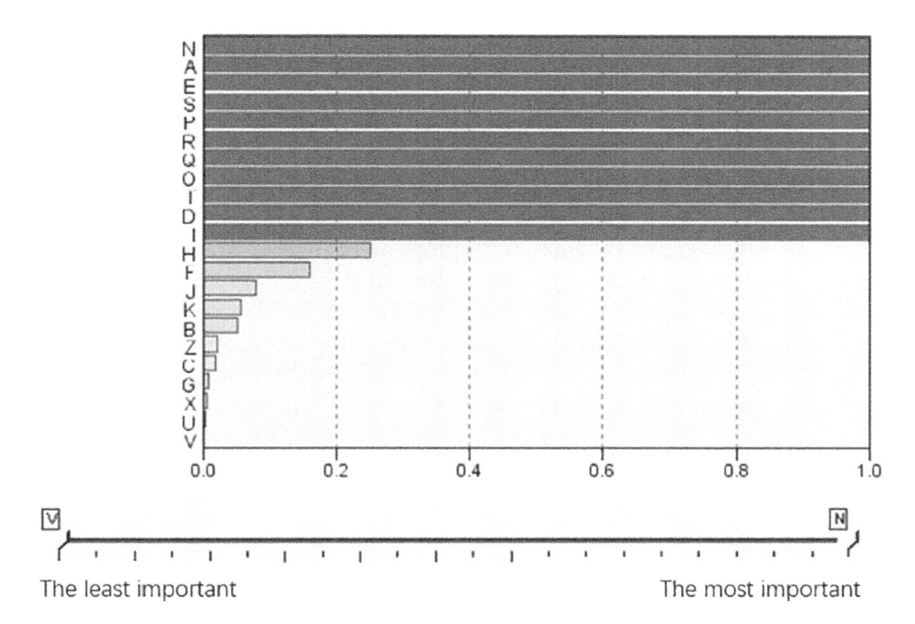

Figure 6.8 Importance of predictive variables.

From Figure 6.8, we can see the importance of 22 types of books to students' borrowing selection. Library managers can appropriately increase the publicity and display of important books according to the importance of different types of books. When the importance of predictive variables is greater than 0.5, they can be regarded as relatively important books. It can be clearly seen from Figure 6.7 that 11 books of N, A, E, S, P, R, Q, O, T, D, and I (see Table 6.2) are the most important books, and the importance of other types is less than 0.5, which is less important for students' choice. Accordingly, teachers can choose appropriate ways to stimulate students' interest in reading without paying attention to book categories, appropriately increase the display and publicity of 11 important books, and arrange personnel with library management experience for supervision and management, so as to promote students' borrowing and learning and further improve students' reading quality.

6.4 CONCLUSION

The main research object of this chapter is the book borrowing volume of college students from 2015 to 2018. The main research method used is the Apriori algorithm of association rule mining, which obtains the general correlation between colleges and book borrowing types. Then, cluster analysis is carried out on the data to study the logical relationship between borrowed books, obtain the distribution of student borrowing data, and predict the figure This chapter puts forward some suggestions on reasonable arrangement of book borrowing.

ACKNOWLEDGMENT

This research was partly financially supported through grants from the Chongqing Science and Technology Bureau Technology Innovation and Application Development Key project (No. cstc2020jscx-dxwtBX0044), the Chongqing Science and Technology Bureau Technology Innovation and Application Development General project (No. cstc2020jscx-msxmX0152), the Scientific Research Project of Chongqing University of Education (No. KY202107B), the University Student Research Project of Chongqing University of Education (No. KY20200144 and No. KY20210166).

REFERENCES

[1] Melentieva, J. P. Reading among young Russians: Some modern tendencies. *Slavic & East European Information Resources*, 2009, 10(4): 304–321.
[2] Dollinger, S J. "You Are as You Read": Do students' reading interests contribute to their individuality? *Reading Psychology*, 2015, (ahead-of-print) 1(1): 1–26.
[3] Yanbin, L. On reader studies. *Research on Publishing and Distribution*, 1987, 1(2): 35–38.
[4] Yixiao, L. Zhiqiang, Z. Analysis of book publishing based on 2014 book ranking. *Library Forum*, 2015, 1(9): 89–94.
[5] Hai, Z. *Research on Job Risk Early Warning based on Association Rule Apriori Algorithm*. Jilin University, Jilin, China, 2014.
[6] Chengyong, W. *Research and Application of Apriori Algorithm for Association Rules*. North China Electric Power University (Beijing), Beijin, China, 2018.
[7] Nan, Y. *Research and Implementation of Web Log Mining based on Association Rule Apriori Algorithm*. Chengdu University of Technology, Sichuan, China, 2012.
[8] Weiguo, Y. *Research and Application of Prediction Method based on Association Rules and Decision Tree*. Dalian Maritime University, Shandong, China, 2012.
[9] Zhixiu, D. *Comparative Study on Readers' Reading Tendency of Different Types of University Libraries in the Digital Age*. Nanjing Agricultural University, Jiangsu, China, 2008.

Chapter 7

Performance evaluation of cargo inspection

Systems with the function of materials recognition

Sergey Osipov, Andrey Batranin, and Sergei Chakhlov
Tomsk Polytechnic University

Sergey Gorbachev
Tomsk State University

CONTENTS

7.1 INTRODUCTION

Inspection systems (ISs) with the material recognition option by the dual and multi-energy methods are the most prominent representatives of technical means for customs and border inspection of large containers and vehicles [1–4]. Without such ISs [5–7], it is impossible to solve most of the problems of ensuring transport safety in passenger and freight transportation. As a result of the material recognition procedure by the dual energy method (DEM), the material of a structurally isolated fragment of the test object (TO) by the value of the recognition parameter (RP) is associated with one of the wide classes of materials. The effective atomic number (EAN) of a material or some of its unique function acts as a RP [8–11]. Each recognition class is associated with its most characteristic representative. As the main materials for fragments of test objects, hydrocarbons with a low EAN value, aluminum alloys, steel, and lead alloys are used [12–14]. The analyzed ISs consist of a bremsstrahlung source (BS), a bremsstrahlung

DOI: 10.1201/9781003381167-7

recorder (BR), and a TO positioning system relative to the BS+BR system. The high-energy DEM implementation is based on highlighting the contributions of the Compton effect and the pair production effect to the bremsstrahlung attenuation; therefore, the studied ISs use BSs with a maximum radiation energy from 4 to 10 MeV.

To create new and modernize existing ISs, it is necessary to develop criteria and algorithms for comparing systems with material recognition option that differ by registrars and bremsstrahlung sources, and experimentally verify the obtained approaches.

7.2 KEY PARAMETERS AND FEATURES OF HIGH ENERGY IS WITH MATERIAL RECOGNITION

When describing the main parameters and characteristics of IS with the material recognition option, we will adhere to the information from the websites of some developers of such complexes [15–17].

The IS manufacturers provide their various parameters and characteristics. From the analysis of the data given in [15–17], a set of basic parameters and characteristics of the systems under consideration was selected and summarized in Table 7.1.

Table 7.1 Key parameters and features for IS with material recognition

Parameters and specifications	*Designation*	*Dimension*
Low maximum energy of bremsstrahlung	E_L	MeV
High maximum energy of bremsstrahlung	E_H	MeV
Radiation dose rate per meter for energy E_L	P_L	G/min
Radiation dose rate per meter for energy E_H	P_H	G/min
Pulse repetition rate	ν	1/s
The number of pulses for energy E_L	k_L	impulse
The number of pulses for energy E_H	k_H	impulse
Distance from BS to BR detector	F	m
Number of arrays in BR	n_{lin}	
RST size in scan direction	A	mm
RST size along the array of detectors	B	mm
The thickness of the RST (scintillator)	h_s	mm
Scintillator density	ρ_s	g/cm^3
Scintillator atomic number	Z_s	
ADC capacity	k_{ADC}	bit
Steel penetration	H_{lim}	mm
Diameter of detectable copper wire without obstruction	d_{Cu}	mm

(*Continued*)

Table 7.1 (Continued) Key parameters and features for IS with material recognition

Parameters and specifications	*Designation*	*Dimension*
Contrast sensitivity	k_X	%
Range of mass thicknesses of material recognition	$(\rho H)_{min}$-$(\rho H)_{max}$	g/cm^2
Inspection performance	Pr	m/s
Number of material recognition classes	K_Z	
EAN of typical representatives of recognition classes	Z_i, $i = 1 \ldots K_Z$	

The expression to estimate the parameters n_{L1} and n_{H1} has the form

$$n_{L,H1} = \left[N_{L,H1}\, ab/F^2 \right] \tag{7.1}$$

The usage of expression (Equation 7.1) greatly simplifies the calculations associated with comparison of various registrars used in IS with the material recognition option.

It is first necessary to briefly describe the corresponding mathematical model to estimate the quality of high-energy ISs with the recognizing materials option by DEM calculation or simulation.

7.3 MATHEMATICAL MODEL OF IS WITH DUAL ENERGY RECOGNITION OF MATERIALS

The method to calculate the required IS characteristics is based on a mathematical model to acquire the primary radiographic images and subsequent recognition of materials by the level line method [8,18].

To build the desired methodology and the corresponding simulation algorithm to acquire radiographic images, we use the discrete model of the integrated registration mode. An analog signal (AS) J_1 for a single bremsstrahlung pulse is the sum of a random number l_1 of identically distributed independent random variables $E_{ab\,i}$, $i = 1 \ldots l_1$

$$J_1 = C_a \sum_{i=1}^{l_1} E_{ab\,i} + A_{dc}, \tag{7.2}$$

where C_a is conversion factor of absorbed photon energy into electrical energy; A_{dc} is random variable of a dark signal.

Random variable l_1 from formula (Equation 7.2) is distributed according to Poisson and represents the number of unreduced photons detected by the detector from one pulse. The relationship of the Poisson distribution parameter $l_{L,H1}$ with the IS parameters is described by the expression

$$\overline{l_{L,H1}(Z, \rho H)} = \left[n_{L,H1} \int_0^{E_{L,H}} f(E, E_{L,H}) \exp(-m(E,Z)\rho H))\varepsilon(Z_s, \rho_s h_s)\, \mathrm{d}E \right], \tag{7.3}$$

where $\varepsilon(Z_s, \rho_s h_s)$ is registration efficiency for RST [19,20].

The random variable E_{ab} is equal to the energy transmitted by the RST photon. The first two initial moments $k = 1; 2$ of random variables $E_{L\,ab}$ and $E_{H\,ab}$ are estimated by the formula

$$\overline{E^k_{L,H\,ab}(Z,\rho H)} = \frac{\int\limits_0^{E_{L,H}} E^k_{ab} f(E, E_{L,H}) \exp(-m(E,Z)\rho H))\varepsilon(Z_s, \rho_s h_s)\, \mathrm{d}\, E}{\int\limits_0^{E_{L,H}} f(E, E_{L,H}) \exp(-m(E,Z)\rho H))\varepsilon(Z_s, \rho_s h_s)\, \mathrm{d}\, E} \tag{7.4}$$

Mean value $\overline{A_{dc}}$ and standard deviation σA_{dc}, mean value A_{dc} describe the analog signal from the RST detector for the switched off BS. They are determined at the stage of preliminary measurements for the prototype of the investigated BR.

The formulas for estimating and $\overline{J_{L,H1}(Z,\rho H)}$ and $\sigma^2 J_{L,H1}(Z, \rho H)$, $J_{L,H1}$ have the form

$$\begin{aligned} \overline{J_{L,H1}(Z,\rho H)} &= C_a \overline{I_{L,H1}(Z,\rho H)}\,\overline{E_{ab}(Z,\rho H)} + \overline{A_{dc}} \\ \sigma^2 J(Z,\rho H) &= C_a^2 \overline{I_{L,H1}(Z,\rho H)}\,\overline{E_{ab}{}^2(Z,\rho H)} + \sigma^2 A_{dc} \end{aligned} \tag{7.5}$$

Analog signals are divided into three classes: speakers without radiation ($\rho H = \infty$); AS for attenuation by air ($\rho H = 0$); measuring AS ($Z \in \{Z_i, i = 1 \ldots K_Z\}$, $\rho H \in [\rho H)_{\min}, (\rho H)_{\max}]$).

The maximum AC value is observed for the energy E_H for attenuation through air with an equal number of integrable pulses; therefore, the conversion of AS $J_{L,H}(Z, \rho H)$ into digital signals (DS) $D_{L,H}(Z, \rho H)$ is described by the expression

$$D_{L,H}(Z,\rho H) = \left[\frac{J_{L,H}(Z,\rho H)}{\Delta_D}\right],\ \Delta_D = \frac{C_{ADC}\overline{J_H(Z,0)}}{\left(2^{k_{ADC}} - 1\right)}, \tag{7.6}$$

where C_{ADC}, $C_{ADC} > 1$, is the coefficient to protect DS from exceeding the maximum level; Δ_D is ADC quantization step.

Digital signals are fed to the processing unit, in which for each TO point an estimate of the thickness $Y_{L,H}(Z, \rho H)$ in the mean-free paths (m-f.p.) is formed. In the processing unit, the signals are calibrated by black and by gain and then taking logarithm.

$$Y_{L,H}(Z,\rho H) = \ln\left(D_{L,H}(Z,0) - D_{L,H}(Z,\infty)\right) - \ln\left(D_{L,H}(Z,\rho H) - D_{L,H}(Z,\infty)\right) \tag{7.7}$$

The resulting sets of estimates $Y_{L,H}(x, y)$ for all points in the image with coordinates (x, y) are images $\mathbf{Y}_L$ and $\mathbf{Y}_H$. Images $\mathbf{Y}_L$ and $\mathbf{Y}_H$ are input to the recognition algorithm and visualized if necessary.

Note 1. To compensate for the instability of the radiation intensity from pulse to pulse, the measured signals and air signals are calibrated in "black" are normalized to a "black" signal calibrated from the reference detector.

Expressions (Equations 7.1–7.7) are a simplified mathematical model for the formation of digital radiographic images in DEM and are easily transformed into a simulation algorithm in any algorithmic language or in any system for mathematical calculations.

In the calculations, theoretical and experimental data on the attenuation of gamma radiation are used [21,22].

The standard deviations $\sigma Y_L(x, y)$ and $\sigma Y_H(x, y)$, random variables $Y_L(x, y)$ and $Y_H(x, y)$ corresponding to the parameters of the fragment $(Z, \rho H)$ are found analytically [18] by simulation or experimentally.

The level line method is the most common material recognition method for high-energy ISs [10,12,23,24].

The recognition parameter Q at each point (x, y) in relation to the method under consideration is calculated as follows:

$$Q(x,y) = Y_H(x,y)/Y_L(x,y) \tag{7.8}$$

In the level line method, level lines are constructed based on preliminary measurements called EAN calibration. Preliminary calibration measurements are carried out for a TO, consisting of a set of fragments. The effective atomic numbers of the TO fragments run through the set $\{Z_1, Z_2, \ldots, Z_{K_Z}\}$, and the mass thickness of the fragments (ρH) varies from $(\rho H)_{\min}$ to $(\rho H)_{\max}$. For all Z_i, $i = 1 \ldots K_Z$ experimental dependencies are formed $Y_L(Z_i, \rho H)$, $Y_H(Z_i, \rho H)$ and $Q(Z_i, \rho H) = Y_H(Z_i, \rho H)/Y_L(Z_i, \rho H)$. For each level Z_i, on the basis of the table $Y_L(Z_i, \rho H)$, $Q(Z_i, \rho H)$, we construct the corresponding interpolating (approximating) function $Q_C(Z_i, Y_L)$ of the variable ρH, which is called the level line Z_i.

We use the rule for assigning material for a point with coordinate (x, y) to i_0 class of materials [24]

$$\min_{i,i=1\ldots K_Z} \left|Q(x,y) - Q_C(Z_i, Y_L(x,y))\right| = \left|Q(x,y) - Q_C(Z_{i_0}, Y_L(x,y))\right| \tag{7.9}$$

Decision rule (Equation 7.9) means that the material corresponds to the EAN corresponding to the nearest level line.

The standard deviations $\sigma Q(x, y)$, random variable $Q(x, y)$ corresponding to the parameters of the fragment $(Z, \rho H)$ are estimated theoretically [18] or experimentally.

The set of expressions (Equations 7.1–7.9) is the basis of the mathematical model and simulation algorithm for the method of recognition of materials and their structural fragments by level line methods.

Note 2. The probability of incorrect recognition of TO material by an experimentally evaluated pair (Y_L, Q) is higher for smaller differences in levels

$$\Delta Q = \left| Q_C(Z_i, Y_L) - Q_C(Z_j, Y_L) \right|, \; i \neq j \tag{7.10}$$

The existence of such areas is confirmed theoretically and experimentally [8,10,12]. The above model provides a methodology to estimate the basic characteristics of IS with the material recognition option listed in Section 7.1.

7.4 PERFORMANCE ESTIMATION OF IS WITH MATERIAL RECOGNITION

A series of calculations, computational, and full-scale experiments were carried out to evaluate the proposed approach for estimation of IS characteristics with the recognition material option by the level line method. The IS of Tomsk Polytechnic University [25] with BR from manufacturers TSNK (Moscow, Russia) [25] and Detection Technologies (Finland) [26] was chosen as a sample.

The small-sized betatron MIB-9 with a pulse repetition rate $\nu = 400$ 1/s was chosen as a source of bremsstrahlung. The experiments were carried out for maximum energies of bremsstrahlung $E_L = 4$ MeV, $E_H = 7.5$ MeV. The distance from the radiation source to the detector array F is approximately 4.2 m. Table 7.2 shows the basic BR parameters used for comparison.

The X-Scan LCS detector provides calibration for signals generated between the beginning and end of bremsstrahlung pulses. This calibration is a black calibration and minimizes the effect of the afterglow of the CsI scintillator on the quality of digital radiographic images. For $CdWO_4$ based detectors, there is no need for such an approach to "black" calibration.

To determine the contrast sensitivity, weakening plates with two thin plates were used. Figure 7.1 shows the corresponding radiographic images. For $E_L = 4$ MeV the maximum thickness of attenuating plates equals 168 mm.

The values of $H_{a\ \min}$ and the corresponding contrast levels k_X were determined by simulation using the same range of digital signals (C_{ADC}) and are given in Table 7.3.

Table 7.2 Basic parameters of bremsstrahlung radiation recorders

No	Name	*Characteristics*					
		Material RST	*a, mm*	*b, mm*	*h_d, mm*	*ADC capacity, k_{ADC}*	*Number detector arrays, n_{lin}*
1	TSNK	$CdWO_4$	6	4	35	16	1
2	X-Scan LCS	CsI	10	5	45	16	2

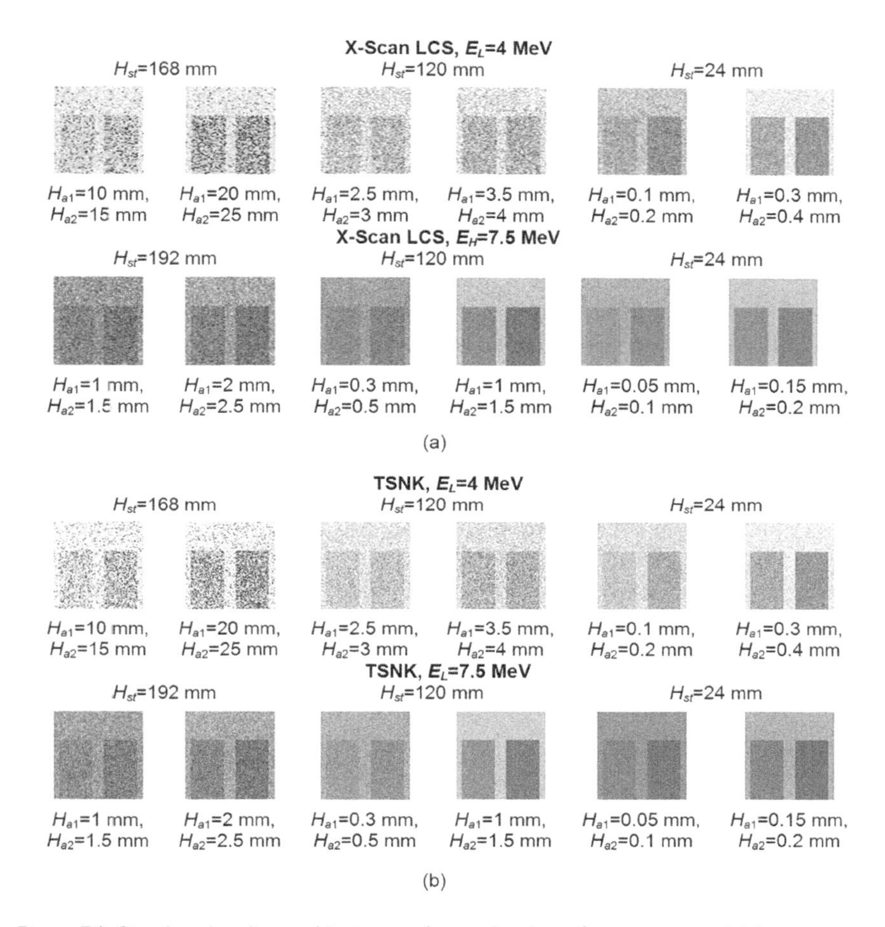

Figure 7.1 Simulated radiographic images for evaluation of contrast sensitivity.

Table 7.3 Simulated estimation of the contrast sensitivity k_X, %

TSNK						*X-scan LCS*					
$E_L = 4$ MeV			$E_H = 7.5$ MeV			$E_L = 4$ MeV			$E_H = 7.5$ MeV		
H_{st}	$H_{a\ min}$	k_X	H_{st}	$H_{a\ min}$	k_X	H_{st}	$H_{a\ min}$	k_X	H_{st}	$H_{a\ min}$	k_X
24	0.2	0.8	24	0.1	0.4	24	0.2	0.8	24	0.1	0.4
120	3	2.5	120	0.5	0.4	120	3	2.5	120	0.5	0.4
168	20	12	192	1.5	0.8	168	20	12	192	1.5	0.8

From the analysis of the simulation results given in Table 7.3, we can conclude that close indicators for the considered BR for the contrast sensitivity. It should be remembered that the calculations were carried out for the same noise level of "dark" signals.

The desired ranges are calculated for a fixed set of recognition classes associated with carbon, aluminum, iron, and lead.

Using the previous set of formulas, we calculate the $Q(Y_L)$ dependencies for different levels Z_i, $Z_i \in \mathbf{Z} = \{6; 13; 26; 82\}$. At the first stage, in accordance with the above approach, $Z = 6$, $\rho = 1$ g/cm³, we determine the range of variation of $Y_L - (Y_{L\ min}(6), Y_{L\ min}(6))$ for $Z = 6$, based on the given range of variation of the mass thickness ρH [g/cm²], $\rho H_{min}(6) \leq \rho H \leq \rho H_{max}(6)$, for example, $\rho H_{min}(6) = 1$ g/cm², $\rho H_{max}(6) = 150$ g/cm². The values of $Y_{L\ min}(6)$ and $Y_{L\ max}(6)$ are easily calculated using the mathematical model described above. This range will be used to determine the mass thicknesses range with the reliable recognition of materials by the level lines method. Figure 7.2 shows the $Q(Y_L)$ dependences calculated for one bremsstrahlung pulse for the considered types of detectors.

The $Q(Y_L)$ dependences for organics, aluminum, and iron are practically the same for the compared types of detectors. The exception is dependencies for lead, however, and they are quite close to each other. The range of reliable recognition of fragments from organic matter, aluminum and iron varies for the analyzed recorders from $Y_L = 1$ m-f.p. to $Y_L = 5.5$ m-f.p. The sawtooth nature of the $Q(Y_L)$ dependences for large values is due to a combination of factors: a low level of the analog signal and insufficient ADC bit capacity.

In Table 7.4, similar to the table from [8], the minimum ρH_- and maximum ρH_+ values of mass thicknesses with reliable material recognition are given.

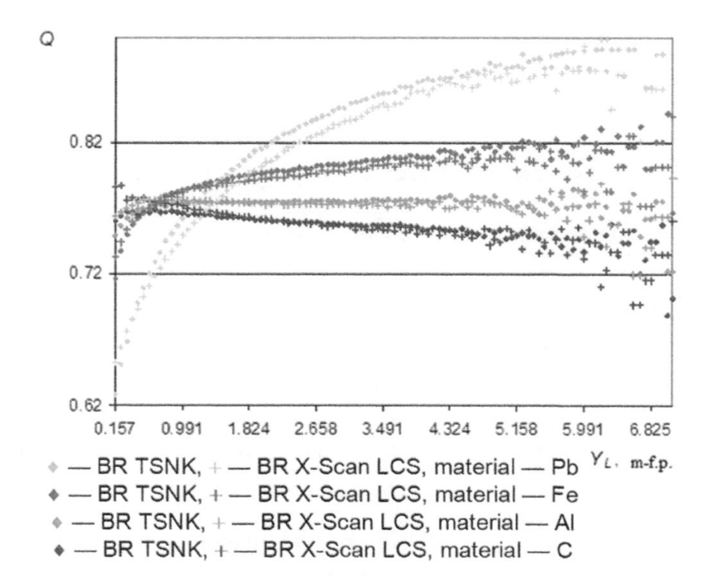

Figure 7.2 Calculated dependencies of recognition parameter on the thickness of material $Q(Y_L)$.

Table 7.4 Reliable material recognition ranges (ρH_{-}, ρH_{+}), g/cm²

Detector	*E_L, MeV*	*E_H, MeV*	*Z_2*	*Z_1*		
				6	*13*	*26*
TSNK	4	7.5	13	(14.6; 100)	(14.9; 100)	(14.6; 100)
			26	(14.6; 100)	(14.9; 100)	(14.6; 100)
			82	(20;) 100	(22; 100)	(30; 100)
X-Scan LCS	4	7.5	13	(14.6); 100)	(14.9; 100)	(14.6; 100)
			26	(14.6; 100)	(14.9; 100)	(14.6; 100)
			82	(22; 100)	(26; 100)	(33; 100)

From an analysis of the data obtained, we can conclude that the proximity of the compared registrars by the studied parameter was significant.

All experimental studies were conducted at IS of Tomsk Polytechnic University [26]. The small-sized betatron MIB-9 was used as the BS, operating in the dual energy mode with energies E_L = 4 MeV and E_H = 7.5 MeV. Test objects were moved in front of the BS+BR systems with a special scanner. The detectors mentioned in Table 7.4 were used as BR.

The expected maximum productivity refers to the number of characteristics that are not evaluated experimentally. The use of the BR X-Scan LCS is a priority for this feature.

Note 3. Some of the IS characteristics described above depend on the number of particles n_{L1} and n_{H1}, falling on the front surface RST. The values of n_{L1} and n_{H1} are determined by the current values of the radiation powers P_L, P_H, the focal length F, and the fraction of the front RST surface irradiated by the bremsstrahlung beam. Ideally, the noted fraction is equal to unity, but in practice this may not be, due to the complexity of the mutual orientations of BS, collimator and BR.

Note 3 leads to the conclusion about the difficulties associated with maintaining uniform measurement conditions for the compared detectors. The noted difficulties increase by a factor in the case of experimental comparison of various IS.

Contrast sensitivity and calibration dependences $Q(Y_L)$ for different levels of Z_i, $Z_i \in \mathbf{Z}$ should be distinguished from the experimental characteristics of IS. These characteristics are relative. They are estimated from large-volume samples, which makes them significantly less dependent on the dose rate of bremsstrahlung.

A braided wire with a central core diameter of 2.5 mm aluminum was selected for testing. This wire is equivalent to a copper wire with a diameter of 1.3 mm in mass maximum thickness.

For illustration Figure 7.3 shows an optical image of a test object with a wire and the corresponding radiographic image for E_H = 7.5 MeV, obtained using BR based on X-Scan LCS detectors. The area of the wire in the image is contrasted. The wire is confidently detected.

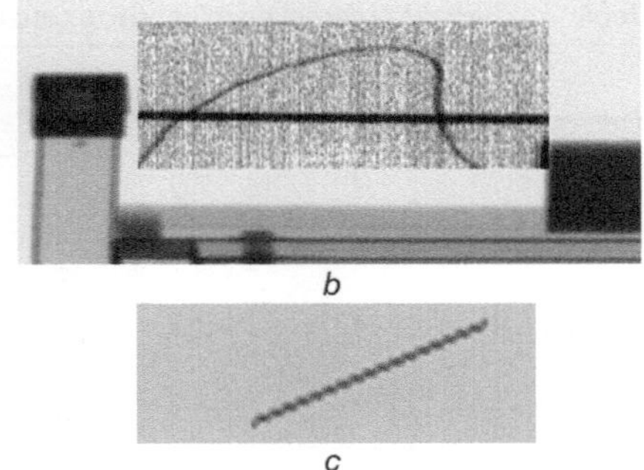

Figure 7.3 Photographic (a), radiographic (b) and simulated (c) images of TO with wire.

For the second type BR, the calculated value of the relative signal caused by the wire was 1.2%, and the experimental value is close to 1.4%. The level of discrepancies between the calculated and experimental estimates is acceptable, which makes it possible to use the above approach to compare BR by this parameter.

A steel standard sample with grooves, similar to that described in [27], was used to estimate the spatial resolution horizontally d_h and vertical d_v. The estimation of the horizontal spatial resolution is determined by the RST size in the scanning direction a and the scanning speed v_x.

Figure 7.4 gives typical radiographic and simulation images of TO for E_H = 7.5 MeV with a spatial resolution standard sample (SRS) for estimating spatial resolutions horizontally and vertically for v_x≈40 mm/s when measuring over the air. The width of the grooves varies from 2 to 8 mm in increments of 2 mm. The wire described above was also present in TO. To improve the perception in Figure 7.4 areas of interest are contrasted.

The experimental estimates of the horizontal spatial resolution d_h and vertical dv for the above-mentioned scanning speed v_x for X-Scan LCS are close to 6 mm. The detection of aluminum wire with a diameter of 2.5 mm is also confirmed.

Comparison of estimates of spatial resolution horizontally d_h and vertical d_v by the simulation method for BR X-Scan LCS and TSNK confirms their proximity to each other and to experimental estimates. This leads to the conclusion about the effectiveness of the approach to comparing IS with the analyzed parameter by the method of simulation.

Note that the experimental horizontal SRS for the TSNK bremsstrahlung recorder was significantly better than for the X-Scan LCS. This is because the size of the detectors in the scanning direction for TSNK BR is smaller. In addition, the actual size of the radiation sensitive element is even smaller; this is due to the detector array displacement relative to a narrow radiation beam.

The penetration capacity of $H_{\lim H,L}$ was measured according to the standard [27]. When measuring for the BR X-Scan LCS, there were certain

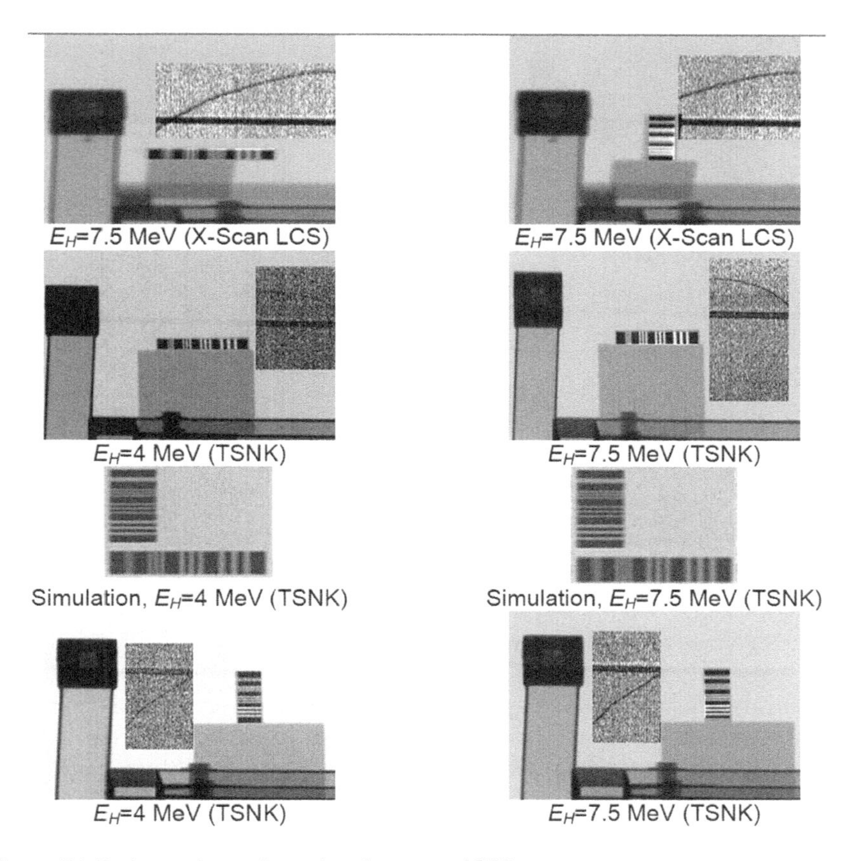

Figure 7.4 Radiographic and simulated images of SRS.

difficulties associated with the limited detector array size, which made it difficult to make gain calibration.

Figure 7.5 shows the photography and radiographic images of STO with a total thickness of 210–220, 232, 250, and 270 mm. The central parts of the radiographic images are highlighted and contrasted.

From the analysis of the data shown in Figure 7.5, we can conclude that the penetration capacity of $H_{\lim L}$ for $E_L = 4$ MeV is about 232 mm, and for $E_H = 7.5$ MeV the penetration capacity of $H_{\lim H}$ exceeds 270 mm. Further refinement of the penetration capacity for the registrar under consideration for energy $E_H = 7.5$ MeV is limited by the length of the BR X-Scan LCS, therefore, for further experiments, we will use $H_{\lim H} = 270$ mm. Comparison of radiographic images obtained by the method of full-scale experiment and simulation, confirms their proximity. This indicates the effectiveness of the simulation algorithms to estimate penetration capacity at IS design stage.

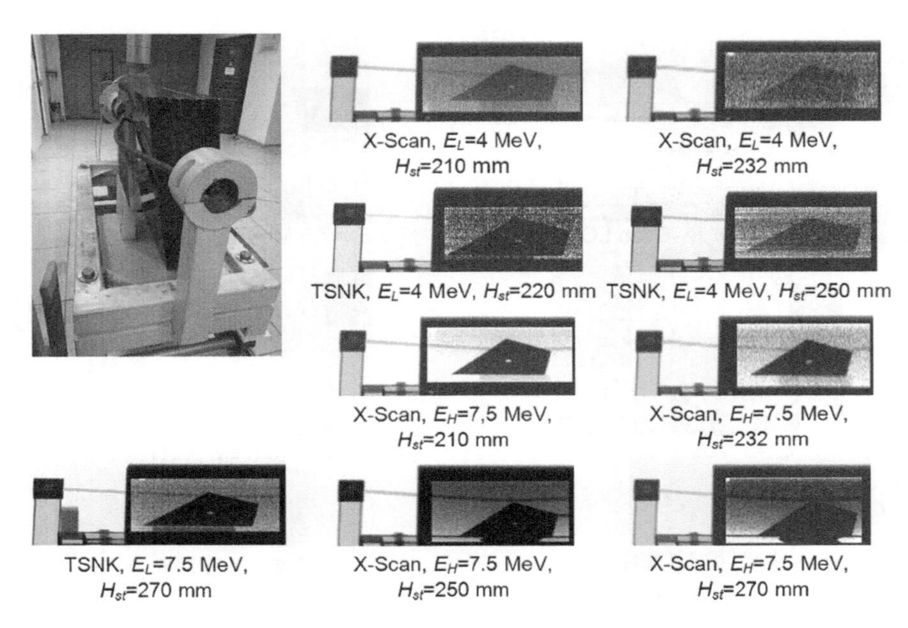

Figure 7.5 Photography and radiographic images of STO.

Contrast sensitivity was experimentally evaluated close to the standard [27], but rectangular steel sheets were used instead of the boom tip.

Table 7.5 shows the characteristics of TO consisting of main plates and pairs of rectangular steel sheets of various thicknesses. Expected level was marked k_X, %.

From an analysis of the results presented, it can be concluded that the experimental images for the considered BR and simulation images of TO are similar to the standards for estimating kX. The boundaries of fragments in the images are quite clear. Estimates kX obtained by the method of full-scale and computational experiments are close to each other and to the expected values from Table 7.5.

Figure 7.6 illustrates optical, radiographic images of the whole TO and the gray marked zone of interest. Images were obtained from X-Scan LCS. Figure 7.7 shows radiographic images of a test object and area of interest for the TSNK detector.

Table 7.5 Test object characteristics

H_{st}, mm	220	220	110	25
H_{a1}, mm	5	5	3	1,5
H_{a2}, mm	1.5	3	1.5	0.35
k_X, %	0.7	1.5	1.4	1.4

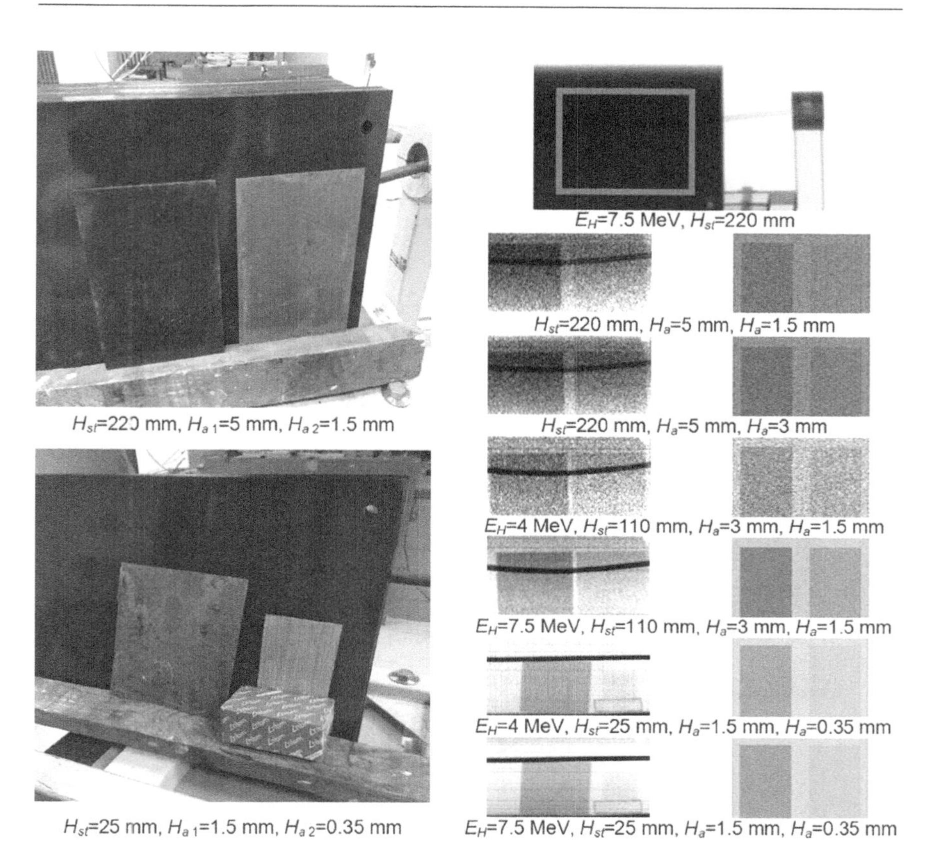

Figure 7.6 Photographic, radiographic and simulated images of TO for the X-Scan LCS system.

The calibration dependences $Q(Y_L)$ intended for recognition by the method of level lines were measured for TO, consisting of fragments with various mass thickness and made from various materials. Fluoroplastic was chosen as a typical representative of organic materials, aluminum represents metals with a small EAN, steel does for metals with an average EAN level, and lead represents heavy metals.

Figure 7.8 shows the experimental dependences $Q(Y_L)$ for the aforementioned classes of materials for the studied bremsstrahlung detectors for maximum energies E_L = 4 MeV, E_H = 7.5 MeV.

From a joint analysis of the data presented in Figures 7.2 and 7.8, we can conclude that the use of linear regressions to approximate experimental dependencies $Q(Y_L)$ is acceptable.

Figure 7.9 shows a top view of a TO with fragments of materials from the classes in question, a calibrated radiographic image and a PR image of the test object in the case of using the BR X-Scan LCS.

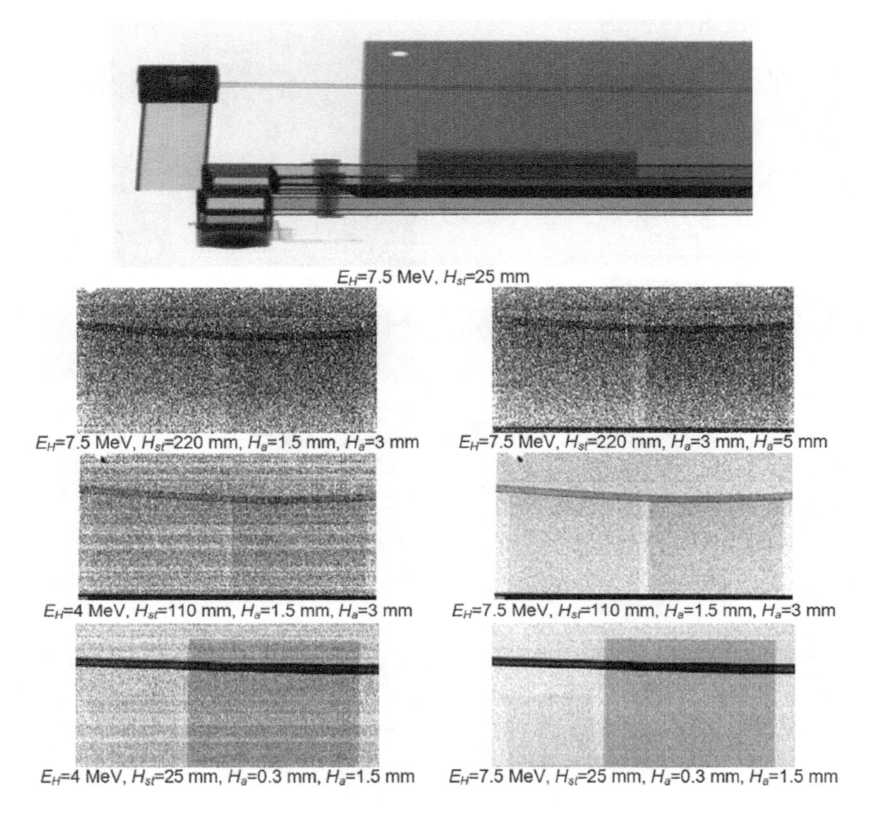

Figure 7.7 Photographic, radiographic and simulated images of TO for the TSNK system.

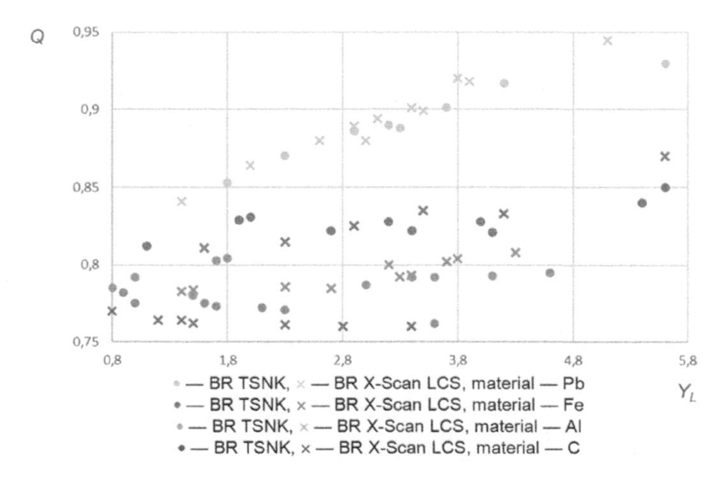

Figure 7.8 Experimental dependencies of recognition parameter on the thickness of material $Q(Y_L)$.

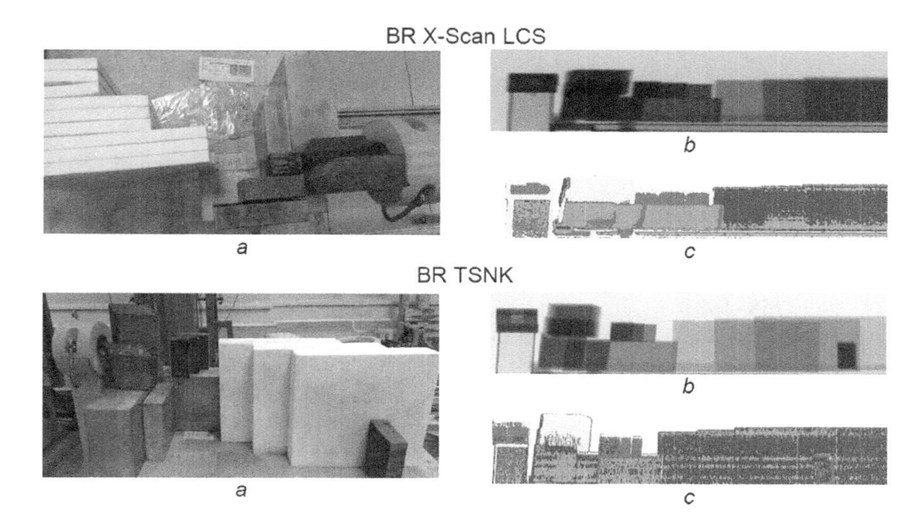

Figure 7.9 Images of the test object. (a) Is top view. (b) Is radiographic image. (c) Images with recognition parameter applied.

As a result of the analysis of the studied dependencies shown in Figure 7.8, and the images in Figure 7.9 for comparable types of rubber goods, it can be concluded that the recognition of materials from the above classes is reliable in the ranges of changes in mass thicknesses noted in Table 7.4.

7.5 CONCLUSIONS AND RESULTS

The methodology to compare the quality of the ISs with various bremsstrahlung detectors was developed based on a simplified model of the formation and processing of information in those systems with an option for recognizing materials. To estimate the feasibility of usage the developed methodology in practice, a series of calculations, computational and full-scale experiments were conducted to compare the bremsstrahlung recorders manufactured by TSNK (Russia) and X-Scan LCS (Finland). The IS characteristics for the studied detectors are close to each other, with the exception of inspection performance, which is 2.33 times higher for the BR X-Scan LCS compared to the BR TSNK. The developed IS comparison methodology takes into account most of their parameters; therefore, it can be used to compare not only BR, but also IS as a whole. The effectiveness of the approaches based on calculations and simulation for the comparison of ISs with the material recognition option is proved, which makes these approaches effective at the design stage of the analyzed systems.

REFERENCES

1. Lee D., Lee J., Min J., Lee B., Lee B., Oh K., Kim J., Cho S. Efficient material decomposition method for dual-energy X-ray cargo inspection system. *Nuclear Instruments and Methods in Physics Research Section A: Accelerators, Spectrometers, Detectors and Associated Equipment*, 2018, 884, 105–112.
2. Shikhaliev P. M. Megavoltage cargo radiography with dual energy material decomposition. *Nuclear Instruments and Methods in Physics Research Section A: Accelerators, Spectrometers, Detectors and Associated Equipment*, 2018, 882, 158–168.
3. Langeveld W.G.J. Comparison of dual-energy, Z-SCAN, and Z-SPEC material separation techniques for high-energy x-ray cargo inspection. *AIP Conference Proceedings. AIP Publishing LLC*, 2019, 2160(1), 050018.
4. Osipov S.P., Usachev E.Y., Chakhlov S.V., et al. Limit capabilities of identifying materials by high dual-and multi-energy methods. *Russian Journal of Nondestructive Testing*, 2019, 55(9), 687–699.
5. Bairashewski D.A., Drobychev G.Y., Karas V.A., Komarov V.V., Protsko M.V. Development of the X-ray security screening systems at ADANI. *International Conference on Engineering of Scintillation Materials and Radiation Technologies*. Springer, Cham, 2018, 249–259.
6. Bendahan J. Vehicle and cargo scanning for contraband. *Physics Procedia*, 2017, 90, 242–255.
7. Cui Y., Oztan B. Automated firearms detection in cargo x-ray images using RetinaNet. Anomaly Detection and Imaging with X-Rays (ADIX) IV. *International Society for Optics and Photonics*, 2019, 10999, 109990P.
8. Osipov S.P., Chakhlov S.V., Osipov O.S., Li S., Sun X., Zheng J., Hu X., Zhang G. Physical and technical restrictions of materials recognition by the dual high energy X-ray imaging. *International Journal of Applied Engineering Research*, 2017, 12(23), 13127–13136.
9. Slavashevich I., Pozdnyakov D., Kasiuk D., Linev V. Optimization of physicotopological parameters of dual energy X-ray. *International Conference on Engineering of Scintillation Materials and Radiation Technologies*, Springer, Cham, 2018, 261–269.
10. Chen Z. Q., Zhao T., Li L. A curve-based material recognition method in MeV dual-energy X-ray imaging system. *Nuclear Science and Techniques*, 2016, 27(1), 25.
11. Wang X.W., Li J., Tang C.X., Chen Z.Q., Zhong H.Q. Material discrimination by high-energy X-ray dual-energy imaging. *High Energy Physics and Nuclear Physics*, 2007, 31, (11), 1076–1081.
12. Ogorodnikov S., Petrunin V. Processing of interlaced images in 4–10 MeV dual energy customs system for material recognition. *Physical Review Special Topics-Accelerators and Beams*, 2002, 5(10), 104701.
13. Nattress J., Nolan T., McGuinness S., et al. High-contrast material identification by energetic multiparticle spectroscopic transmission radiography. *Physical Review Applied*, 2019, 11(4), 044085.
14. Harms J., Maloney L., Erickson A. Low-dose material-specific radiography using monoenergetic photons. *Plos one*, 2019, 14(9), 0222026.

15. http://scantronicsystems.com/en/products/idk_train_st2630t/.
16. http://www.nuctech.com/en/SitePages/ThDetailPage.aspx?nk=PAS&k=AIFEEF.
17. https://www.rapiscansystems.com/en/products/rapiscan-eagle-r60.
18. Osipov S.P., Chakhlov S.V., Osipov O.S., Shtein A.M., Strugovtsev D.V. About accuracy of the discrimination parameter estimation for the dual high-energy method. *Materials Science and Engineering Conference Series*, 2015, 81(1), 2082.
19. Yaffe M.J., Rowlands J.A. X-ray detectors for digital radiography. *Physics in Medicine & Biology*, 1997, 42(1), 1–39.
20. Maddalena F., Tjahjana L., Xie A., et al. Inorganic, organic, and perovskite halides with nanotechnology for high–light yield x-and γ-ray scintillators. *Crystals*, 2019, 9(2), 88.
21. Berger M.J., Hubbell J.H., Seltzer S.M., Chang J., Coursey J.S., Sukumar R., Zucker D.S., Olsen K. *XCOM: Photon Cross Sections Database*, National Institute of Standards and Technology, 2010, http://www.nist.gov/pml/data/xcom/index.cfm/.
22. Chadwick M.B., Obložinský P., Herman M., et al. ENDF/B–VII. 0: Next generation evaluated nuclear data library for nuclear science and technology. *Nuclear Data Sheets*, 2006, 107(12), 2931–3060.
23. Zhang G., Zhang L., Chen Z. An HL curve method for material discrimination of dual energy X-ray inspection systems. *IEEE Nuclear Science Symposium Conference Record, 2005*, IEEE, 2005, 326–328.
24. Osipov S.P., Usachev E.Y., Chakhlov S.V., et al. Identification of materials in fragments of large-sized objects in containers by the dual-energy method. *Russian Journal of Nondestructive Testing*, 2019, 55(9), 672–686.
25. Scientific Educational Cargo Vehicle Inspection System. http://portal.tpu.ru/departments/laboratory/rknl/eng/products/iDK.
26. Detection Technology Plc. https://www.deetee.com/x-scan-d-series.
27. American national standard for determination of the imaging performance of X-ray and gamma-ray systems for cargo and vehicle security screening, Standard ANSI N42.46, 2008.

Chapter 8

Automated medical report generation on chest X-ray

Images using co-attention mechanism

B. K. Tripathy, Rahul Sai R.S, and Sharmila Banu K.
VIT Vellore

CONTENTS

8.1 INTRODUCTION

Every problem is decomposed into several sub-problems by human beings, and further subdivisions are carried out. Deep neural networks (DNNs) [5] are modeled as a combination of ANN [6] and deep learning. As a branch, it has several types that are suitable for different applications [5,7,8,24,25]. It has high classificatory capability if the number of hidden layers is chosen efficiently [6]. Convolutional neural networks (CNNs) are framed by using the convolutional operator instead of matrix multiplication in at least one of the intermediate layers [9]. The convolution operator has the benefits of sparse interactions, parameter sharing, and equivalent representations in order to improve the machine learning (ML) process. Besides the convolutional operator, a concept called "pooling" is used in CNNs. Pooling is a technique used to summarize the information the detected features hold. In some variants of the CNN instead of using the pooling operation, several convolutional operators are used [23]. CNNs are used in localization, detection, semantic segmentation, speech recognition, and visual question answering systems. Introduced in the late 1980s by Dr. Yann LeCun, the first CNN was used to identify handwritten digits (LeNet) and was deployed to read postal codes

DOI: 10.1201/9781003381167-8

and bank cheques in the early 1990s. The working of a CNN is very similar to that of the human visual cortex. A typical convolutional layer will multiply the image (pixel-wise) with multiple predefined filters in order to identify the features. To extract the visual features of the image in this chapter, we make use of CNNs. If we consider the scenario for medical images, say chest X-ray images, which look simple to the naked eye. But highly trained radiologists can mark multiple areas of the images and can write down clear reports for potential abnormalities, which is time-consuming and such radiologists or pathologists are rarely available in rural areas. Therefore, this chapter focuses on generating detailed medical reports using chest X-ray images, which can facilitate the diagnosis of various respiratory and cardiovascular diseases.

The body of work focused on automating the process of medical report generation from images is already highly extensive. The concept of using semantic attention is discussed in reference [1] where the authors first discuss the top-down paradigm and next use the bottom-up paradigm Both paradigms suffer from their own weaknesses, such as lack of attention to fine details in the top-down approach and the lack of end-to-end procedures from individual features to sentences in the bottom-up approach. Their approach involves the detection of semantic attributes using the bottom-up approach, which they call candidates, and then they employ a top-down approach in order to select which candidates require more attention in order to yield better results. A similar concept of caption generation is discussed in reference [2] where the authors discuss the concept of scene understanding with the help of visual attention.

The approach defined by the authors in reference [3] introduces a new technique known as fine-grained label learning, to improve the quality of the generated reports in order to approach the clinically acceptable limit.

8.2 MATERIALS AND METHODS

8.2.1 Dataset

The dataset to be used in this project is the Indiana University chest X-ray (CXR) Image dataset [4]. It is a high-resolution CXR dataset with multiple views. There are 7,470 images accompanying 3,955 well-written reports encoded in Extended Markup language (XML). These XML reports have references to the CXR images, the findings, impressions, and the indication from the CXR images, along with the metadata of the patient and hospital which provided them. These CXR images were obtained from patients diagnosed with tuberculosis, pneumonia, and various heart ailments.

8.2.2 Experimental setup

The model was trained over a Google Colab Pro instance. The processor is a server-grade Intel Xeon processor with support for multithreading, which helps with the initial preprocessing and streaming of the data during the training phase. The GPU is an essential requirement for training deep learning models. The Nvidia Tesla P100 has 16 GB of HBM2 memory along with 3,584 CUDA cores. The Colab instance also has 28 GB of usable RAM and disk storage of 150 GB which provides us plenty of memory in order to retain the model data, weights, and other information. The architecture of the proposed system is shown in Figure 8.1.

8.2.2.1 Natural language processing (NLP) pipeline

The findings, impressions, and indications obtained from the reports are properly cleaned for use in the model. The steps are:

1. Converting all characters into lowercase
2. Removal of contractions from the text e.g., won't – will not, can't – cannot.
3. Removing punctuation from text with the exception of full stop, as the findings from the reports may contain more than one sentence.
4. Removing all numbers and redacted data from the text.
5. Removing smaller words and adverbs with the exception of "no" as it adds significant value. e.g., adverbs such as "there" and "then".
6. Tokenization and addition of identifier tokens such as "_start" and "_end" tokens, which are necessary in the text generation process.

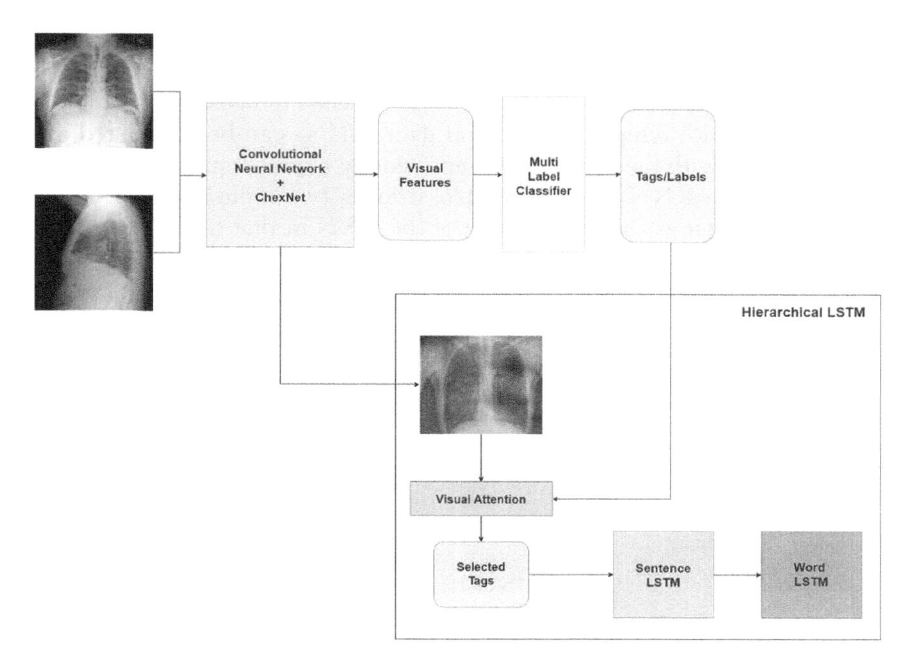

Figure 8.1 Architecture of the proposed system.

8.2.2.2 Convolutional neural network

The CNN acts as the encoder in our model and provides us a feature vector with the visual features of the image, which can be used to predict the tags for that particular X-ray. This is done by considering the problem as a multi-label classification task where the output layer provides with the probability of a set number of tags. These predicted tags help us massively in text generation, which is discussed in the next section. However, the size of our dataset (7,470 images) is not enough to train a CNN properly. To remove this bottleneck, we have to use a transfer learning framework. Most transfer learning frameworks such as VGG16 or InceptionV3 are trained over generic image datasets, which doesn't serve our purpose.

Fortunately, ChexNet [10] is a convolutional neural network especially trained on chest X-ray images. It was trained over 112,120 images and contains 121 layers where the input is a chest X-ray image, and the output is the probability of 14 different diseases along with a localized heatmap that highlights the visual features of the chest x-ray image. However, we do not need to classify the image into one of those 14 categories, so we can remove the final classification layer. From an image of dimensions (224, 224, 3), we get a feature vector with a length of 1,024. An example image and the corresponding features have been provided below in Figure 8.2. We have two images associated with a report, so we concatenate the two feature vectors to get a final feature vector with length of 2,048.

8.2.2.3 Recurrent neural network

The text generation process of the application is handled by recurrent neural networks (RNNs) [11, 22]. RNNs are a special class of neural networks that usually work with temporal/sequential data. RNNs can be considered as a directed graph with the connections representing a time sequence. The working process of a RNN is pretty simple; it takes two input vectors and one of them is known as the hidden state of the RNN in that time step. After a

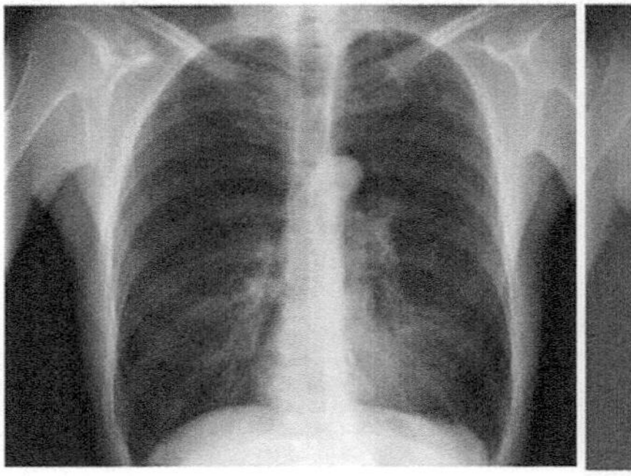
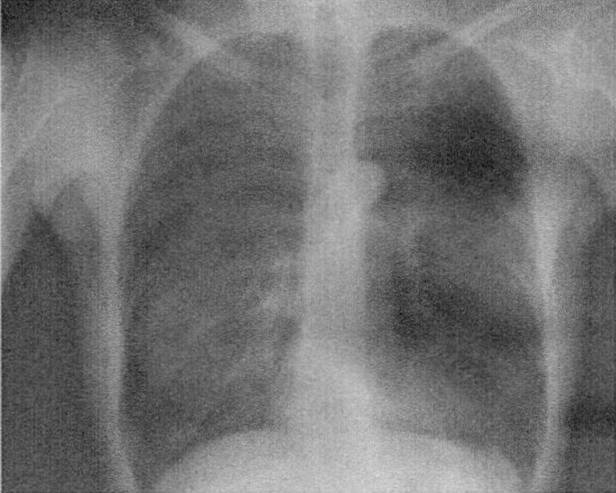

Figure 8.2 Obtaining the image features from the ChexNet.

particular time-step, the hidden state of the RNN is updated and it generates an output vector. Figure 8.3 provides the basic working principle of RNN.

The next timestep will utilize its hidden state (which has the temporal context stored from the previous time step) and the next input vector to generate the next output vector. This is depicted in Figure 8.5. The hidden state opens up a storage mechanism for the RNN, which can be additionally replaced by other networks/cells with feedback mechanisms and time delays. These controlled states form a part of Long Short Term Memory (LSTMs) or Gated Recurrent Unit (GRU), which are also known as feedback neural networks.

However, when there are multiple inputs involved (image and text), a sequence-to-sequence model is better suited for the purpose. A sequence-to-sequence model consists of an encoder and a decoder where both are based on RNNs. A simple encoder and decoder network can be built with a series of LSTM cells. The encoder network works by understanding the input sequence and returns a reduced representation of it, in the form of a feature vector. The decoder network uses this feature vector and generates an output sequence according to its own understanding. However, at each timestep of the decoder, the LSTM cells generate the probability distribution of the next word as the output; therefore, it needs to make a decision on which word should be next in the sequence. A good approach would be to use a beam search (as illustrated in Figure 8.4) which takes the probability of next k words and chooses the best approach to maximize the probability.

The regular seq2seq model passes the final hidden state from the encoder to the decoder, which means the intermediate hidden states are lost, thereby losing the context information from the intermediate states. To fix this, we

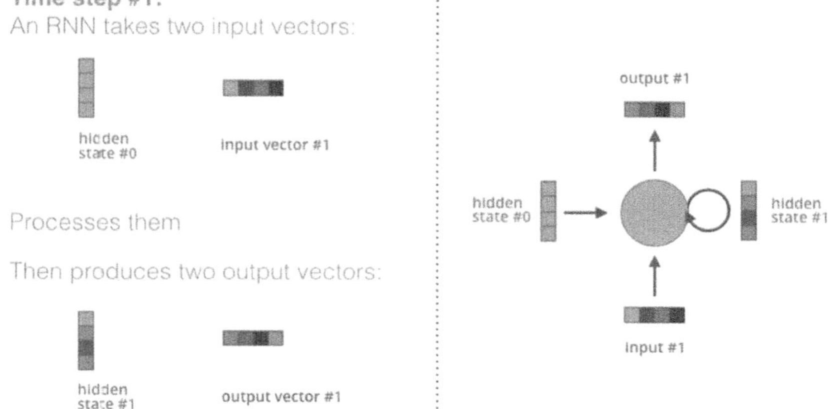

Figure 8.3 Basic working of an RNN.

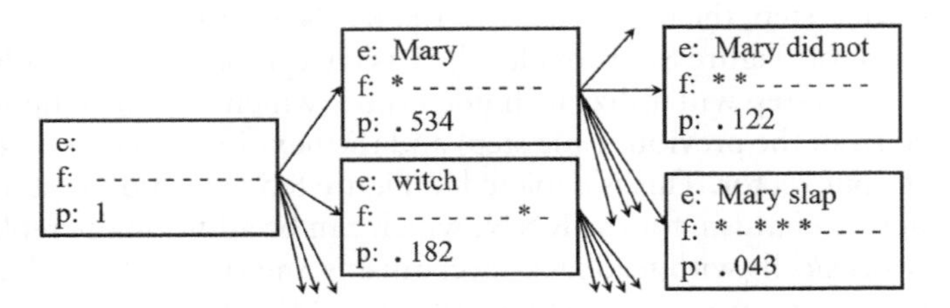

Figure 8.4 Beam search to find k subsequent words.

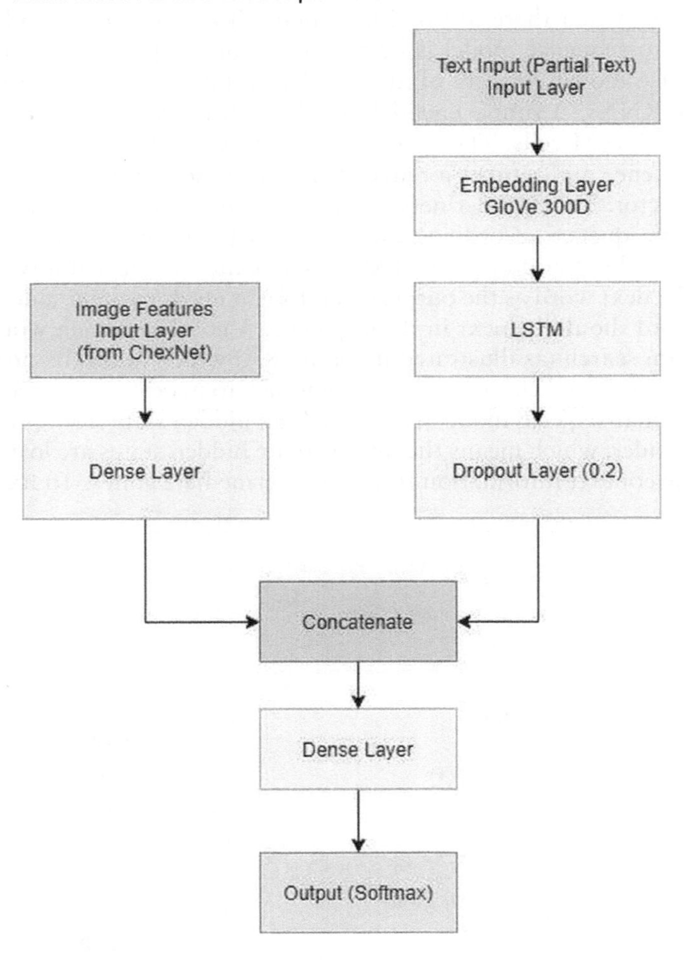

Figure 8.5 Depicting merge architecture for report generation.

can store the intermediate hidden states [12], to greatly improve the quality of the generated caption, as we can recover the lost contextual information. This is known as attention mechanism. It is analogous to how humans pay attention to specific parts of a text or an image to get the most important sections/regions in the text or the image at a particular time. To implement the attention mechanism, the hidden states are assigned weightages according to the importance of the contextual information. Now, that our training data contains images and text, the attention mechanism needs to be adjusted to simultaneously understand both the text content and the visual content of the data. Yu et al. [13] define a co-attention approach which is used for visual question answering. Hierarchical LSTMs [14] are specialized RNNs based on the encoder-decoder architecture which are often used for text generation from images and video frames. Here, it is built to consider both high-level language features from the training text and low-level visual features obtained from the processed image. As the keywords/tags obtained from the image results in a loss of spatial information, an additional mechanism known as co-attention is added. Co-Attention mechanism uses the spatial information from the visual features of the convolutional layers and the semantic features obtained from the tags of the specific image.

At the decoder, the high-level spatial information provided by the localized heatmap helps us focus on the tags that are visually highlighted more, yielding better results. This new context vector with the embeddings of the selected tags is passed on to a sentence LSTM generating multiple sentences as suggestions using beam search which predicts the probability distribution across the given vocabulary and returns the words which have the maximum probability. Beam search selects multiple alternatives for an input sequence, based on conditional probability and a parameter known as beam width. When the sentence vector is successfully produced, it is passed on as a context vector to the Word LSTM, which employs a greedy search mechanism in selecting a single candidate suitable for the input sequence in a time step, which improves the quality of the final sentence.

8.2.2.4 Sequence-to-sequence model with co-attention mechanism

a. During the training phase, the preprocessed reports from the NLP pipeline along with the two associated images (training data) form the inputs to the neural network.
b. The images are then augmented in order to avoid overfitting and increasing the amount of data available to the CNN. This is done using the Image Augmentation utilities which are available in the TensorFlow library. The text is converted into their embedding form using a GloVe embedding (300 dim).
c. The images are then passed through the ChexNet CNN along with a few custom CNN layers which allows us to get a visual feature context vector. The visual features of both images are concatenated and

sent to the visual attention layer. This is the encoded context vector in our sequence-to-sequence model.

d. The visual features are also sent to a multi-label classifier, which predicts the tags/labels from the given images which are sent to the input layer as partial reports.
e. Once the inputs are encoded, they are passed to the hierarchical LSTMs which are specialized RNNs used for image captioning. There are multiple methods for injecting an image in an RNN as mentioned in reference [15], but the merge architecture suits our use-case the best. Using the merge architecture ensures that our RNN is not exposed to the visual features until the partial reports are prefixed. Due to this late binding technique, the RNN does not modify the image representation with every time step. The merge architecture has been illustrated in Figure 8.5.
f. The RNN outputs are regularized using a dropout layer, and the "merged" with the normalized image input vector, which is then passed through the dense layer. The output layer has a SoftMax function, which then generates the probability distribution across the words present in the vocabulary.

Let us consider the inputs Images and Labels (partial text) below. The objective of the decoder is to get the most probable next word in the sentence at each time step. In the training phase, the target tokens are provided to the hierarchical LSTM. This is repeated till the <end> token is obtained.

Image: Visual feature vector from the image encoder (frontal+lateral CXRs)

Labels (partial text): <start>, token1, token2, token3....

Target (output): token1, token2 ...... <end>

Table 8.1 provides the Inputs and Outputs at each timestep of the decoder.

Table 8.1 Inputs and outputs at each timestep of the decoder

Timestep	*Input (partial text)*	*Output word*
1	<start>	the
2	<start>, the	right
3	<start>, the, right	basilar
4	<start>, the, right, basilar	opacities
5	<start>, the, right, basilar, opacities	favored
6	<start>, the, right, basilar, opacities, favored	to
7	<start>, the, right, basilar, opacities, favored, to	represent
8	<start>, the, right, basilar, opacities, favored, to, represent	atelectasis
9	<start>, the, right, basilar, opacities, favored, to, represent atelectasis	<end>

8.3 RESULTS AND DISCUSSION

The generated medical reports can be compared and evaluated using metrics such as BLEU, METEOR, and ROUGE scores. BLEU (Bi-Lingual Evaluation Understudy) scores range from 0 to 1 and compare the n-gram sequence of the generated output sentences and the expected output sentences. It uses a modified precision metric that compares the number of matching words in the candidate text (generated text) and the reference text (expected text) irrespective of their position. It heavily penalizes the repetition of words, which is common in image captioning and machine translation. The METEOR and ROUGE scores also work similarly, but they also measure the ordering of the text sequence and the overlap of words in the sentences. These metrics measure the similarity of language between the reference text and the generated text, but they do not evaluate the clinical correctness of the reports.

From the metrics provided in Table 8.2 and the various correct and incorrect captions in the figures provided in Table 8.3, we can see that the system provides accurate reports for numerous respiratory and cardiovascular diseases. In some cases, it is able to even speculate the diagnosis accurately.

8.4 CONCLUSIONS AND FUTURE WORKS

The proposed deep learning architecture helps us to automate the generation of text reports for chest X-ray images. From the results and the comparison metrics, it can be observed that the system is accurate, efficient, and can assist medical professionals in a productive manner. The multi-step technique with hierarchical LSTMs helps address the missing spatial context in image captioning with the use of co-attention mechanism that identifies both the visual information from the image and the semantic information from the text in order to accurately describe localized regions of interest/abnormality. The diagnosis being not correct in some cases

Table 8.2 Comparison of metrics with state of the art models

Methods	*BLEU-1*	*BLEU-2*	*BLEU-3*	*BLEU-4*	*METEOR*	*ROUGE*
Cascade RNN model [16]	0.399	0.251	0.168	0.118	0.263	0.287
Co-attention [17]	0.300	0.218	0.165	0.113	0.149	0.279
MVH+Attn+MC [18]	0.529	0.372	0.315	0.255	0.343	0.453
Recurrent attention [19]	0.464	0.358	0.270	0.195	0.274	0.366
NLG+CCR [20]	0.313	0.206	0.146	0.103	0.251	0.306
Multi-attention+BG [21]	0.476	0.340	0.238	0.169	0.498	0.347
Simple encoder-decoder	0.198	0.227	0.286	0.314	0.451	0.475
With visual co-attention	0.213	0.258	0.325	0.381	0.516	0.578

Table 8.3 Comparing the true and predicted reports

Image	*True caption*	*Predicted caption*
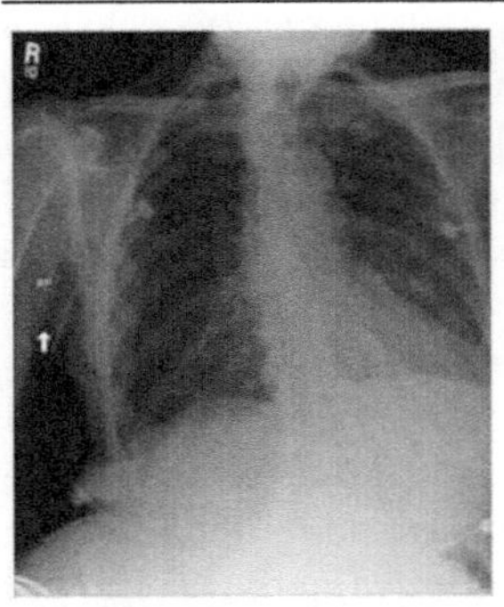	Emphysematous change without acute radiographic pulmonary process	Chronic changes consistent with emphysema. no acute cardiopulmonary abnormality detected
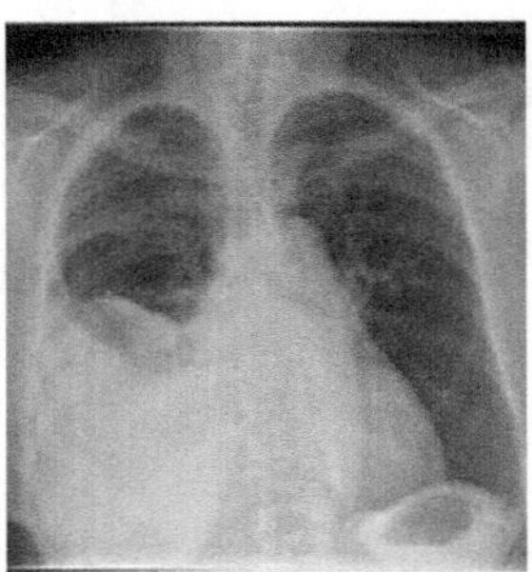	Large right pleural effusion with associated passive atelectasis of the left lung	Partially loculated right pleural effusion with possible atelectasis
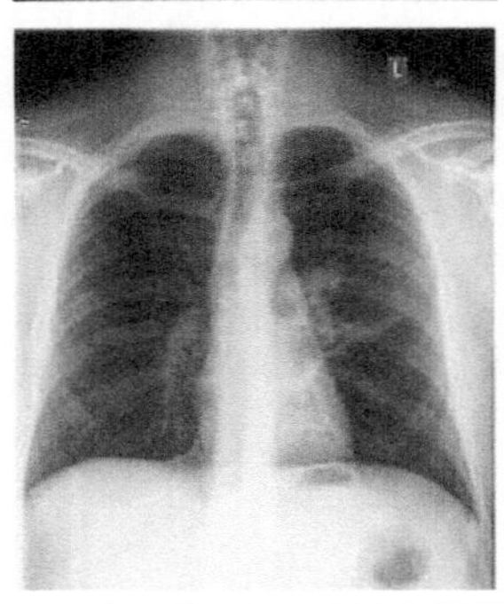	Left mid lung opacity noted most compatible with atelectasis versus infiltrate.	Possible atelectasis in the left lung, opacity noted

indicates that the system has not passed the prototypical phase yet and requires further fine-tuning.

Minor improvements carried out in the preprocessing phase greatly improved the model's accuracy. It is expected that better lip tracking and selection of the frames based on word length may improve the accuracy further. This step may reduce the RAM requirement during data preprocessing and hence improve complexity. This model can be extended to support multiple languages. Also, the fusion of audio in the early training that automates the window selection technique with the help of CNNs [25] can be used.

REFERENCES

1. You, Q., Jin, H., Wang, Z., Fang, C., Luo, J. Image captioning with semantic attention. In: *Proceedings of the IEEE Conference on Computer Vision and Pattern Recognition*, pp. 4651–4659. IEEE, Las Vegas, NV (2016).
2. Xu, K., Ba, J., Kiros, R., Cho, K., Courville, A., Salakhudinov, R., Zemel, R., Bengio, Y. Show, attend and tell: Neural image caption generation with visual attention. In: *International Conference on Machine Learning*, pp. 2048–2057. PMLR, Lille (2015).
3. Syeda-Mahmood, T., Wong, K. C., Gur, Y., Wu, J. T., Jadhav, A., Kashyap, S., Karargyris, A., Pillai, A., Sharma, A., Syed, A. B. Boyko, O. Chest x-ray report generation through fi- ne-grained label learning. In: *International Conference on Medical Image Computing and Computer-Assisted Intervention*, pp. 561–571. Springer, Cham (2020).
4. Demner-Fushman, D., Kohli, M. D., Rosenman, M. B., Shooshan, S. E., Rodriguez, L., Antani, S., Thoma, G. R., McDonald, C. J. Preparing a collection of radiology examinations for distribution and retrieval. *Journal of the American Medical Informatics Association* 23(2), 304–310 (2016).
5. Bhattacharyya, S., Snasel, V., Hassanian, A. E., Saha, S., Tripathy, B. K. *Deep Learning Research with Engineering Applications, De Gruyter Publications.* Volume 7 in the series De Gruyter Frontiers in Computational Intelligence, Berlin (2020). DOI: 10.1515/9783110670905
6. Adate, A., Tripathy, B. K. Deep learning techniques for image processing. In: S. Bhattacharyya, H. Bhaumik, A. Mukherjee & S. De (Eds.). *Machine Learning for Big Data Analysis.* Berlin, Boston: De Gruyter, 69–90 (2018).
7. Adate, A., and Tripathy, B. K. A survey on deep learning methodologies of recent applications. In: D. P. Acharjya, A. Mitra & N. Zaman (Eds.). *Deep Learning in Data Analytics- Recent Techniques, Practices and Applications.* Springer Publications, 145–170. Springer Nature Switzerland AG (2021).
8. Kaul, D., Raju, H., Tripathy, B. K. Deep learning in healthcare, in: deep learning in data analytics. In: D. P. Acharjya, A. Mitra & N. Zaman (Eds). *Deep Learning in Data Analytics- Recent Techniques, Practices and Applications.* Springer Publications, 97–115 (2021).
9. Maheswari, K., Shaha, A., Arya, D., Tripathy, B. K., Rajkumar, R. Convolutional neural networks: A bottom-up approach. In: S. Bhattacharyya, A. E. Hassanian, S. Saha & B. K. Tripathy (Eds.). *Deep Learning Research with Engineering Applications.* De Gruyter Publications, Berlin 21–50 (2020).
10. Rajpurkar, P., Irvin, J., Zhu, K., Yang, B., Mehta, H., Duan, T., Ding, D., Bagul, A., Langlotz, C., Shpanskaya, K., Lungren, M. P. Chexnet: Radiologist-level pneumonia de- tection on chest x-rays with deep learning. *arXiv preprint arXiv:1711.05225* (2017).
11. Tripathy, B.K., Baktha, K. Investigation of recurrent neural networks in the field of sentiment analysis. In: *The Proceedings of IEEE International Conference on Communication and Signal Processing (ICCSP17)*, Melmaruvathur (2017).
12. Bahdanau, D., Cho, K., Bengio, Y. Neural machine translation by jointly learning to align and translate. *arXiv preprint arXiv:1409.0473* (2014).

13. Yu, Z., Yu, J., Cui, Y., Tao, D., Tian, Q. Deep modular co-attention networks for visual question answering. In: *Proceedings of the IEEE/CVF Conference on Computer Vision and Pattern Recognition*, pp. 6281–6290. IEEE, Long Beach, CA (2019).
14. Gao, L., Li, X., Song, J., Shen, H. T. Hierarchical LSTMs with adaptive attention for visual captioning. *IEEE Transactions on Pattern Analysis and Machine Intelligence* 42(5), 1112–1131 (2019).
15. Tanti, M., Gatt, A., Camilleri, K. P. Where to put the image in an image caption generator. *Natural Language Engineering* 24(3), 467–489 (2018).
16. Shin, H. C., Roberts, K., Lu, L., Demner-Fushman, D., Yao, J., Summers, R. M. Learning to read chest x-rays: Recurrent neural cascade model for automated image annotation. In: *Proceedings of the IEEE Conference on Computer Vision and Pattern Recognition*, pp. 2497–2506. IEEE, Las Vegas, NV (2016).
17. Jing, B., Xie, P., Xing, E. On the automatic generation of medical imaging reports. In: *Proceedings of the 56th Annual Meeting of the Association for Computational Linguistics*, vol. 1 (Long Papers), pp. 2577–2586. Melbourne Convention and Exhibition Centre, Melbourne (2018).
18. Yuan, J., Liao, H., Luo, R. Luo, J. Automatic radiology report generation based on multi- view image fusion and medical concept enrichment. In: *International Conference on Medical Image Computing and Computer-Assisted Intervention*, pp. 721–729. Springer, Cham (2019).
19. Xue, Y., Xu, T., Long, L. R., Xue, Z., Antani, S., Thoma, G. R., Huang, X. Multimodal re- current model with attention for automated radiology report generation. In: *International Conference on Medical Image Computing and Computer-Assisted Intervention*, pp. 457–466. 21st International Conference on Medical Image Computing and Computer Assisted Intervention, Granada (2018).
20. Liu, G., Hsu, T. M. H., McDermott, M., Boag, W., Weng, W. H., Szolovits, P., Ghassemi, M. Clinically accurate chest x-ray report generation. In: *Machine Learning for Healthcare Conference*, pp. 249–269. PMLR, University of Michigan North Quadrangle Dining Hall, Ann Arbor, MI (2019).
21. Huang, X., Yan, F., Xu, W., Li, M. Multi-attention and incorporating background information model for chest x-ray image report generation. *IEEE Access* 7, 154808–154817 (2019).
22. Tripathy, B. K., Anuradha, J. *Soft Computing- Advances and Applications.* Cengage Learning Publications, New Delhi (2015).
23. Bhardwaj, P., Guhan, T., Tripathy, B. K. Computational Biology in the lens of CNN, (Chapter 5). In: S.S. Roy & Y.-H. Taguchi (Eds.). *Handbook of Machine Learning Applications for Genomics, Studies in Big Data*, vol. 103. Springer Nature Switzerland AG (2021). ISBN: 978-981-16-9157-7, 496166_1_En.
24. Gupta, P., Bhachawat, S., Dhyani, K., Tripathy, B. K. A study of gene characteristics and their applications using Deep Learning, (Chapter 4). In: S.S. Roy & Y.-H. Taguchi (Eds.). *Handbook of Machine Learning Applications for Genomics, Studies in Big Data*, vol. 103. Springer Nature Switzerland AG (2021). ISBN: 978-981-16-9157-7, 496166_1_En.
25. Bose, A., Tripathy, B. K. Deep learning for audio signal classification. In: S. Bhattacharyya, A. E. Hassanian, S. Saha and B. K. Tripathy (Eds.). *Deep Learning Research and Applications.* De Gruyter Publications, Berlin, pp. 105–136 (2020).

Chapter 9

An energy-efficient secured Arduino-based home automation using android interface

Arghya Dasgupta, Souvik Roy, Sunetra Mukherjee, Shuvam Kabiraj, and Rajib Banerjee
DR. B. C. Roy Engineering College Durgapur

Pulakesh Roy
Kazi Nazrul University Asansol

Arindam Biswas
Kazi Nazrul University Asansol

CONTENTS

DOI: 10.1201/9781003381167-9

9.1 INTRODUCTION

Presently, technology has become an important part of our daily lives and has a continuous influence in areas such as social interaction, entertainment, safety, and the controlling of different types of devices by using devices such as mobile phones, computers, and other communicating devices and gadgets. This simplifies daily technological requirements and brings an immediate solution to different types of problems such as connecting with devices from a remote location, resulting in overall technological advancement superiority and energy saving. There are different types of communication and wireless support technologies available for connectivity issues such as Wi-Fi, Cellular Network, and Bluetooth, etc.

A technological evolution related to home automation systems can be built considering the modern technological aspect, which will be flexible to control the interfacing devices. Presently some of the important applications of home automation involve checking the status of the house, continuous monitoring of the place from any remote location, remotely controlling the home appliances such as air-conditioning, light, motor, heater, etc.

In recent times, IoT devices and its network is the front-runner in this type of application, which also increases the efficiency of the system. The main challenge of IoT-based home automation is the interoperability between heterogeneous systems in terms of systems, standards, and protocols.

The first home automation system was developed in 1966 using a home computer system. The home automation system is the integration of the electrical device in the home to control domestic activities such as lighting, controlling the home entertainment system, and other electrical appliances. The popularity of home automation systems increased day by day due to the introduction of efficient hand-held devices such as smartphones, ease of development, and low cost. The home automation system can be divided into three broad categories, which are:

- Individual device control system.
- Distributed control system.
- Centrally controlled system.

In the individual control system, the device is capable of controlling only one appliance at a time such as the timer, and programmable thermostats. In distributed control systems, appliances can communicate with each other mostly through the wired communication system. In the centrally controlled system, a central unit is responsible to control different devices in wireless/wired mode [1].

There are several protocols available for the home automation system such as Wi-Fi, KNX, ZigBee, Z-Wave, Bluetooth, etc. Among them, Wi-Fi and KNX are less suitable than the other protocols as they are wired base connections and they required a huge power to operate. In the home automation system, Z-Wave and Zigbee are efficient protocols, but the gateway using Z-Wave and Zig-Bee is comparatively expensive. For this, reason the Bluetooth base system is an effective solution for the home automation system. In Bluetooth base system, there is no requirement to design a gateway for setting up device communication with smartphones. Moreover, the advantage is that in modern days all smartphones use Bluetooth, which is energy-efficient. Therefore, it can be stated that Bluetooth is the ideal solution for the home automation system in recent times.

In this chapter, we propose an efficient IoT-based secure home automation base system by using RFID and Bluetooth technology. This system is an effective solution and an alternative to an expensive home automation system. The rest of this chapter consists of Literature Review in Section 9.2. The architecture explains the procedure and functionality of the proposed scheme are described in Section 9.3. The results are shown in Section 9.4. Finally, the work is concluded in Section 9.5.

9.2 LITERATURE REVIEW

Presently IoT devices are massively integrated with our daily life such as in factories, homes, and other public places, as well as in other applications. The research community has focused their work on this aspect. Some of the prominent state-of-the-art work is explicitly described below.

With the growing demand for energy and renewable energy sources [2,3], the smart grid system creates a new revolution [4,5]. In general, the smart grid refers to the transmission and distribution of electricity [6]. It is very much useful in different applications such as advanced meters [7], intelligent control equipment [8], etc. to control the power quality and control different types of equipment. For this, different researchers focus their work on the smart home using smart grids [9,10]. In [11] the author discusses the smart plug by using BLE (Bluetooth Low Energy). Here the author develops a system to control and monitor home appliances by using BLE where the devices are controlled by using Android application. Here the system provides an Application Programming Interface to control the devices based on GATT (Generic Attribute Profile). The system is not only used to control the application but also to monitor power measurement to consume the power by using timer and power limitation.

In another work [12], the author discusses the architecture to manage the smart home by using the BLE-based sensor and actuator. The proposed system is very much efficient for real-time monitoring based on IoT.

In [13] the author proposed the concept of radio communications intrusion detection based on IoT for the intrusions to connect with smart homes, smart factories for profiling and monitoring. The system was implemented by connecting with the object of Wi-Fi, ZigBee, Bluetooth, or other communication protocols. The experimental evaluation is done by deploying an approach in a smart home.

In another work [14], the advanced universal remote controller to control home appliances with maintaining security is proposed. Here all household devices can be easily controlled by URC (Universal Remote Control) by using the internet. Here the author used multiple receivers connected with all the devices in-house by wired or wireless communication method. Here the receivers used multiple IDs for different devices and multiple channels to control them at the same time by using URC. Here the author also developed a PC-based interface for URC connectivity. After the experiment, it is observed that the proposed model is very helpful to control all devices in buildings, offices, hospitals, schools, etc.

In one of the works [15], the author developed a home automation system by using Wi-Fi. Presently, many home appliances that have been developed are controlled remotely. However, it takes a huge cost for this, and integration between systems is very much essential to develop a home automation base system at a very low cost. For this here, the author developed a microcontroller-based home automation system by using NodeMCU, where the server can easily control all the electrical devices by using the web application after completing the authentication process.

In another relevant work [16], the author developed an energy-efficient home automation system by using the internet to control all the home equipment. For this, the system is connected to the internet modem. To make the connection here IP address is used. The entire system has been developed by using Google Assist or web applications according to the needs of the user. Here the authors focus their work to create an efficient home automation system with maintaining security. From the above discussions, it is clear that there are many ways of implementing home automation and this motivates us to design IoT-based home automation systems. However, designing a home automation system consuming very less energy and involving minimum hardware configuration and human intervention is always a challenge and a premium thrust area of research.

In another paper [20], the author discusses the low-cost smart home automation system by using Arduino and ESP32 Wi-Fi module to maintain the current state of home appliances where the devices are controlled by using the web portal. After the experiment, it is observed by the author that the proposed system is very much helpful not only to control the appliances but also to monitor power consumption, and checking the room temperature from anywhere in the world.

In one of the papers [21], the authors develop the home automation system using Raspberry Pi. Here the author introduces the home automation system by using a set of hardware components and software solutions. In this system, the Raspberry Pi is used as a microcomputer device, which is the major hardware element, connected with the sensor node to collect the data. Here the microcontroller runs on the developed software to organize, analyze and process the data, and manage different home appliances by using Wi-Fi. In this work, the system has been tested on different LED lights, and after the experiment, it is observed by the author that the system efficiently controls different appliances remotely.

In another paper [22], the author developed IoT-enabled home automation by using Android application. In this system, the author uses the MQTT broker to control the devices from a remote location. In this system, the author uses the Phidget kit to control the devices. To control the appliances here the author develops a user-friendly application for the mobile device. In this work, the system has been tested to control the appliances and after the experiment, it is observed that the system efficiently works to control the appliances without any difficulty.

9.3 SYSTEM ARCHITECTURE

The main objective of this work is to develop a secured home automation system to control the different home appliances. In this work, we developed a home automation system where the gate opening is secured through RFID. Once the gate security logic is activated, it initializes the Bluetooth unit to control all the devices in the house, office connected to it. As we know, it is very difficult to control all the devices separately, especially for the new user who doesn't have the knowledge of the controller operating which device and this process is also very time-consuming. In this kind of situation, the proposed system is very effective in developing the control through a single unit. The system architecture of the proposed scheme is shown in Figure 9.1.

The proposed system can be used to control all the devices very easily by using the Android application, which creates an easy interface between Android App and smart home devices. The whole system consists of an Arduino UNO board, Bluetooth module, Relay module, RFID module, and servomotor. Initially, to permit a security pass of the system an RFID module is used to identify the Unique ID of the system. Activation of security pass triggers on the servomotor and activates the Bluetooth module. The servomotor is used to unlock the gate, the Bluetooth module is used to create the interface between Android App and home automated devices to control the devices by using the relay switch.

The main purpose of the system is to control all the devices easily by using a single device while maintaining the security of the system and consuming lesser power. When any person wants to enter the house, then, in that case, the system needs the RFID to identify the person and thereafter it triggers on the system. The entire system can be easily handled by using

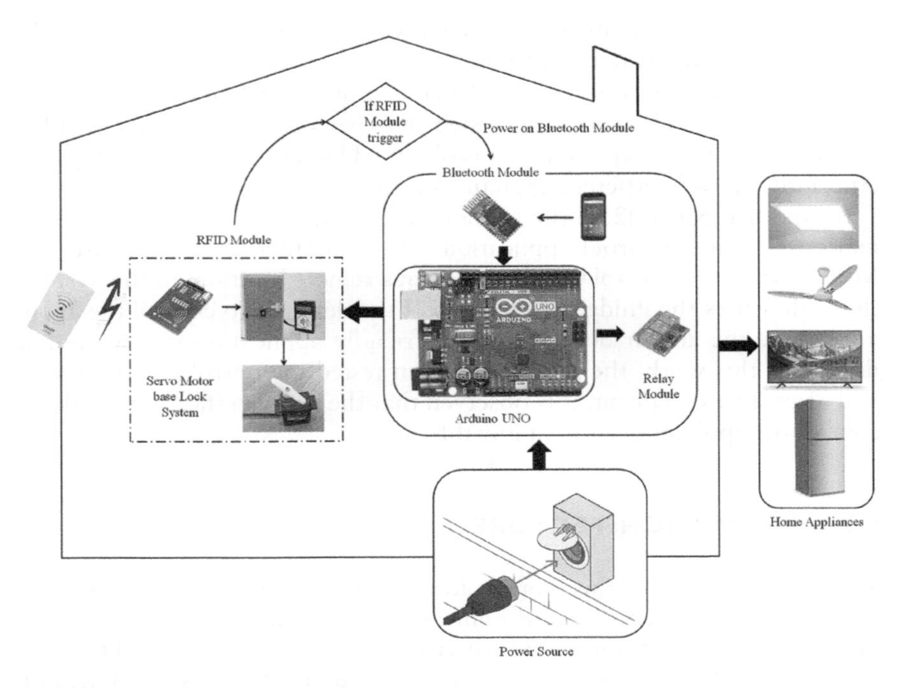

Figure 9.1 Architecture of the proposed system.

Android App with using the Bluetooth system. The whole system is placed inside the building to which all the devices are connected.

Efficiency of the system:

- Power consumption.
- User-friendly control
- High security
- Scalable home appliances
- Hassle-free connection.

9.3.1 Experimental devices

In this system, to control the devices, we use Arduino UNO as the microcontroller board. Moreover, here the RFID RC522 module for the user's unique authentication, Bluetooth module HC05 to create an interface with the smartphone, and the Relay switch to trigger the devices by using the application are also used. A brief discussion of all modules has been given below.

9.3.1.1 RFID-RC522 module

The RFID-RC522 is a 13.56 MHz Radio Frequency Identification module based on the MFRC552 microcontroller, which supports the I2C, and UART (Universal Asynchronous Receiver Transmitter) connection, which operates at a voltage of 3.3 V. Here an RFID card is used for user authentication. This card has a memory of 1 Kb to store the unique identification number where we can do both the read and write operations. This system is very much useful in systems such as user attendance systems, person identification systems, etc.

This system has a range of 5 cm and the data transmission rate is 10 Mbps. Also, it consumes a very low amount of power nearly about 13–26 mA where the minimum power consumption of this module is 10 μA. An illustration replicating the idea of gate unlocking system using RFID is given below (Figure 9.2).

9.3.1.2 Servomotor

A servomotor is one kind of electric motor where the operation is done by using servomechanism. It is the linear or rotary actuator for precise control of the linear position, velocity, and acceleration of the device.

In this system, we use the servomotor as a door lock, which is triggered when the RFID module is activated. Presently the remotely operated door lock becomes very much popular and very much demanding. For this here, we developed a simple circuit by using a simple servomotor for the room lock system. The key benefits of the system are easy to use and easily configurable.

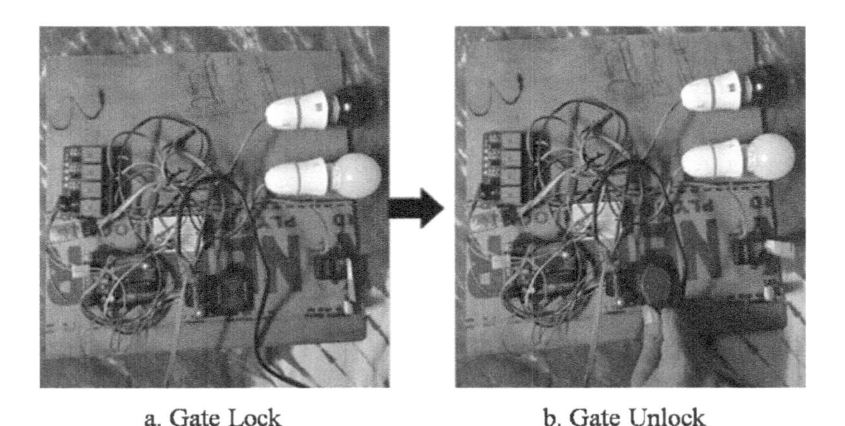

Figure 9.2 Gate unlock system using RFID.

9.3.1.3 Arduino UNO

Arduino UNO is a lightweight microcontroller board based on the ATmega328P chip. It takes an input voltage of 7–12 V and operates at 5 V. It has 6 analog input-output pins and 14 digital input-output pins. It has an EEPROM of 1 KB and SRAM of 2 KB and operates at a frequency of 16 MHz. The main advantage of this is it operates at very low power, and it can be easily programmed through the USB port. Here we developed a home automation system in which when the Arduino UNO received any signal from the mobile through the Bluetooth module, it triggers the Relay [19] to control the devices.

9.3.1.4 Bluetooth HC05 module

In this system, we use the HC05 Bluetooth module, which is a class 2 Bluetooth module, interfaced with Arduino UNO and used to create communications with the Android device, which consumes very low power and is configured to maintain security. Here the Bluetooth device interacts with the microcontroller to receive the data coming from the mobile device. Moreover, the Bluetooth device is triggered only when the RFID is activated.

In this system, the transmission rate is about 3 Mbps. It has an operating voltage of about +5 V and the operating current is about 30 mA where the maximum range is about less than 35 m. It follows the protocol of IEEE 802.15.1 which is a low-power protocol. It has six pins namely Key, VCC, Ground, Transmitter, Receiver, and State pins.

The power consumption of the Bluetooth HC05 module is given below

$$P = I \times V \tag{9.1}$$

$$= 30 \text{ mA} \times 5 \text{ V} = 150 \text{ mW (max)}$$

9.3.1.5 Relay switch

A relay switch is one kind of electrically operated switch, which can be used to control the flow of signals. It can be easily integrated with the microcontroller to switch or select different devices under operation. In the proposed work, relays are used and interfaced with the Arduino to select and control the different devices at home and thus contributing to the efficacy of the home automation system.

9.3.1.6 Android system

A new technology known as Bluetooth Low Energy (BLE) is introduced from Android version 4.3. This Bluetooth-based low energy provides an Application Programming Interface (API) [17,18] to discover, connection, and other operations at a low rate . Presently, almost all smartphone devices use this system. Here the application provides a user-friendly smart home automation control system with consumes less energy. The main challenge of this system is the User Interface without losing connection.

A figure of all the experimental devices has been given below (Figure 9.3).

9.3.2 Process diagram

The process diagram of the entire system has been given below (Figure 9.4).

9.4 RESULT AND DISCUSSION

In this work, a home automation system has been developed to open the gate after security check using RFID and then get access to control the devices of the building such as TV, AC, refrigerator, light, fan, etc. In this work initially, the hardware is set up, and then the performance of the system is judged through experimental analysis.

9.4.1 Experimental setup

The entire experimental setup to operate the system by using the android device has been shown in Figure 9.5. The design of the system is meticulously done to meet the scheme performance requirements.

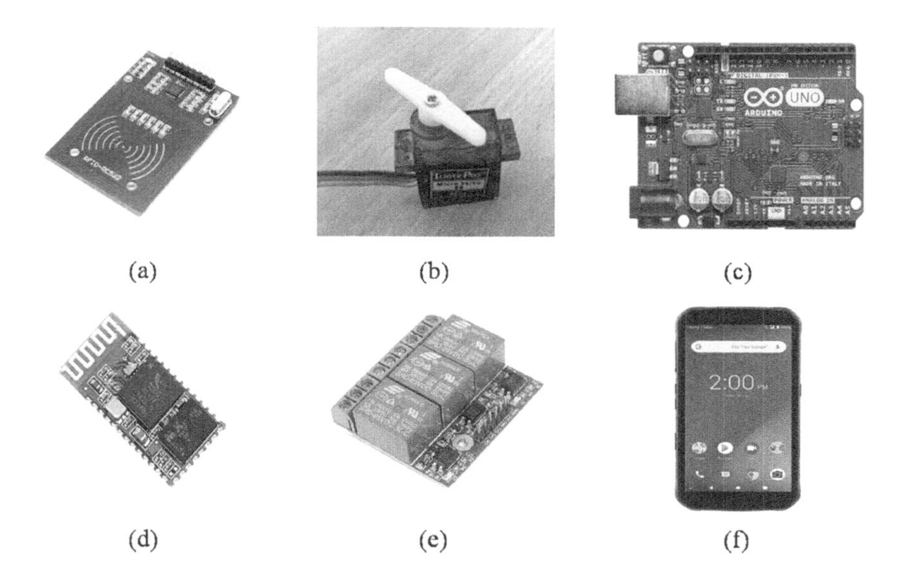

Figure 9.3 Experimental devices. (a) RFID-RC522 Module. (b) Servomotor. (c) Arduino UNO. (d) HC05 Bluetooth module. (e) Relay Switch. (f) Android System.

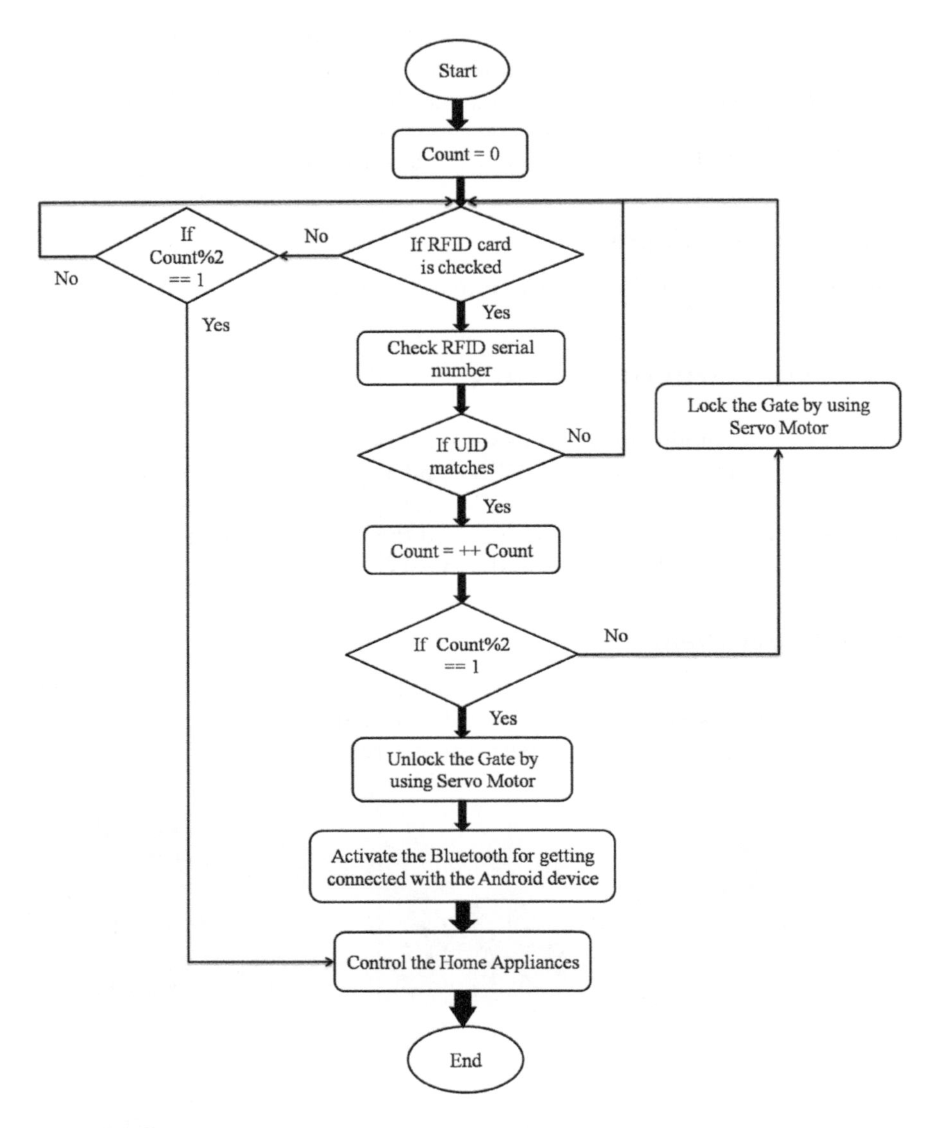

Figure 9.4 Process diagram.

9.4.2 Performance analysis

Three types of experiments are performed to judge the efficacy and effective operation of the system.

In the first set of experiments, the operating time is measured at a different range of the system to know the efficiency of the system. After the

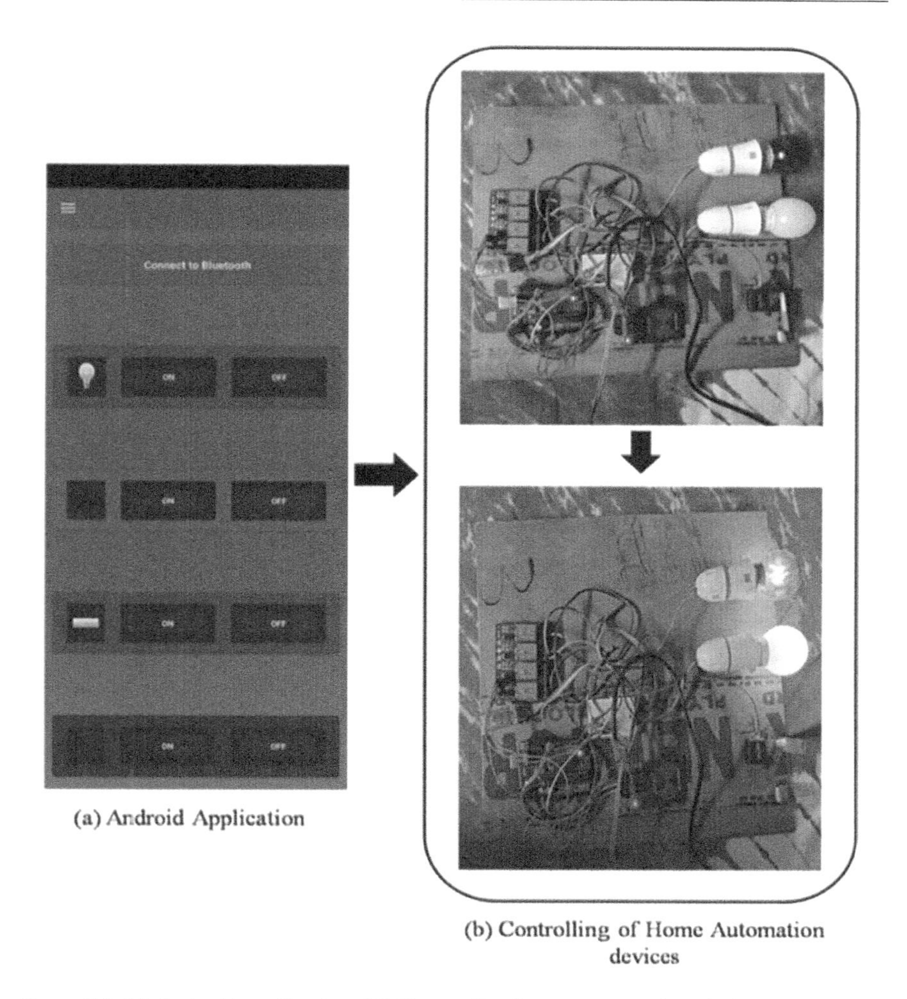

Figure 9.5 (a) Android application. (b) Controlling home automation devices.

experiment, it is observed that the system quickly responds to the response of the Android Device. Here we record the time delay to operate the system at a different distance. In this experiment, 50 observations have been taken at each point for getting the accuracy of the system and the average time delay to operate the system.

The system is developed to get a minimum amount of time delay to operate the system from different positions of the building without any problem. The key benefits of the proposed system are quick response, low power consumption, flexibility, and ease of use. The experimental result of the proposed system is given in Table 9.1.

Table 9.1 Experimental observation

Position	*Distance (m)*	*Time delay (ms)*
P1	1–3	100.1
P2	4–6	128.8
P3	7–9	176.5
P4	10–12	255.6
P5	13–15	334
P6	16–18	468

It is observed that the result shows that the system is highly efficient to operate the household devices from any point of the building. Here we use two different types of security one by using the RFID signal and by using a Bluetooth recognition system. Here the system is enough intelligent to control different household devices. Here P1, P2, P3, P4, P5, and P6 are six different positions where the experiment is conducted to know the efficiency of the system at a different distance. At a different distance, the time delay to operate the system varies depending upon the signal strength to operate the system. It can be seen that as the distance increases the delay in responding increases in terms of milliseconds which is negligible. This is due to the variation in the operating range of the target devices from the access point. However, this delay is negligible within an operating range and does not affect the system performance. A figure of time delay variation at a different distance using a box plot is given in Figure 9.6.

In the second part of the experiment, exhaustive experiments are performed for 50 times at each distance to find the minimum and maximum variation of delay. We have done a statistical analysis using the box plot (Figure 9.6) and from that, it is observed for example the variation of time delay within a 6-m distance is 127–129 ms. On the other hand, if we increase the distance approx 12 m then in that case the variation of time delay will be in the range of 240–260 ms. In the same way for an 18-m distance, this value will be 450–475 ms. This shows the efficiency of the proposed system. However, as in the part of the third experiment, it can be seen that even due to the exhaustive experimentation the average time delay to operate the entire system is the same and found to be consistent as replicated in Table 9.1 and is increasing continuously with distance. This justifies the accuracy and timeliness of our scheme and the observations are shown in Figure 9.7.

Performance-wise the system is also compared in terms of efficiency with other competing schemes and is explicitly demonstrated in Table 9.2.

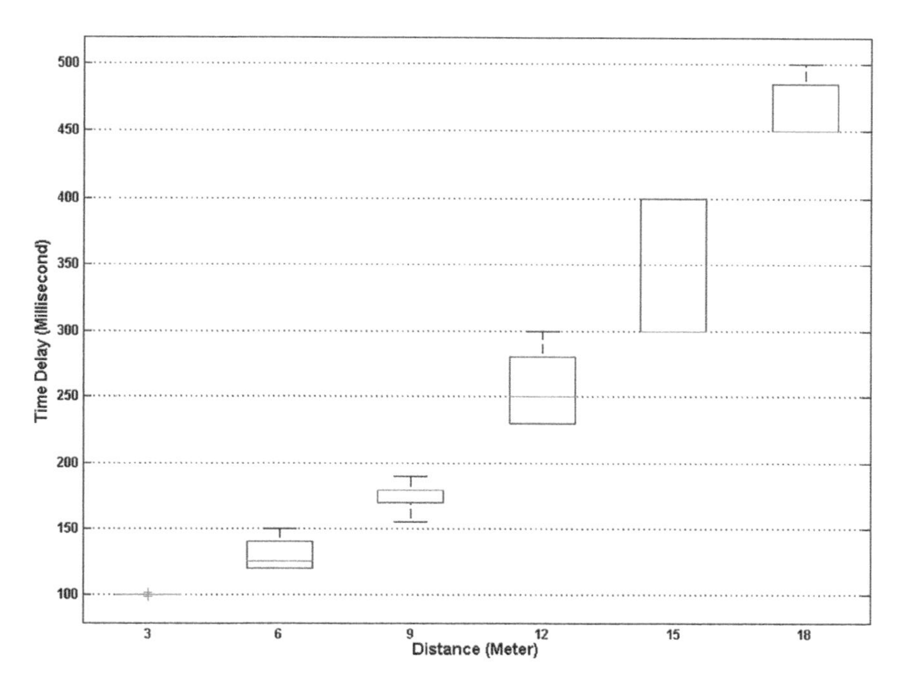

Figure 9.6 Time delay variation at different distance.

Table 9.2 Comparison with established scheme

Scheme	*Central lock using RFID*	*Power consumption*	*Communicating medium*
Harsh Kumar Singh [15]	No	High	Web server
Satyendra K. Vishwakarma [16]	No	High	Web server
Taewan Kim [14]	No	High	GUI
Proposed	Yes	Low	Android application

9.5 CONCLUSION

In this work, a complete solution for household security and in controlling household devices by using the Bluetooth system has been discussed. This IoT-enabled Bluetooth-based communication system makes the system efficient for household applications in terms of cost and it is also energy-efficient. In the proposed system, the communication with the household devices has been done by using the Android application where the user can easily access the devices by using this application.

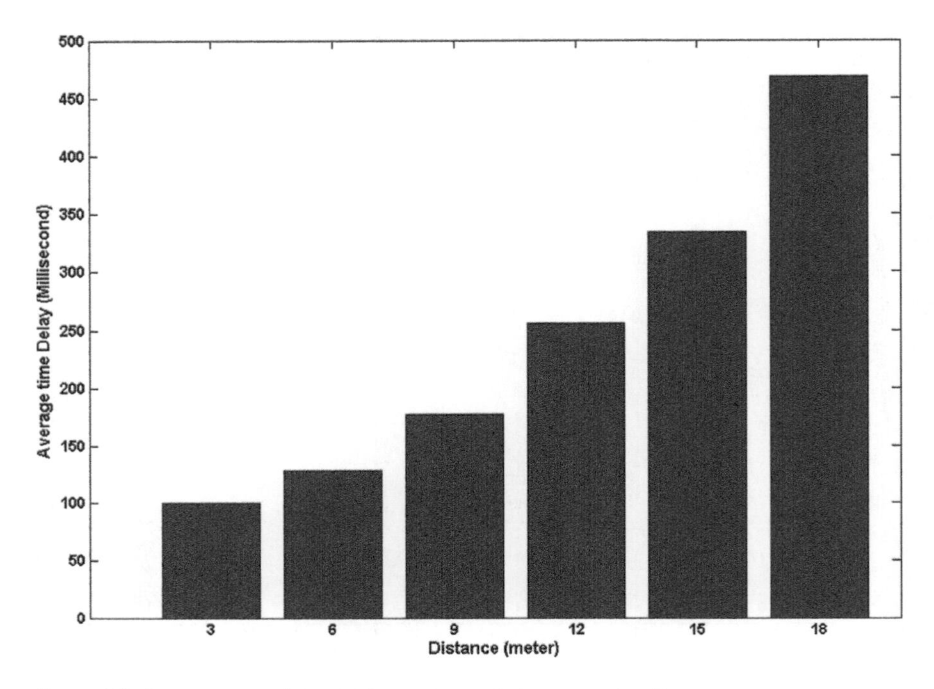

Figure 9.7 Average time delay with continuously increase the distance.

The main goal of the proposed system is to save energy along with maintaining the security of the system. The implementation of the Bluetooth module saves energy while the RFID act as user authentication to authenticate the user and trigger the system to control the devices. The proposed system gives the user freedom to get control all the household devices from any place inside the building.

As a future extension of this work, a broader IoT aspect of clustered network home automation system having a secured connection, communication and storage will be focused on.

REFERENCES

1. V. Vujoic and M. Makshimovic. "Raspberry Pi as a sensor web node for home automation". *Elsevier Transaction on Computer and Engineering* 42, 153–171, 2015.
2. M. Marsan, G. Bucalo, A. Di Caro, M. Meo, and Y. Zhang. "Towards zero grid electricity networking: Powering BSs with renewable energy sources". *2013 IEEE International Conference on Communications Workshops (ICC)*, Budapest, 2013, pp. 596–601.

3. A. Mishra, D. Irwin, P. Shenoy, J. Kurose, and T. Zhu. "Greencharge: Managing renewable energy in smart buildings". *IEEE Journal on Selected Areas in Communications* 31(7), 1281–1293, 2013.
4. H. Nguyen, J. Song, and Z. Han. "Distributed demand side management with energy storage in smart grid". *IEEE Transactions on Parallel and Distributed Systems*, 2014, doi: 10.1109/TPDS.2014.2372781.
5. I. Koutsopoulos and L. Tassiulas, "Optimal control policies for power demand scheduling in the smart grid". *IEEE Journal on Selected Areas in Communications* 30(6), 1049–1060, 2012.
6. Smart Grid Working Group (2003–06). *Challenge and Opportunity: Charting a New Energy Future—Appendix A: Working Group Reports*. Washington, DC: Energy Future Coalition, NW, 2008.
7. J. Zheng, D. Gao, and L. Lin. "Smart meters in smart grid: An overview". *2013 IEEE Green Technologies Conference (GreenTech)*, Denver, CO, 2013, pp. 57–64.
8. Z. Li-ming, L. Bao-cheng, T. Qing-hua, and W. Li-ping. "The development and technological research of intelligent electrical building". *2014 China International Conference on Electricity Distribution (CICED)*, Shenzhen, China, 2014, pp. 88–92.
9. X. Chen, T. Wei, and S. Hu. "Uncertainty-aware household appliance scheduling considering dynamic electricity pricing in smart home". *IEEE Transactions on Smart Grid* 4(2), 932–941, 2013.
10. Y. Hadj-Said, S. Ploix, S. Galmiche, S. Bergeon, and X. Brunotte. "Canopea, an energy-smart home integrable into a smart-grid". *2013 IEEE Grenoble Conference*, Grenoble, 2013, pp. 1–7.
11. I. Horvat, N. Lukac, R. Pavlovic, and D. Starcevic. "Smart plug solution based on bluetooth low energy". *2015 IEEE 5th International Conference on Consumer Electronics Berlin (ICCE-Berlin)*, Berlin, 2015, pp. 435–437.
12. C. Garrido, V. Lopez, T. Olivares, and M. Carmen Ruiz. " Architecture proposal for heterogeneous, BLE-based sensor and actuator networks for easy management of smart homes". *2016 15th ACM/IEEE International Conference on Information Processing in Sensor Networks (IPSN)*, Vienna, 2016, pp. 1–2.
13. J. Roux et al. "RadIoT: Radio communications intrusion detection for IoT - A protocol independent approach". *2018 IEEE 17th International Symposium on Network Computing and Applications (NCA)*, Cambridge, MA, 2018, pp. 1–8.
14. T. Kim, H. Lee, and Y. Chung. "Advanced universal remote controller for home automation and security". *IEEE Transactions on Consumer Electronics* 56(4), 2537–2542, 2010.
15. H. Kumar Singh, S. Verma, S. Pal, and K. Pandey. "A step towards home automation using IOT". *2019 Twelfth International Conference on Contemporary Computing (IC3)*, Noida, 2019, pp. 1–5.
16. S.K. Vishwakarma, P. Upadhyaya, B. Kumari, A. Kumar Mishra. "Smart energy efficient home automation system using IoT". *2019 4th International Conference on Internet of Things: Smart Innovation and Usages (IoT-SIU)*, Ghaziabad, 2019, pp. 1–4.

17. H. Bauer, M. Patel, and J. Veira. "The Internet of Things: Sizing up the opportunity." 2014. Retrieved from: McKinsey at http://www. mckinsey.com/insights/high_tech_telecoms_internet/the_internet_of_things_sizing_up_the_opportunity.
18. C. Stolojescu-Crisan, C. Crisan, and B-P. Butunoi. "An IoT-based smart home automation system." *Sensors* 21(11), 3784, 2021.
19. A. Singh, H. Mehta, A. Nawal, and O.V. Gnana Swathika. "Arduino based home automation control powered by photovoltaic cells". *Second International Conference on Computing Methodologies and Communication*, IEEE, Erode, 2018.
20. Md. S. Mahamud, Md. S. Rahman Zishan, S. I. Ahmad, A. R. Rahman, M. Hasan, and Md. L. Rahman. "Domicile - An IoT based smart home automation". *IEEE 2019 International Conference on Robotics, Electrical and Signal Processing Techniques (ICREST)*, Dhaka, 2019, pp. 493–497.
21. N. Valov and I. Valova. "Home automation system with raspberry Pi". *2020 7th International Conference on Energy Efficiency and Agricultural Engineering (EE&AE)*, IEEE, Ruse, 2020, pp. 1–5.
22. A. Eleyan and J. Fallon. "IoT-based home automation using android application". *2020 International Symposium on Networks, Computers and Communications (ISNCC)*, IEEE, Montreal, QC, 2020, pp. 1–4.

Chapter 10

A multithreaded android app to notify available 'CoWIN' vaccination slots to multiple recipients

Sudipta Saha, Saikat Basu, and Koushik Majumder
Maulana Abul Kalam Azad University of Technology, West Bengal

Debashish Chakravarty
Indian Institute of Technology (IIT), Kharagpur

CONTENTS

10.1 INTRODUCTION

In December 2019, a new coronavirus known as the 2019 novel coronavirus (2019-nCoV) made its first wicked steps in Wuhan, the capital of China's Hubei province. Then it crossed the national border of China and rapidly spread all over the world. Infected persons get flu-like symptoms such as fever, dry cough, sore throat, trouble breathing, and invasive lesions on both lungs. It may lead to viral pneumonia. So far, 562,198,772 people have been diagnosed with novel coronavirus positive worldwide, and 6,375,464 of them have died (Worldometer July 12, 2022). To date, the world is suffering and fighting against the disease. The International Committee on Taxonomy of Viruses (ICTV) renamed the 2019 novel coronavirus 'Severe Acute Respiratory Syndrome Coronavirus 2' (SARS-CoV-2) on 11 February 2020 (Nature microbiology 2020, 536). World Health Organization (WHO)

DOI: 10.1201/9781003381167-10

labelled this disease a 'Pandemic' on 11 March 2020. The United States is the country that has been hit the hardest by this virus. Unfortunately, India is ranked second, with 525,474 Indians who have died out of 43,652,944 positive cases (Worldometers July 12, 2022). The COVID-19 pandemic has put enormous strain on some health systems and pushed others to their limits. COVID-19 healthcare costs in India are too high at private hospitals for the common Indian people. Most countries implemented local or national lockdown policies to combat the spread of the coronavirus. The economic impact of the lockdown has been severe. Except for a few critical services and activities, India's $2.9 trillion economy stayed closed throughout the shutdown (Mangla July 11, 2021). Currently, herd immunity (a large group of people develops immunity against the disease) is the only thing that can save the situation. Herd immunity against COVID-19 can be achieved by immunization through vaccines without causing sickness or other complications. To date, a single dose of the COVID-19 vaccine has been given to almost 68.5% of the world's population (Holder, 2022).

On January 16, 2021, India began its immunization campaign. Since then, India has delivered 1,990,179,793 doses of COVID-19 vaccination. At least one dose of COVID-19 vaccine was given to 1,018,381,042 Indians, while 922,151,680 Indians were fully vaccinated and 49,647,071 got their precautionary dose (CoWIN Dashboard July 13, 2022). Initially, only frontline workers were eligible for vaccination. From March 1, 2021, it will progressively include all residents above 60, individuals aged 45–60 with comorbidities, and ultimately all residents aged 45–60 from April 1, 2021. Until then, all vaccination doses were obtained by the union government from vaccine makers and distributed to government and private hospitals free of cost. From May 1, 2021, vaccinations for those aged 18–44 have begun to be made available. The union government would buy only half of the vaccines manufactured in India for people aged 45 and up, which will be provided free of charge. The remaining 50% was left to state governments and commercial providers, although they had to pay more than the union government. The union government set a COVID vaccine price ceiling on June 8, 2021, eliminating the private vaccination facility's capless fee (Sanyal June 8, 2021). Following much outcry on June 21, 2021, the union government has begun to buy free vaccinations for those aged 18–44, intending to dispense them at no cost in government hospitals. And 25% of vaccinations were reserved for the private sector (Hussain and Suhel September 14, 2021). Although private hospitals can procure up to 25% of manufactured vaccines, it was reported that they had procured only 9.5% of available vaccines. They can only administer 65% of vaccines taken from vaccine manufacturers before August 17, 2021 (Hussain and Suhel September 14, 2021). After huge income losses during the pandemic, it is unrealistic to expect that people living in rural areas or small towns will bear the cost of such expensive vaccines. The Ministry of Health and Family Welfare of India owns and operates 'CoWIN' ('CoWIN'. January 16, 2021), an Indian government web platform/app for COVID-19 vaccine booking. It shows available COVID-19 vaccination slots in the nearest location and allows to book them on the Internet. Finding available slots at government centres is a time-consuming and challenging process due to the huge demand for the free-of-cost COVID-19 vaccine.

The nationwide COVID-19 immunization campaign for recipients between the ages of 15 and 18 started on January 3, 2022. The precautionary third dose for the vulnerable categories (frontline workers and people above 60) commenced on January 10, 2022. India began immunizing children between 12 and 14 against COVID-19 on March 16. From April 10, 2022, the precautionary dose for 18+ population groups started at private vaccination centres. Meanwhile, manufacturers of 'Covishield' and 'Covaxin' (two most popular COVID-19 vaccines of India) lowered the selling prices of single vaccine dosages to private facilities (Sharma, April 9, 2022). Till July 14, 2022, less than 1% of the 77-crore target population in the 18–59 age range received the precautionary dosage, although most of the Indian population received their second dose more than 9 months earlier. The government of India declared on July 13, 2022, that people in 18–59 age range would get free precautionary doses of the COVID-19 vaccine at government vaccination centres under a special 75-day drive commencing on July 15, 2022 (ZeeBusiness July 13, 2022). So, it would be challenging for anyone from these age groups to find free vaccination slots at the government centre within just 75 days.

The current work involves creating an Android application called 'Amar Protishedhak Bondhu' for ordinary smartphone users. The user needs to provide vaccination preferences such as vaccine name, charge kind, dosage type, and desired age group. Then the user needs to schedule a notifications/SMS service by touching a button. This application will be able to create and execute more than one separate background thread. The threads will search for open vaccination slots with different (user-preferred) criteria and can alert users periodically. Users can close the app after scheduling, but the background threads will continue to work.

10.2 RELATED WORKS

The COVID-19 vaccination programme in India is linked to the 'CoWIN' portal ('CoWIN'. January 16, 2021). It is the Indian COVID Vaccine Intelligence Network. It helps the government keep track of vaccination coverage, COVID-19 vials in cold storage, and evaluate the vaccination programme's effectiveness, and develop future strategies to deal with the challenges given by a changing COVID environment. For Indian citizens, 'CoWIN' serves as a conduit for preregistration, appointment confirmation, record upkeep, certificate production after immunization, and reporting unfavourable reactions (Thakur 2021). Citizens of India can use this service to find the nearest available vaccination slots and book vaccination appointments. 'Search By District', 'Search By PIN', and 'Search On the Map' are three types of vaccine search options provided by the 'CoWIN.' While searching for vaccination slots using 'CoWIN', the user can use filters on age group, vaccine type, dose type, and fee type. However, 'CoWIN' does not offer any notification/SMS/email services to its registered members to make them aware of newly available slots in their desired area. Free-of-cost government vaccination slots are in such high demand that all free-of-cost slots will be filled within minutes of their opening. The majority of regular Indians discover that all of the free-of-cost vaccination slots are always filled.

'Getjab.in' (Co-WIN Slot Notifier October 17, 2021) was a website that asked for information like a name, email address, district, and, phone number. It sent out an email to users when COVID-19 vaccines became available in their districts. However, the website's owner has discontinued providing this service due to the high operational expenses.

The 'Vaccine Finder' section of the 'MyJio' app assists the 'Jio' subscribers in finding the vaccination slots by the pincode and by the district for the next seven days. Filters related to dosages type, vaccine type, cost type, and age group. However, this app does not have any notification/SMS/email facilities to inform users about newly available vacant slots.

The 'Get Vaccine Alerts' module of the 'Airtel Thanks' app alerts 'Airtel' users about the availability of COVID-19 vaccination slots. It only allows users to search by the district for the next seven days. Filters for vaccine kind, cost type, and age group can be specified while searching. If slots are unavailable according to user-specified criteria, this app allows setting a notification service using only the age-related filter to inform the user upon slot availability.

'COVID-19 Vaccine Slot Finder' section of the 'Paytm' app notifies users about newly available vaccination slots. This app provides the facilities of vaccination slot searching by pincode and by the district after logging in using a mobile number and one-time password. While searching, users can set filters on age group, dose type, vaccine type, and fee type. If vaccination slots are unavailable according to intended the age group and dose type, this app offers the option of setting notification services for newly open slots using these two filters.

'https://under45.in/' and 'https://above45.in/' send COVID-19 vaccination slot alerts on the telegram messaging app when slots open up for the intended district for the age group 18+ and 45+ only. Users can request an alert service by visiting this website, selecting their state and district, and finally joining the telegram channel. The telegram app must be installed on the user's mobile device in this scenario. However, any filter related to the vaccine, dosage, or cost type cannot be specified while requesting an alert. Table 10.1 compares different COVID-19 vaccination slot searching and notification-sending facilities in India.

Table 10.1 Comparison of COVID-19 vaccination slot searching and notification-sending facilities in India

App/Portal	*Searching facilities*	*Available search filters*	*Booking facility available?*	*Notification/Email facility available?*
CoWIN' app/portal	By district, By pincode, and On Map	Vaccine type, dose type, cost type, and age group	Yes	NO
'Getjab.in' (Not operational)	By district	No filter facility	No	Email if slot available in specified district
'Vaccine Finder' section of the 'MyJio' app	By district, and By pincode	Dosages type, vaccine type, fee type, and age group	No	No
'Get Vaccine Alerts' feature of the 'Airtel Thanks' app	By district	Vaccine type, fee type, and age group	No, but it gives a link to the 'CoWIN' app for booking	Yes, allows setting a notification service using only the age-related filter if vaccination slots are unavailable according to user-specified criteria
'COVID-19 Vaccine Slot Finder' section of the 'Paytm' app	By district, and By pincode	Dosages type, vaccine type, fee type, and age group	Yes	Yes, allows setting a notification service if vaccination slots are unavailable according to user-specified age group and dose type
'https://under45.in/' portal & 'https://above45.in/'	By district	No filter facility, only available for age group 18+ and 45+	No	Yes

10.3 DETAILED DESCRIPTION OF APPLICATION

10.3.1 Functional features

In the present work, an Android application called 'Amar Protishedhak Bondhu' (APB) is developed. It is a Bengali name, meaning My Vaccine Friend. This application can search for available COVID-19 vaccination slot details for any place in India like the 'CoWIN' platform. The application 'APB' provides search by pincode, search by district, and search on map facility, just like 'CoWIN'. Additionally, users can search for any COVID-19 vaccination centre within a 15-km radius of a provided pincode. This new facility is included because often, there may not be a vaccination centre inside a certain pincode, but it may be very close to that particular pincode. When searching, users can set filters on the vaccine type, dose type, fee type, and age group (using drop-down lists) like the 'CoWIN' app. The application 'Amar Protishedhak Bondhu' makes the 'Search on the Map' facility provided by the 'CoWIN' more convenient and detailed. This application implements a pincode-wise search inside the map view. When a certain pincode is entered in the map view's search field, the 'CoWIN' app places marks on vaccination centres both inside and outside of that pincode. This feature is quite misleading. The map view of 'CoWIN' also does not give relevant details about vaccination centres. For example, the 'CoWIN' map view does not provide information about available precaution doses. In 'CoWIN', information of all vaccination slots for a week is displayed together in one place when a user clicks on a centre's marker. The software 'APB' can provide a date-wise separate list of vaccination slots under a vaccination centre in map view. This app place markings on all vaccination centres under only this pincode. The user can get this centre's slot details on the selected date by clicking on a marker. This software also has a facility; it can set markers within a 15-km radius of a provided pincode if the user wants. Clicking any marker in the map view, the users can get the details of vaccination slots, including vaccine name, available doses (including the precaution doses), fee type, and age group. At the time of vaccination slot searching using 'CoWIN', users cannot see the weekdays of the search date. Weekdays are sometimes necessary for proper daily planning. 'APB' uses a popup calendar at the time of the 'date' input taken from a user, so weekdays information is always available to a user while searching for vaccination slots using this application.

People can more realistically plan for a vaccination with weather forecasts. Depending on the weather, we could decide not to go outside to get our vaccinations, or we might take the appropriate safety measures while visiting a far-off vaccination centre. 'APB' can provide a brief weather forecast for the current and following seven days during vaccination slot searching. The weather forecast includes maximum temperature, minimum temperature, humidity, and a short weather description. When searching inside a pincode or within a 15-km radius of a pincode, 'APB' provides the weather forecast of that particular pincode according to the search date. At the time of the district-wise search, 'APB' provides weather forecasts for the whole district at once.

One attractive feature of the application 'APB' is that it can send notifications and SMS if slots are available under the searching location for the next 1–15 days. This application allows scheduling these notifications/SMS sending work by accepting necessary details from users. It performs these slots searching and notifications/SMS sending jobs in the background every 15 minutes. A user can schedule notifications/SMS services for the vaccination slots under a specific pincode, under one particular district, or within 15 km radius of a specific pincode. Several vaccination-related options can be set when scheduling the notifications/SMS services (using radio buttons). Users can configure the application to send notifications and SMS only for free-of-cost slots, paid slots, or both, for a specific age group, or the age-neutral way, for a particular vaccine or any vaccine, for a specific dose, or any doses. This app allows users to put several restrictions on the device's battery condition and network connection while scheduling (using radio buttons and switches). This notification/SMS sending service may be scheduled for a 'Wi-Fi' connection, or a metered connection, or any connection type, for just battery charging conditions, or only when the device's charge is high, or at any time. The notifications/SMS sending service can be set for the next 1–15 days using a drop-down box. Because each SMS sent by this app deducts money from the user's existing balance, a user may only choose between a notification service with no SMS or a notification service with SMS if he is willing to pay. A user must enter a phone number into the app if he/she wishes to receive SMSes. Finally, the user must press the 'SCHEDULE SMS/NOTIFICATION' button to schedule searching and notifications/SMS. A user can cancel the already scheduled searching and notifications/SMS sending service at any time by touching the 'CANCEL SMS/NOTIFICATION' button.

The application interfaces of the following figures are described here. Figures 10.1–10.10 shows the user interfaces of the application 'Amar Protishedhak Bondhu'. The notifications and SMS delivered by the app 'Amar Protishedhak Bondhu' are depicted in Figures 10.11 and 10.12.

The most notable aspect of this software is that households may use it without access to a smartphone. Although this software requires an Android smartphone to be installed, a single app installation may be used to schedule numerous searching and notification/SMS services with various vaccination preferences and phone numbers. If a family does not possess a smartphone, but a friend, relative, or another known person of the family has. In that case, he or she can schedule vaccination slot searching and notification/SMS service for that family using the vaccination preferences and mobile number of that family's ordinary phone. This app can deliver SMS to this regular phone based on the vaccination preferences that have been set. Thus, a single smartphone may serve the needs of multiple families.

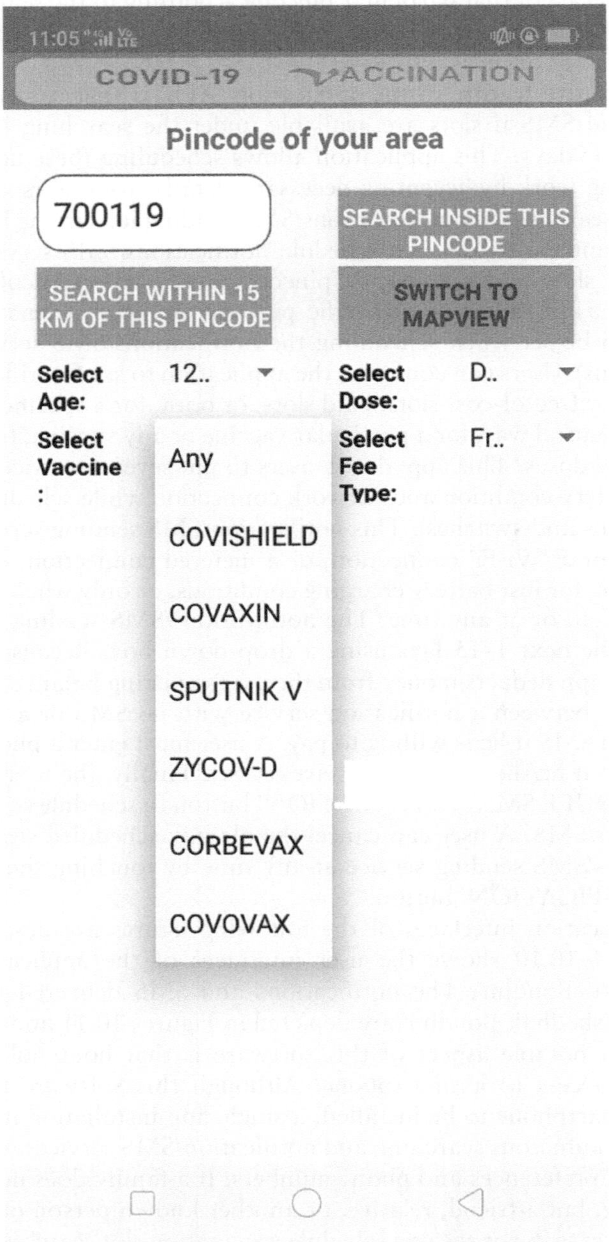

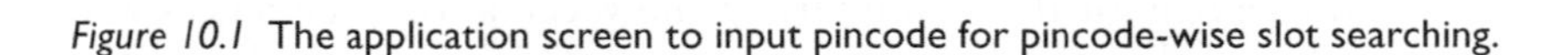

Figure 10.1 The application screen to input pincode for pincode-wise slot searching.

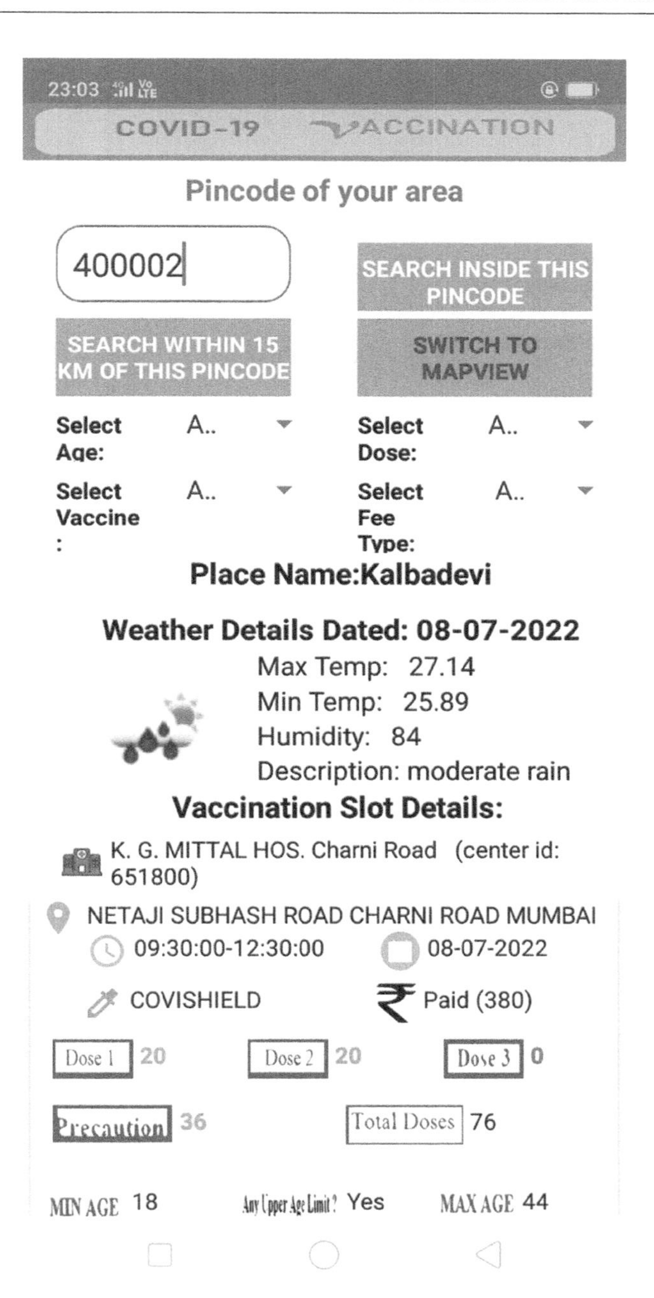

Figure 10.2 The application screen to display result of pincode-wise slot searching.

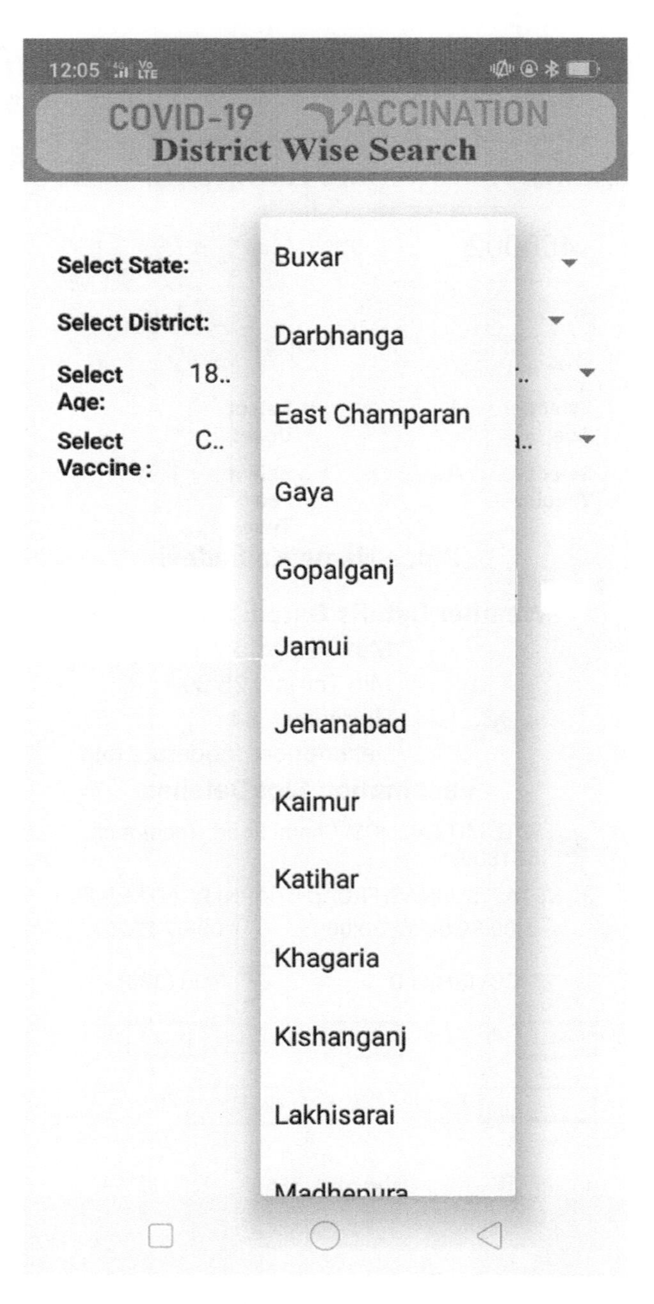

Figure 10.3 The application screen to input district for district-wise slot searching.

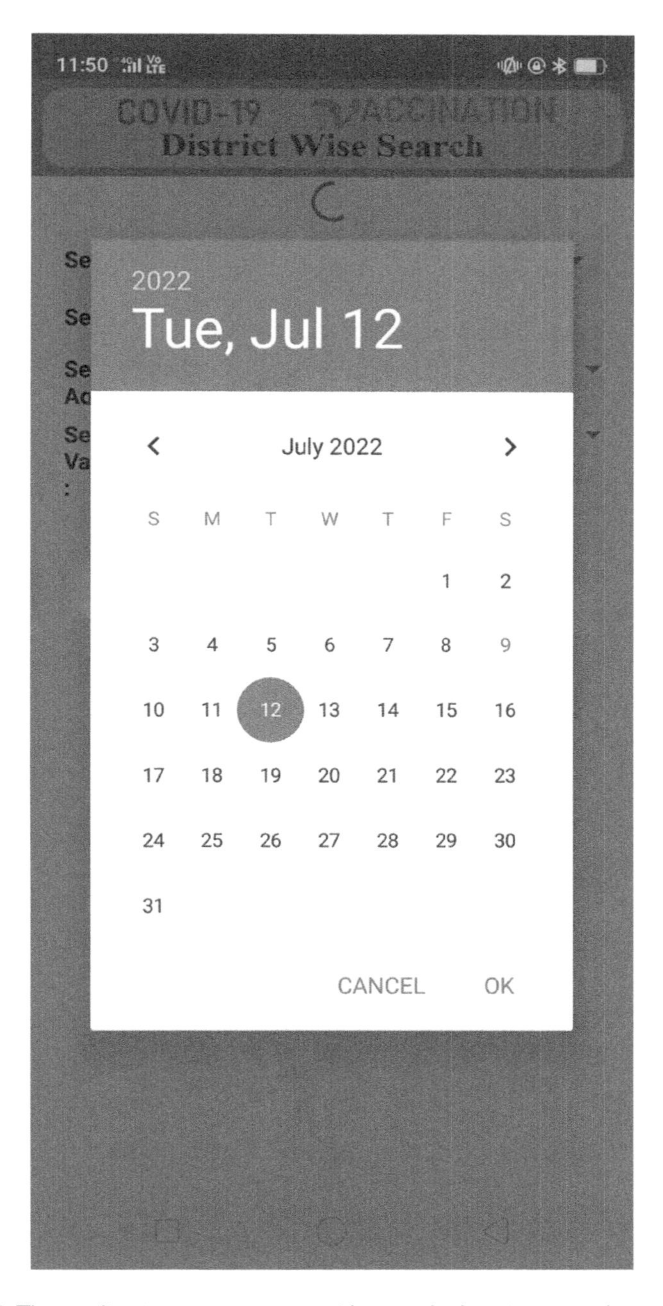

Figure 10.4 The application screen to provide search date input at the time of slot searching.

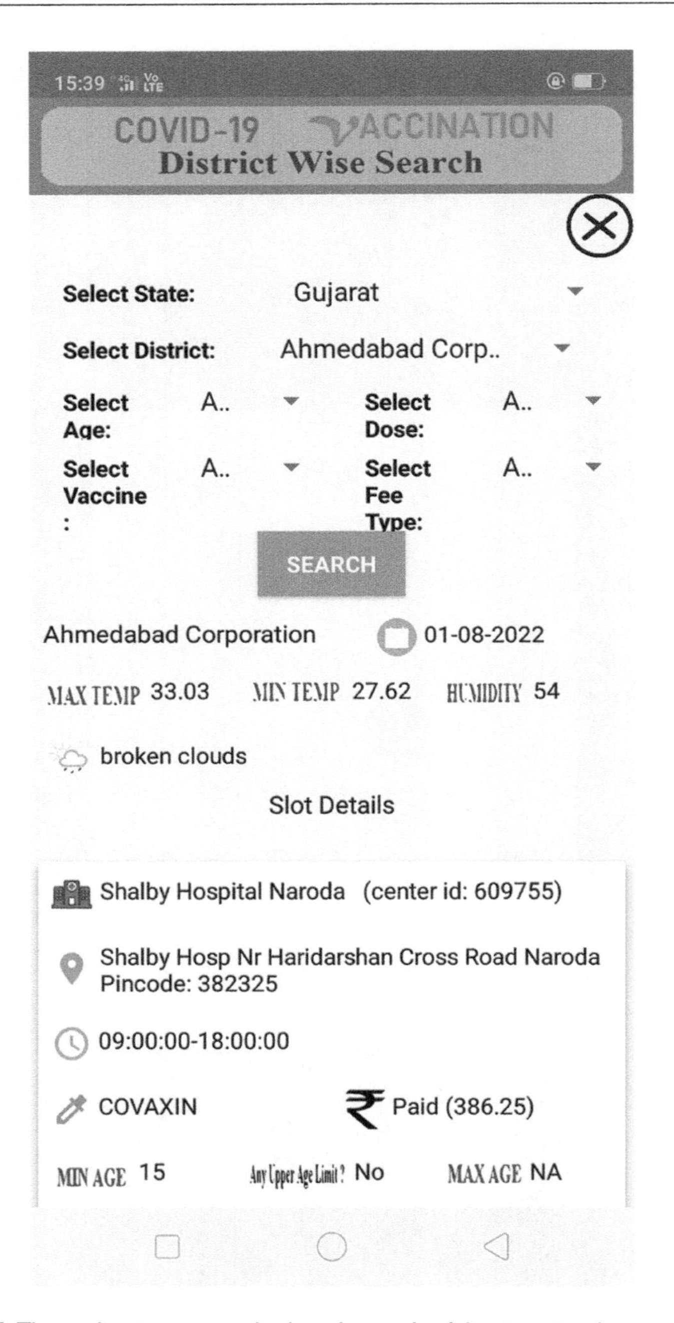

Figure 10.5 The application screen displays the result of district-wise slot searching.

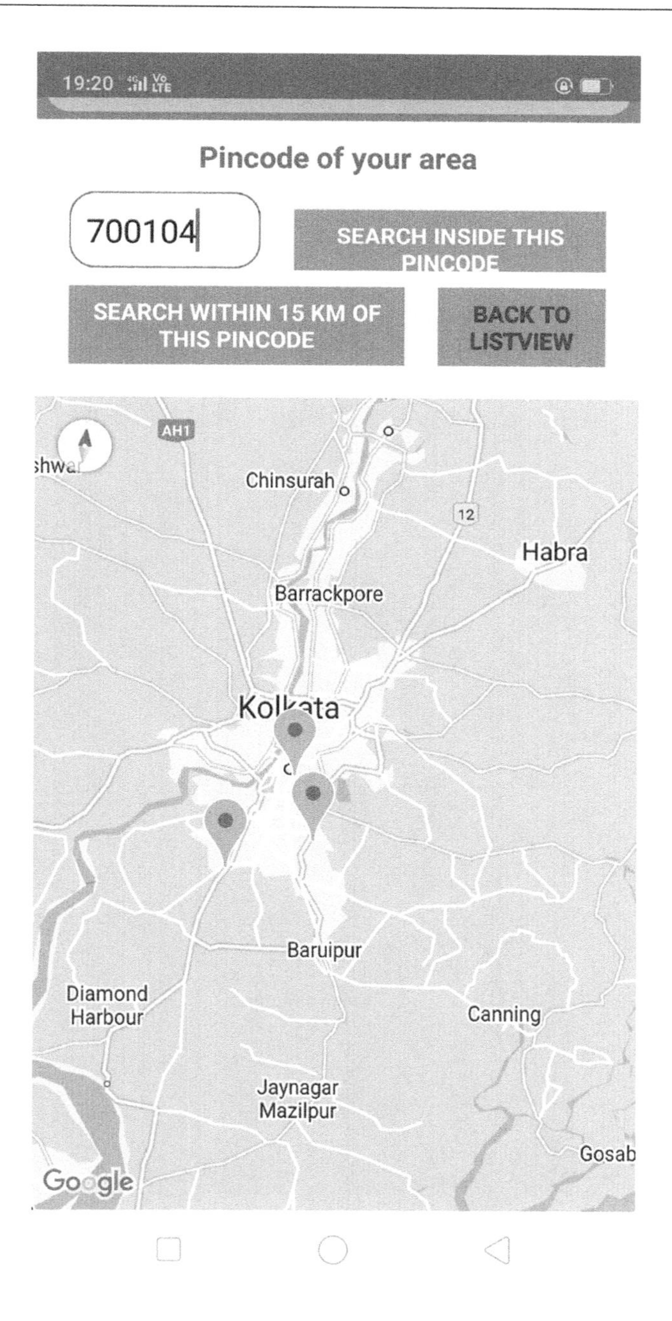

Figure 10.6 The application screen places markings on vaccination centres under a pin-code in map view.

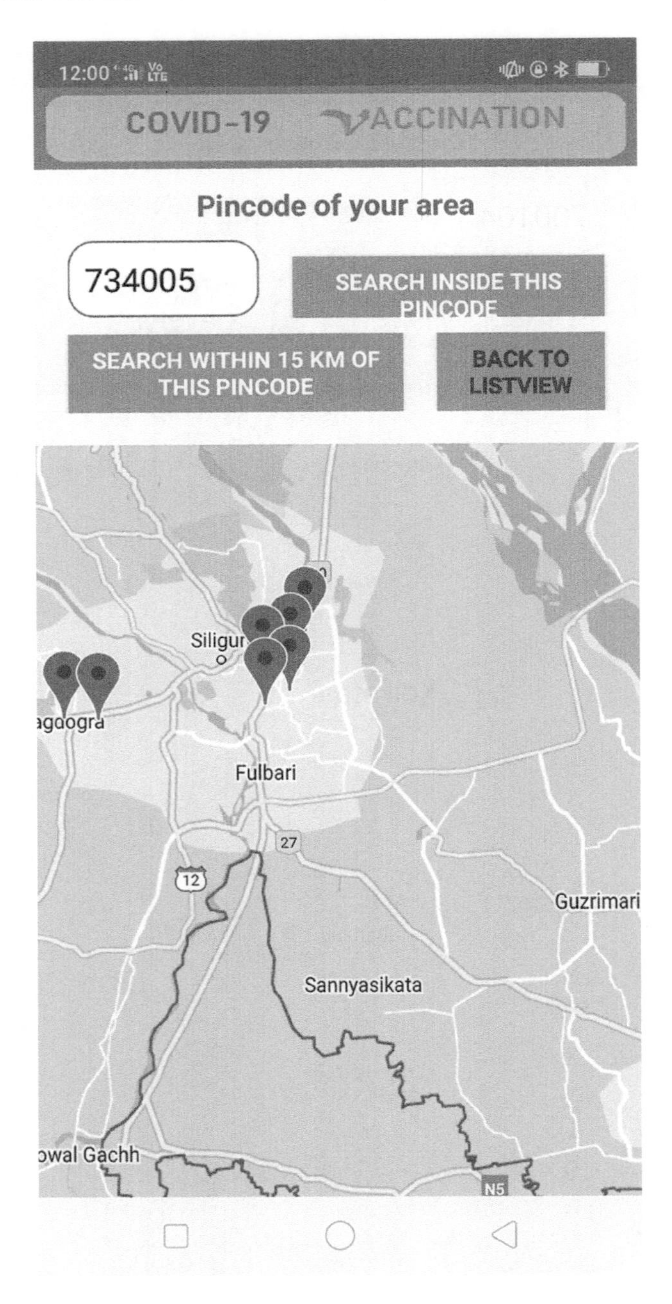

Figure 10.7 The application screen places markings on all vaccination centres within a 15-km radius of a pincode in map view.

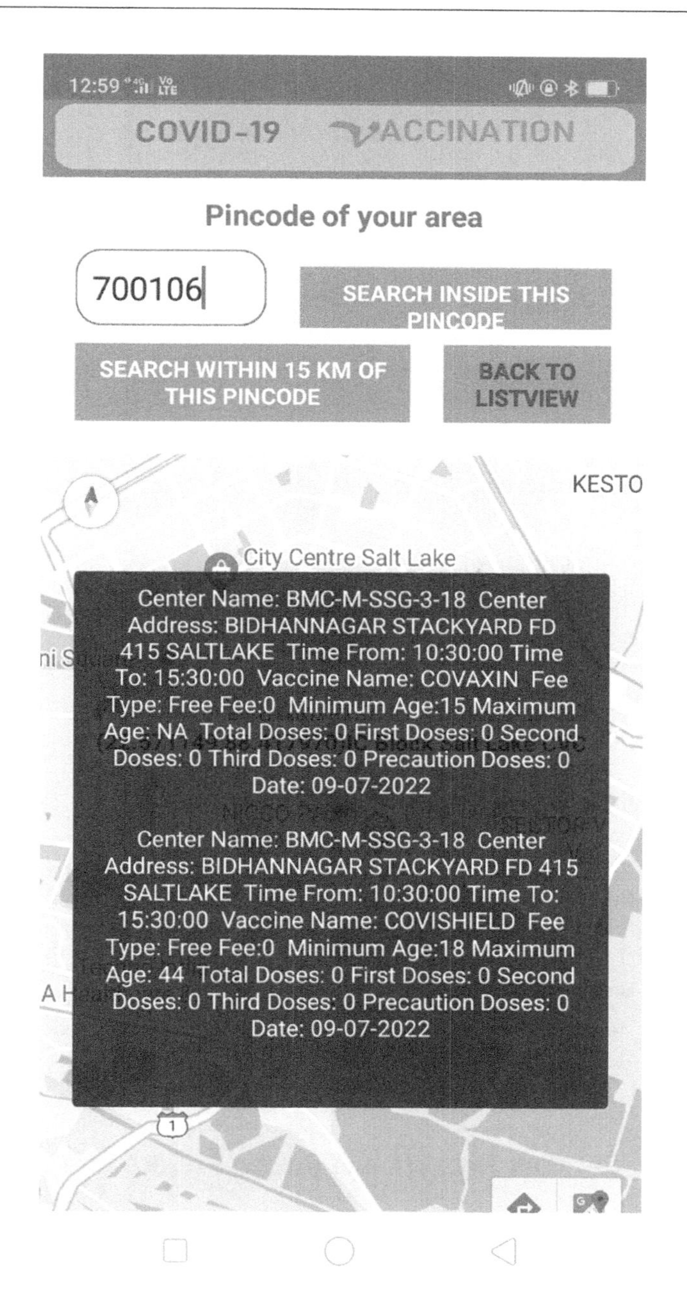

Figure 10.8 Application screen that displays centre-wise slot details in toast view.

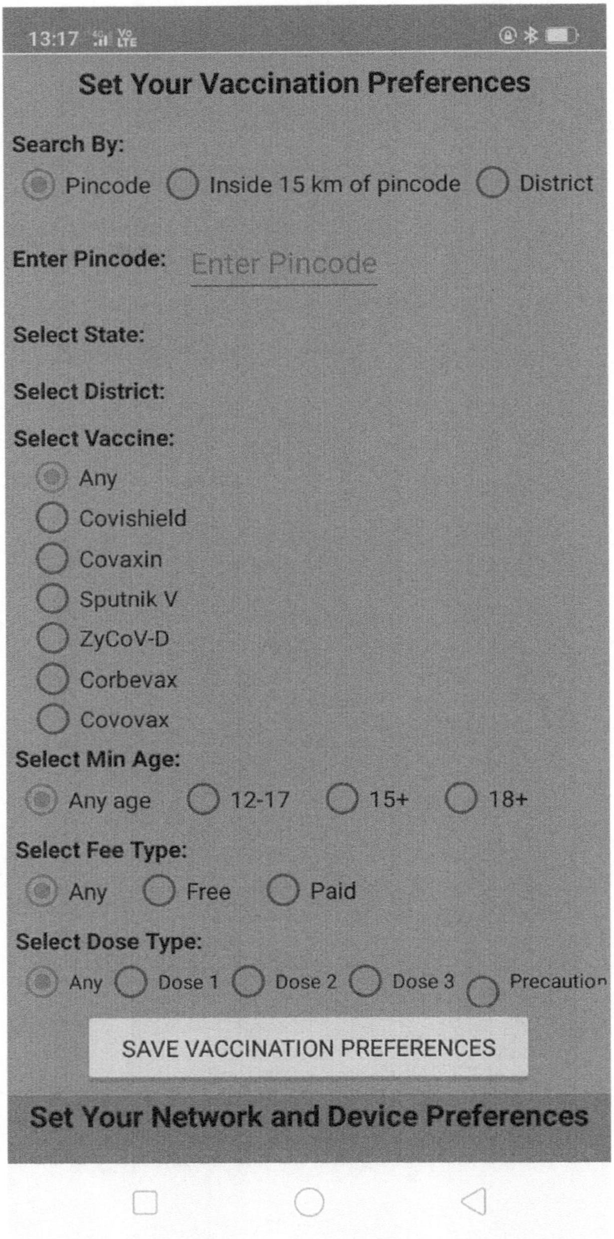

Figure 10.9 The application screen to set the vaccination preferences to schedule notification.

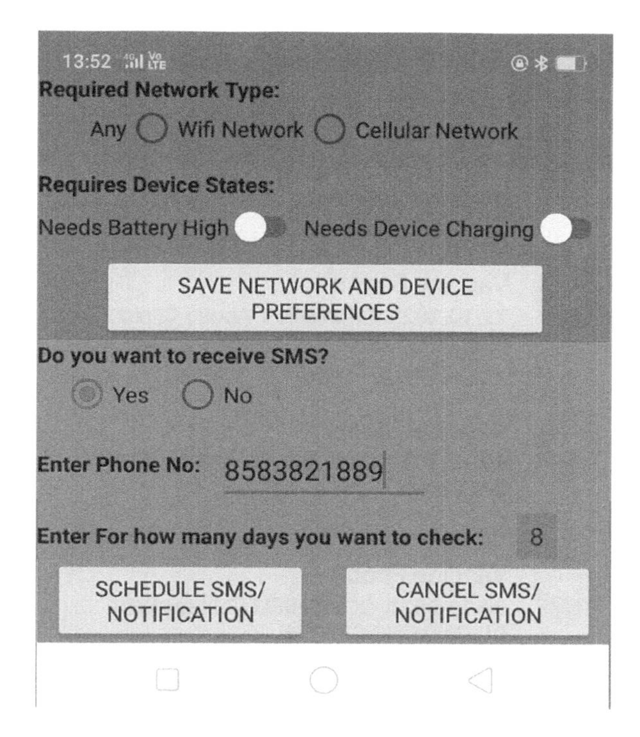

Figure 10.10 The application screen to set network, device preferences, and schedule notifications/SMS service.

10.3.2 User feedback module

Besides the vaccination slot searching and notification-sending modules, the application APB also possesses a user feedback module to collect insightful user feedback. An analysis of this feedback is necessary to assess this application's utility. This application presents a feedback form to a user as a long scrolling screen. Users are asked to fill out the form and submit it with their name and mobile phone number. The feedback form is designed with English statements and its translation (using a regional language) to make it completely understandable for every user. This app is initially targeted at the state of West Bengal only. So, after every statement/question in English, a Bengali (main language of West Bengal) translation is provided. In the user feedback form, the developers are primarily interested to know how users perceive the application, how users benefit from using the application, and what is the usefulness of included features to users. The feedback form is depicted in Figures 10.13–10.16 (partly in each figure).

Figure 10.11 The notifications sent by Amar Protishedhak Bondhu.

Figure 10.12 The SMS sent by Amar Protishedhak Bondhu.

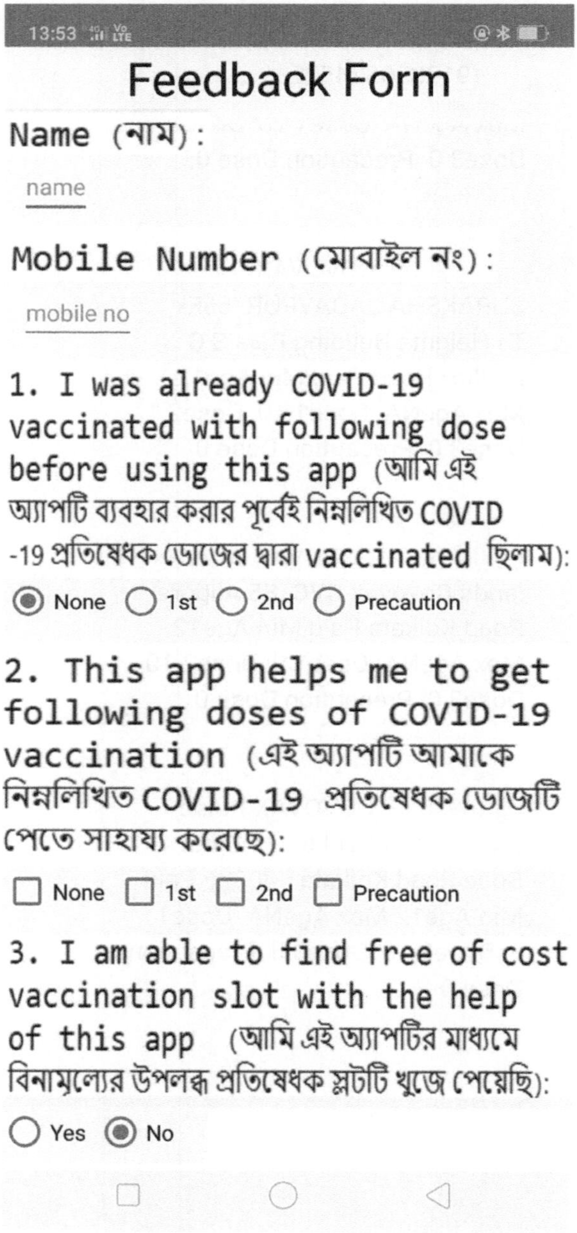

13:53

Feedback Form

Name (নাম):

name

Mobile Number (মোবাইল নং):

mobile no

1. I was already COVID-19 vaccinated with following dose before using this app (আমি এই অ্যাপটি ব্যবহার করার পূর্বেই নিম্নলিখিত COVID -19 প্রতিষেধক ডোজের দ্বারা vaccinated ছিলাম):

◉ None ○ 1st ○ 2nd ○ Precaution

2. This app helps me to get following doses of COVID-19 vaccination (এই অ্যাপটি আমাকে নিম্নলিখিত COVID-19 প্রতিষেধক ডোজটি পেতে সাহায্য করেছে):

☐ None ☐ 1st ☐ 2nd ☐ Precaution

3. I am able to find free of cost vaccination slot with the help of this app (আমি এই অ্যাপটির মাধ্যমে বিনামূল্যের উপলব্ধ প্রতিষেধক স্লটটি খুজে পেয়েছি):

○ Yes ◉ No

Figure 10.13 User feedback form part 1.

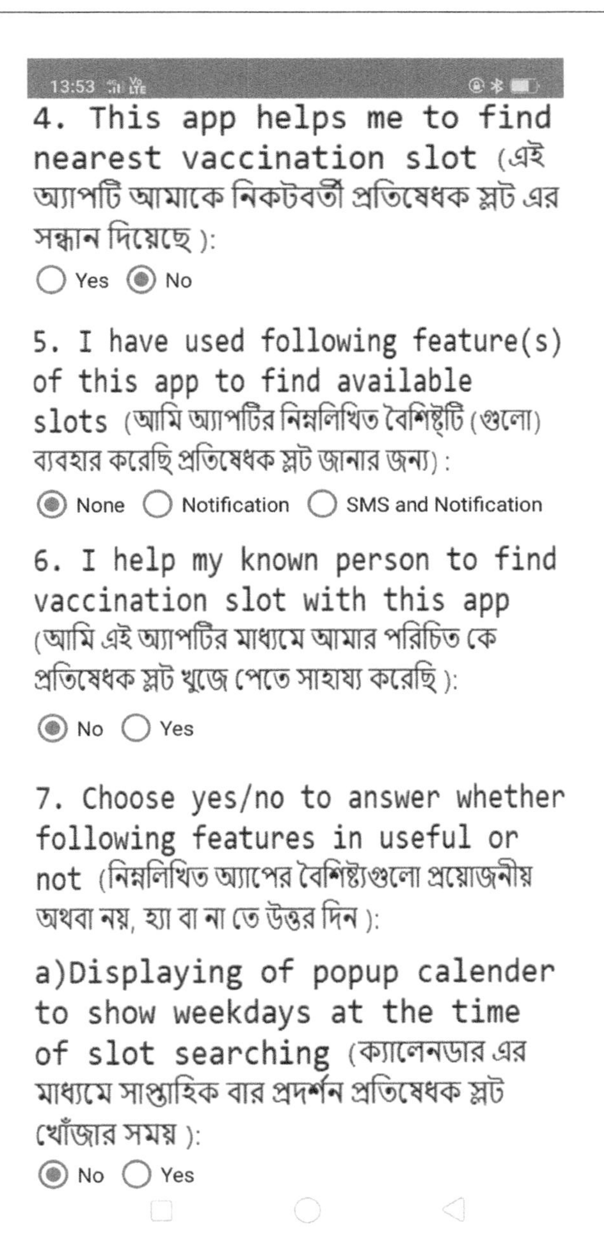

13:53

4. This app helps me to find nearest vaccination slot (এই অ্যাপটি আমাকে নিকটবর্তী প্রতিষেধক স্লট এর সন্ধান দিয়েছে):

Yes No

5. I have used following feature(s) of this app to find available slots (আমি অ্যাপটির নিম্নলিখিত বৈশিষ্টটি (গুলো) ব্যবহার করেছি প্রতিষেধক স্লট জানার জন্য) :

None Notification SMS and Notification

6. I help my known person to find vaccination slot with this app (আমি এই অ্যাপটির মাধ্যমে আমার পরিচিত কে প্রতিষেধক স্লট খুজে পেতে সাহায্য করেছি):

No Yes

7. Choose yes/no to answer whether following features in useful or not (নিম্নলিখিত অ্যাপের বৈশিষ্ট্যগুলো প্রয়োজনীয় অথবা নয়, হ্যা বা না তে উত্তর দিন):

a)Displaying of popup calender to show weekdays at the time of slot searching (ক্যালেন্ডার এর মাধ্যমে সাপ্তাহিক বার প্রদর্শন প্রতিষেধক স্লট খোঁজার সময়):

No Yes

Figure 10.14 User feedback form part 2.

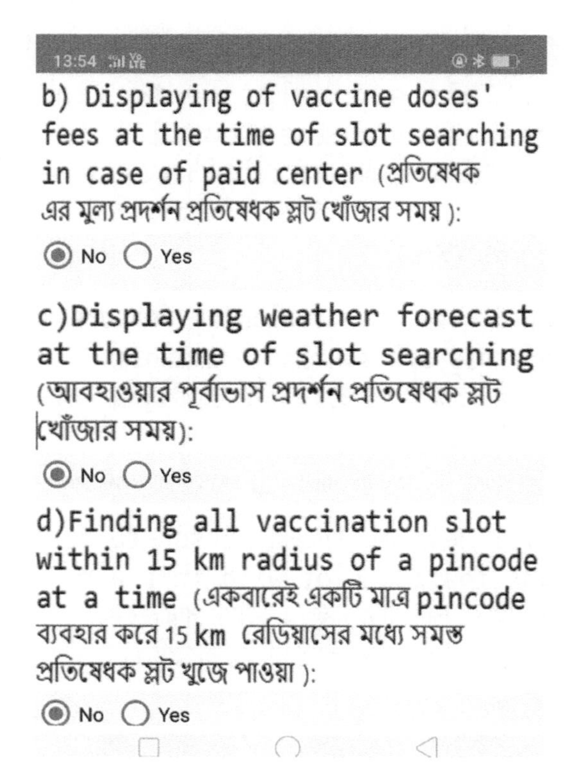

13:54

b) Displaying of vaccine doses' fees at the time of slot searching in case of paid center (প্রতিষেধক এর মুল্য প্রদর্শন প্রতিষেধক স্লট খোঁজার সময়):

(●) No ( ) Yes

c)Displaying weather forecast at the time of slot searching (আবহাওয়ার পূর্বাভাস প্রদর্শন প্রতিষেধক স্লট খোঁজার সময়):

(●) No ( ) Yes

d)Finding all vaccination slot within 15 km radius of a pincode at a time (একবারেই একটি মাত্র pincode ব্যবহার করে 15 km রেডিয়াসের মধ্যে সমস্ত প্রতিষেধক স্লট খুজে পাওয়া):

(●) No ( ) Yes

Figure 10.15 User feedback form part 3.

10.3.3 APIs used in this application

This application uses the 'CoWIN' portal's four public Application Programming Interfaces (APIs) to find available slots. All the public APIs of the web portal 'CoWIN' are listed in the 'https://apisetu.gov.in/api/cowin/cowin-public-v2' (Co-WIN Public APIs 2022) section of 'https://apisetu.gov.in/'. This application uses two appointment availability APIs and two metadata APIs of 'CoWIN'. The appointment availability APIs collect vaccination slot information under a particular pincode or a district. For fetching available slots under a district, the application needs to provide a district ID to appointment availability APIs. The application uses two metadata APIs of 'CoWIN' to know the required district's ID.

Two free weather APIs are used by the application 'Amar Protishedhak Bondhu' to gather weather forecasts. These two are provided by the web portal 'openweathermap.org'. The application first fetches the pincode's precise latitude and longitude are first fetched using a 'Current Weather Data' API. After that, this application uses the fetched latitude, longitude, and current date as parameters to call the 'One call' API and retrieves the

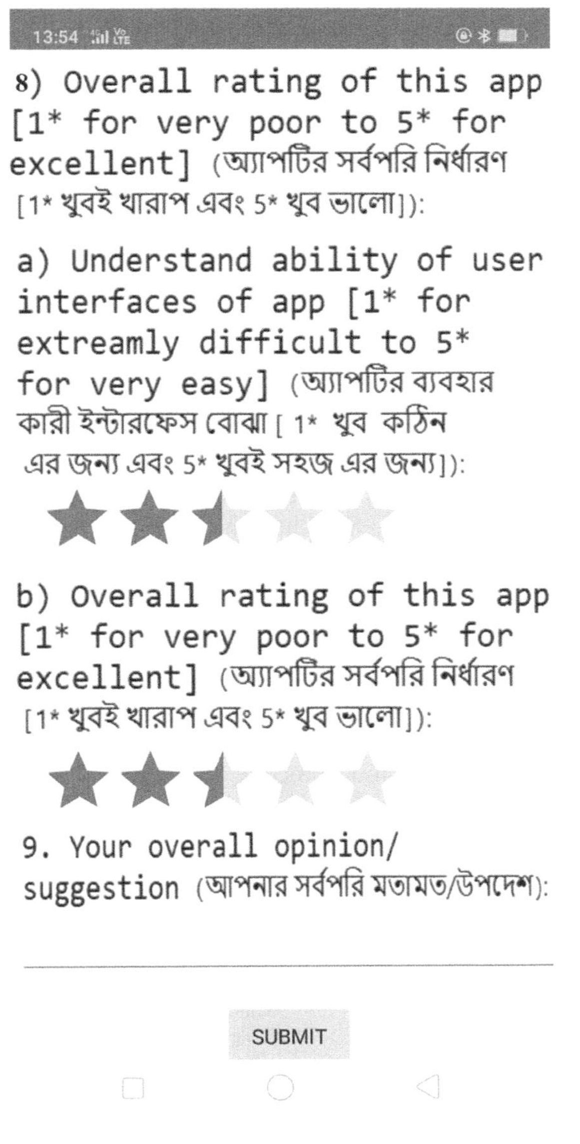

Figure 10.16 User feedback form part 4.

weather prediction for the next eight days, including the current date. For district-wise searches also, the application uses the district's latitude and longitude and the current date as parameters to call the 'One call' API and retrieves the weather prediction.

Two APIs of Google Maps Platform are used by the application 'APB' to provide 'Search by Map' facility. API called 'Maps SDK for Android' is required to display google maps. For setting markers on vaccination centres in map view, the precise latitude and longitude of the centres are needed. Precise latitude and longitude of the centres are not obtained from 'CoWIN' APIs. To overcome this, the 'Geocoding' API of Google Maps is used. It is capable of fetching the precise latitude and longitude of the vaccination centre using the centre name. The centres' precise latitude and longitude data are then used to establish marks on the corresponding Google Map locations.

10.3.4 Design descriptions

The version of Android Studio used to create the application 'Amar Protishedhak Bondhu' is 4.2.2. It can run on any Android device with API level 19 (Android KitKat) or above.

This application offers four different COVID-19 vaccination slot search options: 'Search by Pincode', 'Search within a 15-km radius of a specified Pincode', 'Search by District', and 'Search by Map'. The modular structure of the application 'Amar Protishedhak Bondhu' is shown in Figure 10.17. It is now possible to search within 15 km of a specific pincode only in the state of West Bengal. An adjacency matrix is created to provide this service. Every row of this matrix represents a unique pincode of the state of West Bengal; 1st column of this matrix contains all the unique pincodes of the state of West Bengal. The remaining columns of this matrix listed all the pincodes inside 15 km of the first column's pincode in this row. The website 'https://www.prokerala.com' is used by the developers of 'Amar Protishedhak

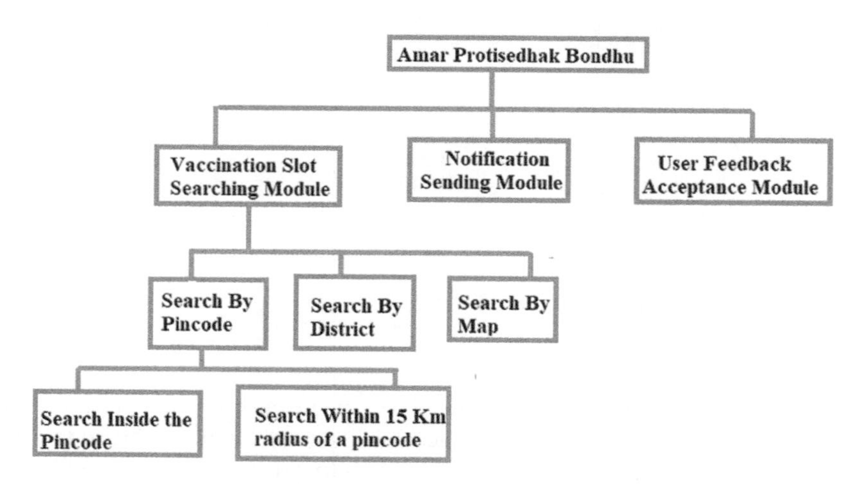

Figure 10.17 The modular structure of app 'Amar Protishedhak Bondhu'.

Bondhu' to gather all 'West Bengal's pincode. Developers also have gathered the information about pincodes within a 15-km radius of a given pincode from the website https://sharedbooks.in/distance-between-pincodes.

The App 'Amar Protishedhak Bondhu' displays vaccination slots with all doses, highlighting the user's requested dose. If a user searches for any dose, the app displays vaccination slots with all doses without highlighting anyone. This app creates a Jab_Model object to store the 'CoWIN' appointment availability API call's results. It then calls a method named Match_Making1() with each Jab_Model object and user-specified vaccine, dose, age group, and fee type choices. Match_Making1() method returns one if the Jab_ Model object matches user-specified criteria; otherwise, it returns zero. Depending on the returned result, this app adds this particular Jab_Model object to an array named Centre_List array. Finally, it displays the contents of the Centre_List array. Algorithm 10.1 is the pseudocode to display the vaccination slot to users. Algorithm 10.2 is partial pseudocode of the Match_Making1() method, which matches the Jab_Model object with user-specified criteria. This pseudocode covers any combination of dose type, age group, or fee type with only the 'Covishield' vaccine). For other vaccines, similar logic to Algorithm 10.2 was implemented.

Algorithm 10.1	*Algorithm to display the vaccination slot to users*
Input :	VaccineName, MinAge, FeeType, DoseType, Pincode/DistrictID, Date
Output :	Vaccination Slot Details according to user-specified criteria
Step 1 :	Create a Jab Model class with data members like JabCenterID, JabCenterName, JabCenterAddress, Timing, Jab_Name, Fee_Type, Fee_Amount, Min_Age, Max_Age, Age Limit Exist, Total_Available_Doses, Dose1, Dose2, Dose3, DosePrecaution
Step 2 :	Create an empty Array List named Center_List of type Jab Model
Step 3 :	Call 'CoWIN' appointment availability API with parameter Date and (Pincode or DistrictID) to fetch vaccination slot details
Step 4 :	For each vaccination slot fetched, create a Jab Model object with the fetched data
Step 4.1 :	Matching =Call Match_Making1(Jab Model object, DoseType, VaccineName, MinAge, FeeType)
Step 4.2 :	If matching=1
Step 4.2.1 :	Insert this Jab Model object to the Array List Center_List
Step 5 :	End For
Step 6 :	Display Elements of Center_List with highlighting doses according to DoseType

Algorithm 10.2	*Partial Algorithm of Match_Making1() method (Covers any combination of dose type, age group or fee type with only 'Covishield' vaccine)*
Input :	Jab_Model object, VaccineName, MinAge, FeeType, DoseType
Output :	1, 0
Step 1 :	If VaccineName='Any' AND MinAge='Any' AND FeeType='Any'
Step 1.1 :	Return 1
Step 2 :	Else If VaccineName='COVISHIELD' AND DoseType='Three'
Step 2.1 :	Return 0
Step 3 :	Else If VaccineName= 'COVISHIELD' AND MinAge='Any' AND FeeType='Any'
Step 3.1 :	If Jab_Model.Jab_Name='COVISHIELD'
Step 3.1.1 :	Return 1
Step 3.2 :	Else Return 0
Step 4 :	Else If VaccineName='COVISHIELD' AND MinAge= '18' AND FeeType ='Any'
Step 4.1 :	If Jab_Model.Jab_Name='COVISHIELD' AND Jab Model_Min Age='18'
Step 4.1.1 :	Return 1
Step 4.2 :	Else Return 0
Step 5 :	Else If VaccineName='COVISHIELD' AND (MinAge='15' OR MinAge='12')
Step 5.1 :	Return 0
Step 6 :	Else If VaccineName='COVISHIELD' AND MinAge='Any' AND FeeType='Free'
Step 6.1 :	If Jab_Model.Jab_Name='COVISHIELD' AND Jab_Model.Fee_Type='Free'
Step 6.1.1 :	Return 1
Step 6.2 :	Else Return 0
Step 7 :	Else If VaccineName='COVISHIELD' AND MinAge='Any' AND FeeType='Paid'
Step 7.1 :	If Jab_Model.Jab_Name='COVISHIELD' AND Jab_Model.Fee_Type ='Paid'
Step 7.1.1 :	Return 1
Step 7.2 :	Else Return 0
Step 8 :	Else If VaccineName='COVISHIELD' AND MinAge='18' AND FeeType='Free'
Step 8.1 :	If Jab_Model.Jab_Name ='COVISHIELD' AND Jab_Model.Min_Age='18' AND Jab_Model.Fee_Type='Free'
Step 8.1.1 :	Return 1
Step 8.2 :	Else Return 0
Step 9 :	Else If VaccineName='COVISHIELD' AND MinAge='18' AND FeeType='Paid'
Step 9.1 :	If Jab_Model.Jab_Name='COVISHIELD' AND Jab_Model.Min_Age ='18' AND Jab_Model.Fee_Type='Paid'
Step 9.1.1 :	Return 1
Step 9.2 :	Else Return 0

The Notification-sending module of this app captures the user-preferred vaccine type, age, dose type, and payment option using integers. For example, it uses vaccine id 0, 1, 2, 3, 4, 5, 6 for any vaccine, 'COVISHIELD', 'COVAXIN', 'SPUTNIK V', 'ZYCOV-D', 'CORBEVAX' and 'COVOVAX', respectively. This application implements notifications/SMS sending works using separate background threads that will execute once every 15 minutes (the minimum timing interval provided by Android studio). Multiple threads can be created, capable of searching and sending notifications/SMS with their own vaccination, network, and device preferences. Different threads are also capable of sending SMS to different mobile numbers. This application uses the abstract class 'Worker' of Android Studio to implement thread scheduling. 'Worker' class enables scheduling for any asynchronous task so that it is capable of running in the background even if the application exits or the device restart. This application extends the 'Worker' class into a class named 'MyWorker'. At the time of this background thread/task scheduling, all the constraints related to network and device (input by the user) are also set to the task by the main programme. The main programme also sent all vaccination preferences, phone no, etc., to the 'MyWorker' class instance as arguments at the time of scheduling. 'MyWorker' class overrides the 'doWork()' method of the base class 'Worker'. In this application, all the necessary tasks of vaccination slot searching and notification/SMS scheduling are performed by the 'doWork()' method of the 'MyWorker' class. To match user-specified vaccination criteria with fetched slot details, the 'doWork()' method calls the 'Match Making2()' method with user-specified vaccination preferences and currently fetched vaccination slot details using the 'CoWIN' API. If matches are found, the 'Match Making2()' returns one; otherwise, it returns zero. Depending on the returned value sent by the 'Match Making2()' method, the 'doWork()' method of the 'MyWorker' class sends SMS and notification.

One of the objectives of the application 'Amar Protishedhak Bondhu' is to provide the user with a mechanism to send their insight, suggestions, and comments about 'APB' to the developers. It is expected that the user will use the application for a few months, get benefit from its features and ultimately give feedback. This app uses Firebase's real-time database to store user's feedback. App's multithreaded structure is depicted in Figure 10.18.

10.4 ADVANTAGES OF 'AMAR PROTISHEDHAK BONDHU'

1. The app 'APB' enhances the COVID-19 vaccination slot search facilities of the Indian government web portal/app 'CoWIN'. Like 'CoWIN', it includes pincode and district-based search facilities with all the relevant filters. Additionally, it improves the accuracy and information of the 'CoWIN' 'Search on Map' facility. This software offers a unique

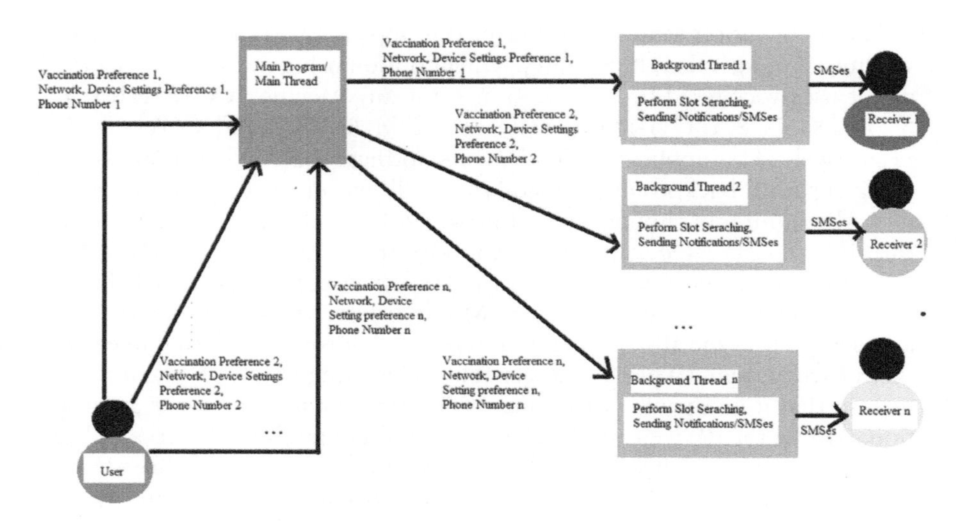

Figure 10.18 The multithreaded structure of notification module of app 'Amar Protishedhak Bondhu'.

search feature that allows users to instantly obtain information for all vaccination slots within a 15-km radius of the selected pincode.

2. The app offers a brief weather prediction according to the search date for the immunization centre's location.
3. This app can send notifications/SMS upon the availability of slots according to the user's preferred vaccine type, dosage type, age group, and charge kind.
4. Every 15 minutes, whether a user is using another app on the same smartphone or when the smartphone is inactive, this app can carry out the tasks of searching and delivering notifications in the background.
5. Users can schedule the vaccination slots searching and notifications/ SMS sending services according to their preferred battery and network conditions.
6. A single instance of an installed app can send SMS to many users with ordinary mobile phones informing them of available slots based on their preferences.

10.5 USABILITY ANALYSIS

Ten users were surveyed to gather data for a usability study. The application 'Amar Protishedhak Bondhu' was distributed to six adults personally and a Non-Government Organization (NGO). The NGO is situated in the North 24 Parganas district, West Bengal. It provides free private tuition to poor children. In the state of West Bengal, vaccination camps were organized by government and government-aided schools for their students free of cost. Some students of this NGO drop out of school and will give board exams in private. The poor dropout students did not have this opportunity to vaccinate themselves in school. The app 'Amar Protishedhak Bondhu' is distributed to four such students free of cost. All ten persons (six adults and four children) submitted the feedback form after filling it out. Figures 10.19–10.28 depict user feedback using pie charts.

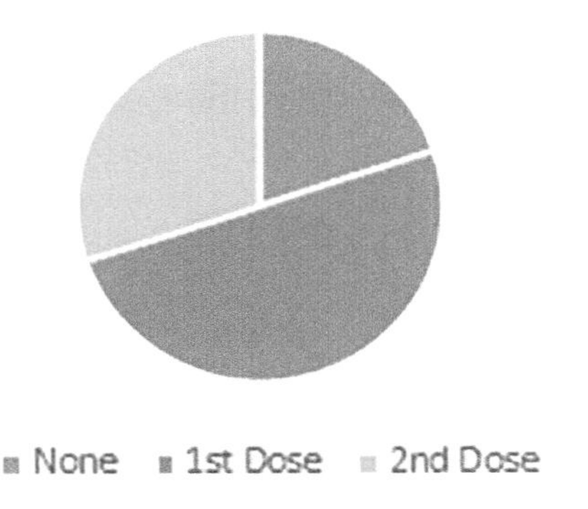

Figure 10.19 Pie chart depicts how many were vaccinated before using this app.

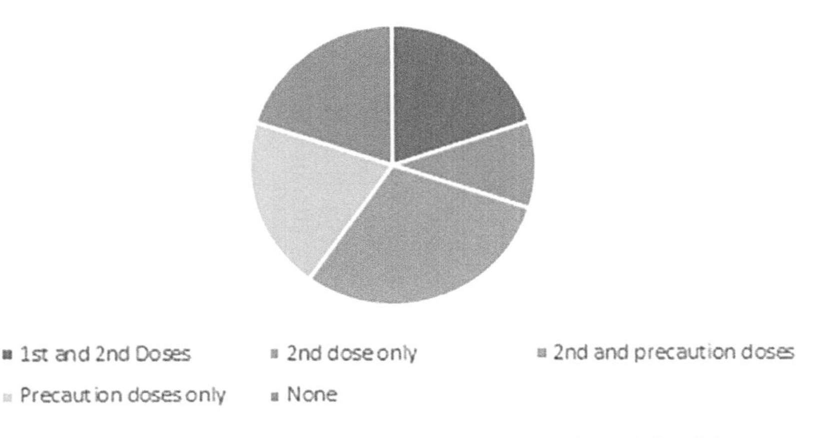

Figure 10.20 Pie chart depicts how many were vaccinated with the help of this app.

I am able to find free of cost vaccination slot with the help of this app

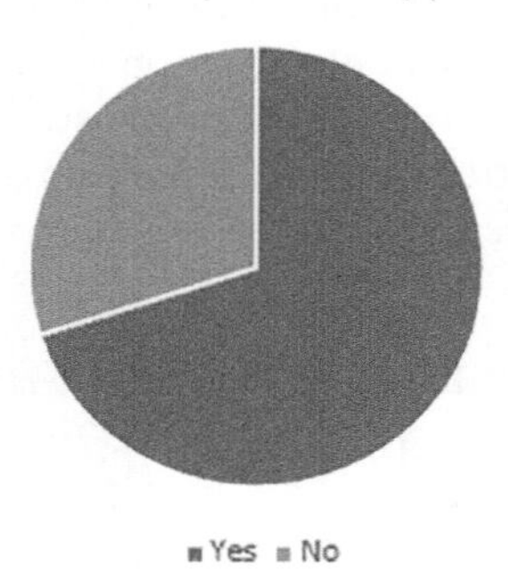

Figure 10.21 Pie chart depicts how many found free vaccination slots using this app.

This app helps me to find the nearest vaccination slot

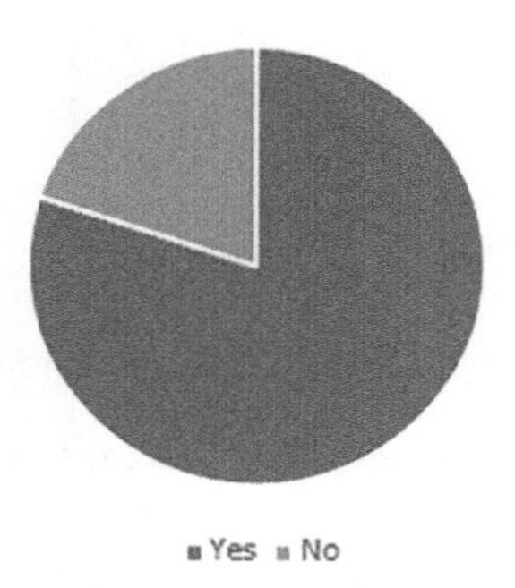

Figure 10.22 Pie chart depicts how many found nearest vaccination slots using this app.

I have used the following features of this app to find available slots

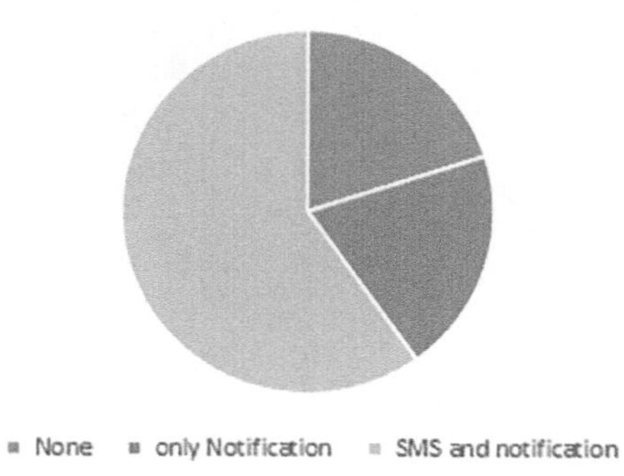

Figure 10.23 Pie chart depicts how many use the notification/SMS feature of this app.

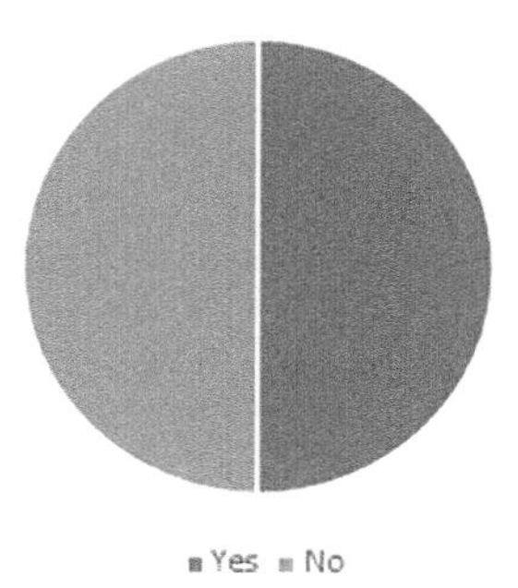

Figure 10.24 Pie chart depicts how many people help any known person using this app.

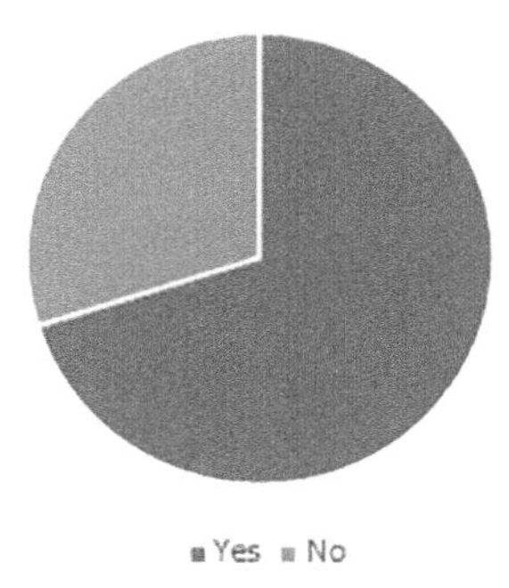

Figure 10.25 Pie chart depicts how many think displaying weekdays is helpful while searching.

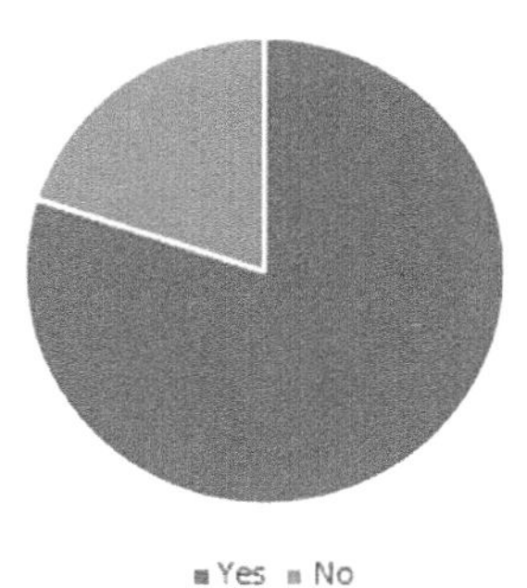

Figure 10.26 Pie chart depicts how many think displaying fees is helpful while searching.

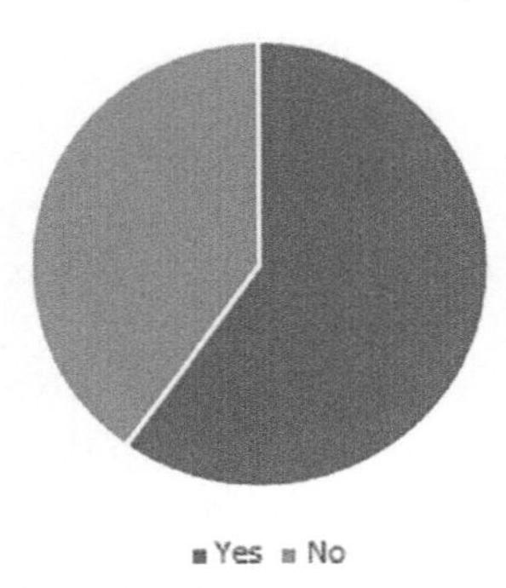

Figure 10.27 Pie chart depicts how many think displaying weather details is helpful while searching.

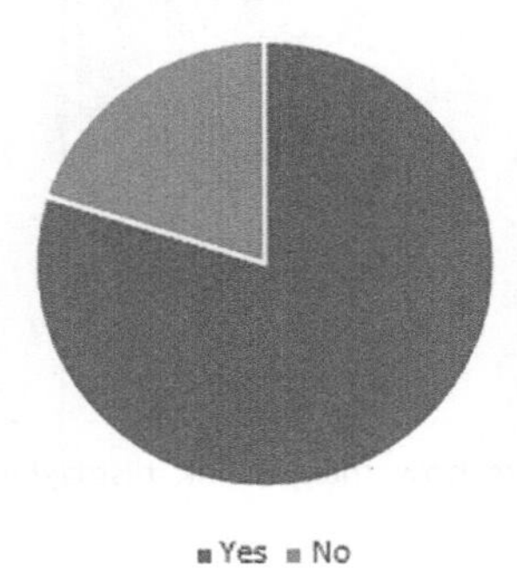

Figure 10.28 Pie chart depicts how many think the search facility to display slots within a 15-km radius is useful.

10.6 CONCLUSIONS AND FUTURE SCOPES

COVID-19 vaccinations are widely acknowledged as the most effective strategy to halt the spread of COVID-19. The covid-19 vaccine costs a lot at private vaccination clinics in India. For most Indian families, the entire cost of three immunization doses for all family members is too expensive. As a result, free-of-cost government vaccination slots are constantly in high demand. Using the Indian Government's COVID-19 vaccination portal/app 'CoWIN', it is pretty challenging to obtain free-of-cost vaccination slots in a convenient location.

Android multithreaded app is created in the present work. This app can search for vaccination slots every 15 minutes and sends notifications/SMS to users when slots become available. When the Indian government begins providing free precautionary dosages for the 18–59 age group for 75 days, it is anticipated that this app will reach the peak of its usability. Like 'CoWIN', this app provides pincode and district-wise search for all over India. This app can also search for vaccination slots under a 15 km radius of a given pincode for the state of West Bengal Only. Non-smartphone users can also benefit from this software by receiving SMS on their ordinary mobile phones. This is because a single instance of the installed app may generate several threads, each of which can search for vaccination slots using separate locations, separate search parameters, and send SMS to different telephone numbers. Soon, this app will include the ability to search for vaccination slots within a 15-km radius of a given pincode across India. When the Government of India begins immunizing toddlers, pre-schoolers, grade-schoolers, or preteens (under 12), this app will be updated to reflect the newly introduced vaccine type, dosage type, and age group.

REFERENCES

CoWIN Dashboard. "CoWIN dashboard". Accessed July 13, 2022. https://dashboard.cowin.gov.in/.

Co-WIN Public APIs. Accessed July 10, 2022. https://apisetu.gov.in/api/cowin.

Co-WIN Slot Notifier. Accessed October 17, 2021. http://getjab.in/.

CoWIN. "Search your nearest vaccination center". Accessed January 16, 2021. https://www.cowin.gov.in/.

Holder, J. "Covid world vaccination tracker - The New York Times". Tracking coronavirus vaccinations around the world. Accessed July 11, 2022. https://www.nytimes.com/interactive/2021/world/covid-vaccinations-tracker.html.

Hussain, S., and F. Suhel. "Why the private sector hasn't been able to fulfil its 25% COVID vaccine quota – The wire science". Why the private sector hasn't been able to fulfil its 25% COVID vaccine quota – The wire science. Accessed September 14, 2021. https://science.thewire.in/health/private-sector-not-able-fulfil-25-percent-covid-19-vaccine-quota/.

Mangla, S. "Impact of Covid-19 on Indian Economy." *Times of India Blog*. Accessed July 11, 2021. https://timesofindia.indiatimes.com/readersblog/shreyansh-mangla/impact-of-covid-19-on-indian-economy-2-35042/.

Sanyal, A. "Covishield at 780, Covaxin at 1,410: Maximum price for private hospitals". Accessed June 8, 2021. https://www.ndtv.com/india-news/covishield-at-780-covaxin-at-1-410-sputnik-v-at-1-145-maximum-price-that-can-be-charged-for-vaccines-by-private-hospitals-2459353.

Sharma, M. "Covishield, Covaxin to now cost private hospitals rs 225 per dose - Coronavirus outbreak news". India Today. Accessed April 9, 2022. https://www.indiatoday.in/coronavirus-outbreak/vaccine-updates/story/covishield-sii-covid-vaccine-price-dose-private-hospitals-1935466-2022-04-09.

Thakur, J. S., and H. Kaur. "Vaccine distribution for COVID-19 and equity issues in India". *International Journal of Noncommunicable Diseases* 6, no. 5 (2021): 98.
Coronaviridae Study Group of the International Committee on Taxonomy of Viruses. "The species severe acute respiratory syndrome-related coronavirus: Classifying 2019-nCoV and naming it SARS-CoV-2". *Nature Microbiology* 5, no. 4 (2020): 536–544.
Worldometer. "Countries where coronavirus has spread". Accessed July 12, 2022. https://www.worldometers.info/coronavirus/countries-where-coronavirus-has-spread/?rel=outbound.
ZeeBusiness. "All 18+ citizens to get free covid-19 booster doses from July 15, say officials | Zee Business". Zee Business. Accessed July 13, 2022. https://www.zeebiz.com/india/news-all-18-citizens-to-get-free-covid-19-booster-doses-from-july-15-say-officials-189438.

Chapter 11

Binary MMBAIS for feature selection problem

Hema Banati
Dyal Singh College, University of Delhi

Asha Yadav
University of Delhi

CONTENTS

11.1 INTRODUCTION: FEATURE SELECTION AND METAHEURISTIC ALGORITHM

Feature selection is an optimization problem that seeks a subset of "x" original features where "x<n". The complete search space for the problem is all feasible "2^n" subsets, given time, memory, and computing cost limitations [1]. Feature selection reduces data dimensions to increase knowledge and prediction accuracy [2]. Selecting important features reduces classifier complexity, speeding convergence to optimal results [3]. Filter-based, wrapper-based, and embedding approaches are feature selection categories. Filter-based approaches provide independent, cheap feature evaluation. Correlation, chi-squared, mutual information, etc. can determine their relationship to the prediction class [4]. Wrapper determines the most relevant feature subset to maximize learning. Such computationally intensive methods may provide the most valuable feature subset [5]. The embedded

DOI: 10.1201/9781003381167-11

learning model includes feature selection. They're better computationally but not as fast as wrappers [6].

Metaheuristic algorithms are widely employed to address the challenging issue of feature selection [7]. Evolutionary algorithms use an approximate, non-deterministic approach to efficiently search for Pareto-optimal solutions. They can be combined with other optimization approaches to solve objectively formulated problems quickly and with acceptable inaccuracy [8].

These algorithms start with random solutions and enhance them with each repetition. Individual solutions are evaluated using explicit criteria. Best solutions are kept, and iterations are repeated till termination. This allows maximum search space for optimal solutions [9]. Exploration or randomization breaks out of local optima, while exploitation or local search improves a good answer [10].

Xin She Yang et al. proposed Bat Algorithm (BA) in 2010 [11]. It uses microbat's echolocation to hunt prey. Echolocation lets bats detect the distance, size, orientation, and speed of their prey. BA portrays bats foraging by flying randomly in search space at position, velocity, and frequency. Bats hunt by varying their pulse, volume, and wavelength. The Bat Algorithm exhibits pre-mature convergence and is caught in local optimal while working around global best. Banati and Chaudhary's [10] MMBAIS improves the Bat Algorithm. In this method, bats move toward their best neighbor. This prevents early convergence and improves exploration. Bats are multimodal because they share information locally and travel toward their best neighbor. A parameter step size helps BA's convergence while the algorithm runs.

Each bat in MMBAIS represents a solution and moves in search space. Feature selection requires discrete solutions. Binary evolutionary algorithms describe solutions as n-dimensional binary vectors of features, where 1 or 0 signifies existence or absence. This research proposes binary MMBAIS to optimize classification feature selection. The algorithm's multimodal nature and precise step size allow for improved convergence. This study tests binary MMBAIS for real-world feature selection. This chapter explores the following objectives:

- Capability of binary MMBAIS to provide improvement over existing binary Evolutionary Algorithms for the task of feature selection.
- Measure the stability of the features subset selected by binary MMBAIS for existing classification algorithms with an m-fold cross-validation.

For the objective stated above, the experimental setup for binary MMBAIS is taken same as provided by Ibrahim and Tawhid [12]. Evaluation of binary MMBAIS is done over 15 different types of benchmark dataset from UCI [13] repository and one dataset from [14] and results are compared with Ibrahim and Tawhid [12]. The stability of binary MMBAIS is measured using the relative weighted consistency proposed by Somol and Novovicova [15] for each dataset. Further, the feature subset with best accuracy of each dataset is used to run other standard machine learning classifiers and their accuracy score is presented.

The structure of the chapter is as follows: Section 11.2 discusses the literature background of various evolutionary algorithms for feature selection problems. The proposed algorithm binary MMBAIS is explained in Section 11.3. Section 11.4 details the experimental setup and dataset used for the study. Section 11.5 provides a detailed analysis of the results obtained from experimentation. The last section provides the conclusion followed by the future scope of the work.

11.2 RELATED WORK

Eberhart and Kennedy [16] presented particle swarm optimization (PSO) for feature selection. Qasim and Algamal [17] proposed PSO-LRBIC, which combines PSO intelligence with logistic regression with a Bayesian information criteria fitness function on a medical dataset. Pashaei et al. [18] hybridized BBHA and BPSO to classify cancer. With fewer attributes, the BPSO-BBHA algorithm achieved 99.33% accuracy. Xue et al. [19] proposed SPS-PSO to reduce high-dimensional datasets. It used a self-adaptive parameter and numerous classifiers to improve accuracy and reduce computing cost.

Tawhid and Dsouza [20] proposed a BA and PSO hybrid for feature selection to overcome BA's slow convergence. This uses Bat's search and PSO's convergence to increase dimensionality reduction. Image steganalysis was proposed by Liu et al. [21]. It discovers the best subset of original attributes to identify stego from cover photos. Al-Betar et al. [22] proposed rMRMR-MBA, a hybrid filter and wrapper on BA, to optimize diagnosis and prognosis. This rMRMR method improves the accuracy of gene expression datasets. BA is utilized as a search methodology in the wrapper method to generate a final feature subset for gene dataset.

Grey wolf optimization (GWO) replicates grey wolf hunting. Emary et al. [23] developed a binary GWO for optimal feature subset selection. Al-tashi et al. [24] presented a hybrid GWO and PSO in binary. Al-Tashi et al. [25] Best disease identification dataset features are obtained using GWO and assessed using SVM.

11.3 MMBAIS ALGORITHM

MMBAIS is an enhancement over the original Bat Algorithm. It allows bats to move toward the best neighbor instead of global best thus dividing the population into non-overlapping groups and each subgroup explores independently by randomly walking around their local best. This multimodal nature of bats leads to proper exploration of a large search space and avoids premature convergence. The random movement of the original Bat Algorithm is controlled by the step-size parameter of MMBAIS. The authors have proved superiority of MMBAIS over BA on 30 Benchmark functions.

11.4 BINARY MMBAIS

To deal with the optimization problem of feature selection binary version of the MMBAIS algorithm is proposed. Feature selection aims to enhance classification accuracy by employing a subset of features instead of all features. The FS problem requires searching an exponentially 2^n-large feature space of n, which becomes exhaustive as the number of features increases. Binary MMBAIS's large exploration capabilities make it a candidate for feature selection.

In MMBAIS, each bat searches continuously. The search space for feature selection is an n-dimensional Boolean vector, where each dimension represents a feature, whose value is 0 or 1 depending on whether the feature is included for the particular bat. So, a binary MMBAIS must be modified. Applying sigmoid function to bat positions restricts them to binary numbers. Sigmoid Equation (11.1) is given below.

$$S(x) = \frac{1}{1+e^{x}} \tag{11.1}$$

11.4.1 Objective function

The objective of the FS problem is twofold: Maximizing the prediction accuracy measure while minimizing the features subset. Both of them are conflicting goals. This multi-objective is defined as single objective function as given by Equation (11.2).

$$\begin{aligned} \text{Fitness}(X) &= \mu * \text{Accuracy Score}(1-\mu) * \frac{s}{n} \\ \text{AccuracyScore} &= \frac{\text{Correctly Classified Instances}}{\text{Total Number of Instances}} \end{aligned} \tag{11.2}$$

Where $\mu \varepsilon$ [0,1] is a parameter that defines weight of classification accuracy as the measure of performance of X having features subset s out of total features n on training dataset D.

11.4.2 Structure of binary MMBAIS

The various Equations (11.3–11.12) used to calculate bat movement and reaching the global best are given below. The bat scans its environment and finds net energy gain obtained by prey. Equation (11.3) gives possible energy gain of bat, and Equation (11.4) calculates the distance between "bat i" and "bat j" i.e., the difference between information gain obtained by selected features of both bats respectively. Equation (11.5) p energy spent in chasing the prey and Equation (11.6) gives net energy gain. Equations (11.7 and 11.8) calculate the frequency and velocity of bat. Equations (11.9–11.11) define positional change of bat via random walk for exploration and exploitation. Equation (11.12) gives step size parameter tuning.

$$eG_{i,j} = \left|E_j - E_i\right| \tag{11.3}$$

$$\gamma_{i,j} = x_i - x_j = \sqrt{\sum_{k=1}^{n} (x_{i,k} - x_{j,k})^2} \tag{11.4}$$

$$eS_{i,j} = r_{i,j} * \delta \tag{11.5}$$

$$\Upsilon_{i,j} = eG_{i,j} - eS_{i,j} \tag{11.6}$$

where E_i and E_j represent objective function value of ith and j^{th} bats.

$$f_i = f_{\min} + (f_{\max} - f_{\min}) * \beta \tag{11.7}$$

$$\upsilon_i^t = \upsilon_i^{t-1} + (x_i^{t-1} - x_j^t) * f_i \tag{11.8}$$

$$x_i^t = x_i^{t-1} + \Upsilon_{i,j} * \upsilon_i^t \tag{11.9}$$

Equation (11.9) is modified for binary positions

$$x_i^t = \begin{cases} 1, \text{ if } S(\Upsilon_{i,j}) * \upsilon_i^t \geq \sigma \\ 0, \text{ otherwise} \end{cases} \tag{11.9a}$$

where σ is a random value ε [0,1] and rand ε [−1,1] and $A_t = < A_i^t >$ is the average loudness of all the bats at this iteration.

$$x_i = x_j^t + \text{rand} * A^t \tag{11.10}$$

Equation (11.10) is modified as follows to attain binary characteristics

$$x_i = x_j^t \tag{11.10a}$$

Generate random number of features using f=int (rand * A^t) and get f random positions to set the feature value in array k.

$$x_i^t = \begin{cases} 1, \text{ if } i \; \varepsilon \; k \\ x_i^t, \text{ otherwise} \end{cases} \tag{11.10b}$$

To control the random movement of bat around itself to explore a new solution to the bat position is calculated by Equation (11.11) as follows, where rand ε [−1,1] and $\mathfrak{R}$ ε [0.1,1] the value of $\mathfrak{R}$ is reduced using ω if global best is not improved over a number of generations G.

$$x_i = x_i + \mathfrak{R} * \text{rand} \tag{11.11}$$

$$\mathfrak{R} = \mathfrak{R} * \omega \text{ where } \mathfrak{R} \rightarrow 0 \text{ as } t \rightarrow \infty \tag{11.12}$$

Equation (11.11) is modified to generate random number of features using f=int ($\mathfrak{R}$ * rand[1,a]) where a is number of features of current best and get f random positions among the selected features of current best to set the feature value in array k f random positions to set the feature value in array k

$$x_i^t = \begin{cases} 0, & \text{if } i \; \varepsilon \; k \\ 1, & \text{otherwise} \end{cases} \tag{11.11a}$$

11.5 METHODOLOGY

In the following section, we will discuss how binary MMBAIS can be used to solve problems involving feature selection. The following is an explanation of the methodology that was used to carry out this study:

- For the purpose of the experiment, 15 datasets, including 14 benchmark datasets from UCI [13] and the Olive dataset [14], are taken.
- The same experimental setup is used for binary MMBAIS in order to facilitate a comparison of our findings with those of the other five binary evolutionary algorithms presented by Ibrahim and Tawhid [12]. Specifically, the population size is maintained at 20, and the maximum number of iterations is set to 50. As classifiers, the K-Nearest Neighbor (KNN) algorithm with k=1 and the Decision Tree for Classification (DT) algorithm are implemented.
- The algorithm is carried out using 10-fold cross-validation for every classifier applied to each dataset, with loss of classification accuracy serving as the objective criterion.

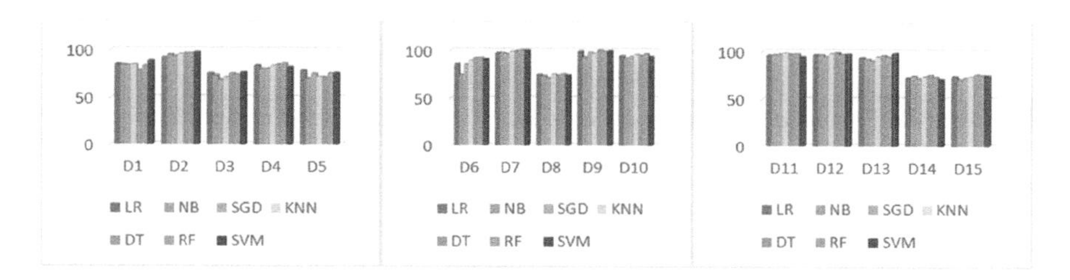

Figure 11.1 Classification accuracy of various classifiers.

- Table 11.4 presents the findings of a comparison between the mean classification error of binary MMBAIS and that of various other binary evolutionary algorithms.
- Table 11.5 contains a presentation of the consistency measure that was used to evaluate the algorithm's stability.
- The results of running other classifiers using the best feature set found through experimentation are displayed in Figure 11.1.
- The detailed binary MMBAIS algorithm is given in Table 11.1 below.

11.6 EXPERIMENTAL SETUP

This section describes the technique used to apply binary MMBAIS to solve feature selection problems. The algorithm is executed on 15 datasets, 14 UCI [13] benchmark datasets and Olive dataset [14] as mentioned in [12]. Ibrahim and Tawhid [12] have proposed a hybrid version of Binary Bat and Differential Evolution algorithm. This binary version was compared with four existing states of the art evolutionary feature selection algorithms. To compare the performance of binary MMBAIS with other algorithms the same experimental setup is used as mentioned in [12]. The experiments are done on Intel® Core™ i5 CPU with 8 GB RAM. In addition, the population size is kept 20 while maximum iterations are 50. The classifiers KNN with $k=1$ and Decision Tree for Classification (DT) are used for measuring the accuracy of the features subset.

11.6.1 Datasets

Table 11.2 contains information about the dataset, and all of the datasets have had their records pre-processed in order to get rid of any missing values. Datasets are taken same as in Ibrahim and Tawhid [13].

Table 11.1 Algorithm

Algorithm : Binary MMBAIS for feature selection
Require: Bat Population N, Total features n, Generations G, Labeled dataset D for training and testing
Ensure: Subset S having features s ε [1,n]; where, s$\leq n$ and n=Total features of dataset D that gives the maximum accuracy in evaluation
1: Define objective function $F(x, y)$
2: Initialize Auxiliary Variables : $\alpha, \gamma, \delta, \omega, \mathfrak{R}$, fmin, fmax
$X_{bati,j}$ (i = 1, 2, 3...N)(j = 1, 2, 3...n) $\leftarrow$ 0
For each *bat* in $X_{bati,j}$ (i = 1, 2, 3...N):
Generate a random number k between (Min, Max):
For i = 1 : k:
Generate random number j between (1, n)
$X_{bat,j} \leftarrow$ 1:
4: Initialize velocity $V_{i,j}$ per bat per dimension
5: Initialize r_i(pulse rate) and A_i (loudness)
6: Evaluate objective function and assign Energy value E_i for each bat
7: while counter $\leq$ total iterations do
8: Rank all bats on Energy Value
9: for $i \leftarrow$ 1 to N do
10: for $j \leftarrow$ 1 to i do
11: Calculate and save highest $Y_{i,j}$ using Equations (11.3–11.6)
12: end for
13: Generate new solutions using Equations (11.7–11.9a)
14: Calculate transfer function value for velocity using Equation (11.1)
15: Update positions using Equation (11.9a)
16: if rand > r_i then
17: Generate a solution around local best using Equation (11.10a)
18: Change position vector dimension with local best using Equation (11.10b)
19: end if
20: Generate a new solution using Equation (11.11a)
21: Evaluate objective function for new solution
22: if (rand < A_i) and (Fitness(X_i) < Fitness(X_{old})) then
23: Update the new solution to bat position
24: $r_i = r^0{}_i$ [1 – exp(–γt)], where γ is constant
25: $A^{t+1}{}_i = \alpha A^t{}_i$, where α is constant
26: end if
27: end for
28: Update and save best
29: Update $\mathfrak{R}$ by Equation (11.12)
30: end while

Table 11.2 Dataset

S. no	*Dataset*	*(Features, instances)*	*Classes*
D1	(Balance Scale)	(4,625)	3
D2	(Olive)	(8,572)	9
D3	(Breast Cancer)	(9,683)	2
D4	(Heart)	(13,270)	2
D5	(Vehicle)	(18,946)	4
D6	(Lymphography)	(18,148)	4
D7	(Zoo)	(18,101)	2
D8	(German credit)	(24,1000)	2
D9	(Wincosin diagnosis breast cancer)	(32,596)	2
D10	(Ionosphere)	(34,351)	2
D11	(Chess)	(36,3196)	2
D12	(Cancer)	(56,32)	3
D13	(Sonar)	(60,20)	2
D14	(Hill valley)	(101,606)	2
D15	(LSVT voice rehabilitation)	(309,126)	2

11.6.2 Classification methods

In [13] the author has used two popular supervised classifiers to measure the fitness of a solution based on classification accuracy namely, KNN and Decision Tree for Classification (DT). The same classifiers are used for experiments in this chapter.

The purpose of the fitness function in this context is to optimize the accuracy of the prediction while simultaneously minimizing the number of features that are employed in the classification process. At each iteration, both of the classifiers are applied to each solution and analyzed. The loss of classification accuracy serves as the objective criterion for this situation. To verify that the results of an evaluation of fitness can be trusted, the data are subjected to 10-fold cross-validation.

The value of $\mu=0.9$ in Equation (11.2) for fitness function. In this experiment, the generation and number of bats are chosen to be 50 and 20 respectively to have a fair comparison with other algorithms as given by Ibrahim and Tawhid [12]. Other parameters of binary MMBAIS have been configured in accordance with the authors in Banati and Chaudhary [10]. The values of various parameters are presented in Table 11.3. The mean of the classification error that is produced from an average of 50 runs that is computed across each dataset is used as the comparison criterion for determining how well binary MMBAIS performs in comparison to other state-of-the-art algorithms, such as those that are listed in [12].

Table 11.3 Parameter values

Bat population size (*N*)	20[b]	Runs (*R*)	50[b]
Generation (*G*)	50[b]	Pulse rate (*r*)	0.85[a]
Loudness (*A*)	0.85[a]	Search weight (ω)	0.5[a]
Gamma (γ)	0.9[a]	Step size ($\Re$)	0.8[a]
Energy consumption coefficient (δ)	0.8[a]	Alpha (α)	0.9[a]

[a] Values are taken from Banati and Chaudhary [10].
[b] Values are taken from Ibrahim and Tawhid [12].

11.7 RESULT AND PERFORMANCE ANALYSIS

In this work, the mean of classification error was attained on an average of 50 runs on the various datasets by utilizing both KNN and DT. This mean was then computed and compared to several other binary evolutionary functions.

Table 11.4 presents the comparison results of mean classification error using KNN and Decision Tree classifier for six evolutionary algorithms Binary Bat, BPSOGSA (binary PSO integrating with the gravitational search algorithm), Binary Crow Search Algorithm, Binary Bat and Differential Evolution, Binary Gray Wolf Optimization Algorithm and binary MMBAIS for feature selection. As it is shown that for KNN classifier binary MMBAIS outperforms for 11 datasets D1, D2, D4, D6, D8, D10, D11, D12, D13, D14, and D15 among datasets mentioned by having low mean classification error for all mentioned evolutionary algorithm. Binary MMBIAS is performing well for D5 dataset against Binary Bat, Crow Search, Binary Gray Wolf and comparable equal to BPSOGSA, Binary Bat and Differential Evolution. BPSOGSA outperforms for D3 datasets. Binary MMBAIS is in second place for D7 and gives more accurate result than Binary Bat, Crow Search, Binary Gray Wolf, BPSOGSA and Binary Bat and Differential Evolution.

Furthermore, for Decision Tree classifier binary MMBAIS has performed well for 12 datasets (D2, D4, D6, D8, D9, D10, D11, D12, D13, D14, D15) among 15 datasets for Binary Bat, Crow Search, Binary Gray Wolf, BPSOGSA, Binary Bat and Differential Evolution. Binary MMBAIS outperform for dataset D1 as compared to other algorithms except Binary Bat and Differential Evolution also achieved the same performance to Binary Bat, Crow Search for D3. For D5 dataset, binary MMBAIS gives better results than Binary Bat, Crow Search, Binary Gray Wolf, Binary Bat, and Differential Evolution. Binary Bat and Differential Evolution perform best for D7 and binary MMBAIS gives more accuracy compared to Crow Search, Binary Gray Wolf. Deep analysis of Table 11.4 indicates that features subset obtained by binary MMBAIS is able to classify various datasets more accurately. Thus, it can be concluded that binary MMBAIS can

Table 11.4 Mean classification error (for 50 runs)

Dataset	*Classification technique*	*BBA*[a]	*BPSOGSA*[a]	*BCS*[a]	*BBADE*[a]	*BGW*[a]	*BMMBAIS*
D1	KNN	0.343	0.358	0.388	0.221	0.348	**0.176**
	DT	0.268	0.281	0.277	**0.2**	0.265	0.204
D2	KNN	0.087	0.097	0.106	0.081	0.087	**0.065**
	DT	0.077	0.085	0.091	0.074	0.078	**0.062**
D3	KNN	0.054	**0.052**	0.059	0.053	0.053	0.257
	DT	0.042	**0.04**	0.042	0.041	**0.04**	0.042
D4	KNN	0.169	0.172	0.198	0.163	0.178	**0.16**
	DT	0.159	0.157	0.166	0.155	0.157	**0.088**
D5	KNN	0.27	0.267	0.298	**0.266**	0.276	0.269
	DT	0.257	**0.248**	0.262	0.252	0.249	0.25
D6	KNN	0.127	0.12	0.162	0.123	0.128	**0.078**
	DT	0.139	0.133	0.148	0.132	0.139	**0.077**
D7	KNN	0.007	0.009	0.022	**0**	0.012	0.005
	DT	0.029	0.035	0.054	**0.025**	0.044	0.03
D8	KNN	0.277	0.271	0.3	0.275	0.2807	**0.242**
	DT	0.251	0.241	0.26	0.238	0.247	**0.229**
D9	KNN	**0**	**0**	**0**	**0**	**0**	0.004
	DT	0.015	0.016	0.016	0.015	0.015	**0.009**
D10	KNN	0.066	0.059	0.085	0.062	0.074	**0.034**
	DT	0.061	0.057	0.066	0.06	0.062	**0.048**
D11	KNN	0.026	0.027	0.069	0.024	0.03	**0.023**
	DT	**0.008**	0.009	0.038	**0.008**	0.008	**0.008**
D12	KNN	0.034	0.053	0.084	0.047	0.066	**0.018**
	DT	0.123	0.125	0.125	0.125	0.125	**0.097**
D13	KNN	0.075	0.057	0.098	0.063	0.075	**0.038**
	DT	0.163	0.157	0.163	0.165	0.175	**0.142**
D14	KNN	0.375	0.363	0.384	0.372	0.373	**0.244**
	DT	0.377	0.373	0.376	0.372	0.375	**0.357**
D15	KNN	0.341	0.318	0.344	0.31	0.348	**0.25**
	DT	0.138	0.133	0.137	0.137	0.141	**0.072**

[a] Values of the column are taken from Ibrahim and Tawhid [13]
Bold values depicts the best result

provide better subset of features for classification than other binary evolutionary methods.

To measure the stability of the proposed algorithm the relative weighted consistency measure proposed in [16] is chosen and calculated for each dataset. This measure is independent of the size of the subset selected. As 50 runs are executed for each dataset and there is no restriction on the subset size selection. Each run of binary MMBAIS can select different-sized subsets, which give maximum accuracy; thus, the relative weighted consistency can be used as an appropriate stability measure for this experiment. This measure calculates the relative degree of randomness of the system (S), where S depicts a selected feature subset for r runs. The value of CW_{rel} (S, n) can be calculated from Equation (11.13).

$$CW_{rel} = \frac{|n|(Y - D + \sum f \in YF_f(F_f - 1)) - Y^2 + D^2}{n(H^2 + r(Y - H) - D) - Y^2 + D^2} \quad (11.13)$$

where

for a given system S=set of feature vector of each run on a given data set for r runs

n=Total features of dataset

Y=Sum of frequency of all selected features in S

r=Number of runs

D=Y mod |n|

H=Y mod r

In Equation (11.13) parameters D, H are used to balance the selection size bias i.e., higher consistency values are obtained when the sizes of feature subsets in system approach the total number of features "n". The value of CW_{rel} (S, n) € [0,1], where higher value indicates more stability and lower value indicates more randomness.

Table 11.5 presents the total features, mean subset size, and stability measure CW_{rel} for 50 runs over each of the 15 Datasets. The value of CW_{rel} is greater than 0.80 in 5 Datasets (D9, D11, D12, D14, D15) while three

Table 11.5 Stability measure of binary MMBAIS

Dataset	*Total features*	*Mean selected features*	*Consistency measure*
D1	4	3	0.649
D2	8	5	0.672
D3	9	4	0.495
D4	13	5	0.731
D5	18	8	0.337
D6	18	7	0.61
D7	18	6	0.522
D8	24	10	0.755
D9	32	12	0.816
D10	34	13	0.743
D11	36	16	0.881
D12	56	28	0.851
D13	60	27	0.52
D14	101	52	0.857
D15	309	142	0.891

datasets (D4, D8, D10) have achieved value greater than 0.70 and datasets (D1, D2, D6) have obtained value greater than 0.60, thus 11 datasets are able to achieve stability value above 0.60. This indicates that the binary MMBAIS is able to give stable feature subsets. The main concern for applying evolutionary algorithms on feature selection is the randomness in the subsets generated from different runs. Binary MMBAIS gives stable results due to its multimodal ability to explore the search space and step size exploitation to further narrow down on the best selected feature subset till the same accuracy can be obtained with a smaller number of features.

Experiments indicate that the feature subset selected by binary MMBAIS gives a stable performance that can be seen by evaluation of feature subset over six different classifiers using 10-fold cross-validation providing almost equal accuracy. Figure 11.1 presents the performance analysis of six standard machine learning classifier LR (Logistic Regression), NB (Naïve Bayes), SGT (Stochastic Gradient Descent), KNN, DT (decision tree), RF (Random Forest), SVM (Support Vector Machine) using features selected by binary MMBIAS for given 15 datasets. It has been observed that more than 80% accuracy is attained in D1, D2, D4, D6, D7, D9, D10, D11, D12, and D13 datasets among 15 datasets. The remaining datasets D3, D5, D8, D14, and D15 have achieved almost 70% accuracy. Binary MMBAIS is able to find minimal subset for almost all the datasets due to its ability of exploitation.

11.8 CONCLUSION

In this chapter, a binary version of the MMBAIS algorithm has been proposed to solve the problem of feature selection. The multimodal improved search over Bat Algorithm helps binary MMBAIS to achieve better results over other five state-of-the-art evolutionary algorithms for feature selection proposed earlier. An extensive experimental study was made on the performance of binary MMBAIS over 15 benchmark datasets and the result demonstrate the superiority of algorithm over other five binary evolutionary algorithm used for comparison. Binary MMBAIS performed significantly better on 73.33% of instances used in experiments. The ability of binary MMBAIS to focus on refining the best solution obtained in iterations by taking the precise steps to omit the irrelevant features further till accuracy is improved, allows it to select minimal relevant subset. Hence, the feature subset obtained by binary MMBAIS during training were able to achieve an at par accuracy by various classifiers in test data also. The accuracy of more than 80% was achieved in 66.67% of instances out of 15 datasets. The stability of any evolutionary features selection is very important as it measures the effectiveness of algorithm to regenerate the same set of features over different runs. The related weighted consistency measure applied to quantify the stability of binary MMBAIS shows a confidence level of more than 60% for 11 out of 15 datasets.

The findings indicate that binary MMBAIS is capable of performing well when it comes to the problem of feature selection. This demonstrates that the binary MMBAIS is an effective algorithm for binary optimization. It can be further adapted for discrete variants and used to find solutions to problems that occur in the real world, such as gaming and scheduling.

REFERENCES

1. Rao, Haidi, Xianzhang Shi, Ahoussou Kouassi Rodrigue, Juanjuan Feng, Yingchun Xia, Mohamed Elhoseny, Xiaohui Yuan, and Lichuan Gu. "Feature selection based on artificial bee colony and gradient boosting decision tree." *Applied Soft Computing* 74 (2019): 634–642.
2. Jia, Zhen, Zhenbao Liu, Chi-Man Vong, and Michael Pecht. "A rotating machinery fault diagnosis method based on feature learning of thermal images." *IEEE Access* 7 (2019): 12348–12359.
3. da Silva, Diogo L., Leticia M. Seijas, and Carmelo J. A. Bastos-Filho. "Artificial bee colony optimization for feature selection of traffic sign recognition." *International Journal of Swarm Intelligence Research (IJSIR)* 8, no. 2 (2017): 50–66.
4. Miao, Jianyu, and Lingfeng Niu. "A survey on feature selection." *Procedia Computer Science* 91 (2016): 919–926.
5. Srivastava, Shweta, Nikita Joshi, and Madhvi Gaur. "A review paper on feature selection methodologies and their applications." *IJCSNS* 14, no. 5 (2014): 78.
6. Bach, Francis R. "Bolasso: model consistent lasso estimation through the bootstrap." In *Proceedings of the 25th International Conference on Machine Learning*, pp. 33–40. Helsinki, 2008.
7. Arora, Sankalap, and Priyanka Anand. "Binary butterfly optimization approaches for feature selection." *Expert Systems with Applications* 116 (2019): 147–160.
8. Heidari, Ali Asghar, Seyedali Mirjalili, Hossam Faris, Ibrahim Aljarah, Majdi Mafarja, and Huiling Chen. "Harris hawks optimization: Algorithm and applications." *Future Generation Computer Systems* 97 (2019): 849–872.
9. Wang, Zhichun, Minqiang Li, and Juanzi Li. "A multi-objective evolutionary algorithm for feature selection based on mutual information with a new redundancy measure." *Information Sciences* 307 (2015): 73–88.
10. Banati, Hema, and Reshu Chaudhary. "Multi-modal bat algorithm with improved search (MMBAIS)." *Journal of Computational Science* 23 (2017): 130–144.
11. Yang, Xin-She. "A new metaheuristic bat-inspired algorithm." In: Janusz Kacprzyk (ed.), *Nature Inspired Cooperative Strategies for Optimization (NICSO 2010)*, pp. 65–74. Springer, Berlin, Heidelberg, 2010.

12. Ibrahim, Abdelmonem M., and Mohamed A. Tawhid. "A new hybrid binary algorithm of bat algorithm and differential evolution for feature selection and classification." In: Nilanjan Dey and V. Rajinikanth (eds.), *Applications of Bat Algorithm and Its Variants*, pp. 1–18. Springer, Singapore, 2021.
13. Dua, Dheeru, and Casey Graff. *UCI Machine Learning Repository* (2017).
14. https://rdrr.io/cran/zenplots/man/olive.html.
15. Somol, Petr, and Jana Novovičová. "Evaluating the stability of feature selectors that optimize feature subset cardinality." In: Ronan Nugent (ed.), *Joint IAPR International Workshops on Statistical Techniques in Pattern Recognition (SPR) and Structural and Syntactic Pattern Recognition (SSPR)*, pp. 956–966. Springer, Berlin, Heidelberg, 2008.
16. Eberhart, Russell, and James Kennedy. "A new optimizer using particle swarm theory." In *MHS'95. Proceedings of the Sixth International Symposium on Micro Machine and Human Science*, pp. 39–43. IEEE, 1995.
17. Qasim, Omar Saber, and Zakariya Yahya Algamal. "Feature selection using particle swarm optimization-based logistic regression model." *Chemometrics and Intelligent Laboratory Systems* 182 (2018): 41–46.
18. Pashaei, Elnaz, Elham Pashaei, and Nizamettin Aydin. "Gene selection using hybrid binary black hole algorithm and modified binary particle swarm optimization." *Genomics* 111, no. 4 (2019): 669–686.
19. Xue, Yu, Tao Tang, Wei Pang, and Alex X. Liu. "Self-adaptive parameter and strategy based particle swarm optimization for large-scale feature selection problems with multiple classifiers." *Applied Soft Computing* 88 (2020): 106031.
20. Tawhid, Mohamed A., and Kevin B. Dsouza. "Hybrid binary bat enhanced particle swarm optimization algorithm for solving feature selection problems." *Applied Computing and Informatics* 16, no. 1/2 (2018): 117–136.
21. Liu, Feng, Xuehu Yan, and Yuliang Lu. "Feature selection for image steganalysis using binary bat algorithm." *IEEE Access* 8 (2019): 4244–4249.
22. Al-Betar, Mohammed Azmi, Osama Ahmad Alomari, and Saeid M. Abu-Romman. "A TRIZ-inspired bat algorithm for gene selection in cancer classification." *Genomics* 112, no. 1 (2020): 114–126.
23. Emary, Eid, Hossam M. Zawbaa, and Aboul Ella Hassanien. "Binary grey wolf optimization approaches for feature selection." *Neurocomputing* 172 (2016): 371–381.
24. Al-Tashi, Q., S. J. A. Kadir, H. M. Rais, S. Mirjalili, and H. Alhussian. "Binary optimization using hybrid grey wolf optimization for feature selection." *IEEE Access* 7 (2019): 39496–39508.
25. Al-Tashi, Qasem, Helmi Rais, and Said Jadid. "Feature selection method based on grey wolf optimization for coronary artery disease classification." In *International Conference of Reliable Information and Communication Technology*, pp. 257–266. Springer, Cham, 2018

Chapter 12

Audio to Indian Sign Language Interpreter (AISLI) using machine translation and NLP techniques

B. K. Tripathy and Nivedita
VIT Vellore

CONTENTS

12.1 INTRODUCTION

There are millions of people on earth with hearing disabilities, and it is difficult for people to interact with them comfortably without using sign languages (SLs). For people in India, the main method of communication is Indian Sign Language (ISL) [1]. It becomes very difficult to interact with people with hearing disabilities if we don't know SL. Learning SL is expensive due to its verities. The business market is coming up with programs for efficient communication in such situations. But a single such program accessible makes it monetarily. This is where Audio to ISL Interpreter (AISLI), which is proposed in this chapter, provides a solution. AISLI is compatible with all available platforms as the Python used here is platform-independent. AISLI will be easy to use and readily available bridging the communication gap between the hearing-impaired and the rest of the world. This application takes speech as input through microphone using PyAudio speech is recognized and translated to text using Google speech recognition API and Natural Language Processing (NLP). If this is matched with the GIFs, it gives ISL GIFs as output; if not matched, it gives letters as output according to the phrase or word recognized.

DOI: 10.1201/9781003381167-12

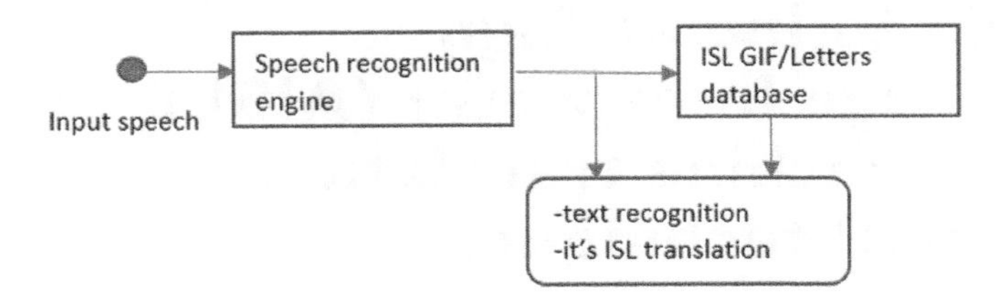

Figure 12.1 Architecture of AISLI.

NLP empowers computers to understand spoken and textual words as efficiently as human beings. It is a branch of AI that combines computational linguistics with deep learning (DL), machine learning (ML), and statistics. As a result, computers can understand the full meaning, completely with the speakers or the writers while keeping their sentiment and intention intact. The programs used by computers to translate languages are taken care by NLP. In addition, NLP adds the capability to respond to commands in speech and rapid summarization of large volumes of text quickly in real time. NLP is also used in GPS systems operated by voice, digital assistants, and dictation using software for the conversion of speech into texts, and chatbots provide customer services. Even though business operations are streamlined, the productivity of the employees is enhanced, and in general critical business processes are simplified using NLP. In Figure 12.1, we provide the architecture of our proposed system AISLI.

12.2 LITERATURE REVIEW

Natural languages are filled with words with uncertain meanings, which require suitable models to represent them. Fuzzy set [2,3] is one such model. Crisp C-means algorithm was put forth by Lloyd in 1957 [2,4], which generates disjoint clusters and restricts its applications. This led to the development of uncertainty-based C-means algorithms and the fuzzy C-means algorithm (FCM) [2]. This mean has been extended in several directions [5,6]. Due to the advancement of big data, many suitable algorithms were proposed [7,8]. There are several applications of clustering algorithms in different diversified areas [9]. A SL system for identification depending on FCM was developed in [10], where picture frames are resized to ensure a level playing field between all images using OpenCV. This process of extraction and classification of video for gesture labeling to identify the expressions of ISL had an accuracy of 75%, which is much higher than that of other methods, which was at 40%. A 3D animated SL interpreter (Deaf Talk) was developed in [11] with 84% sensitivity. It was achieved using the Continuous Gesture Builder from Microsoft Kinect V2. An effective text sentiment analysis model based upon the convolutional neural networks (CNNs) model and support vector machine (SVM) was proposed in [12], with improved text classification. Microsoft Kinect is used to take the input, where the images were contoured using convex hull process, fingertip, and joint position extracting the three fundamental features using artificial neural networks [13] and SVM, to achieve better results. Double-handed ISL was used by taking a series of images, and MATLAB was used to convert these images to speech, which was further converted to text in [14]. However, the limitation of this method lies in the use of data gloves. Another method using MATLAB was proposed in [15] for the recognition of ISL, which is based on the K-nearest neighborhood (KNN) technique, with 100% accuracy when $K=1$. Using the Zardoz system, TEAM system, and ViSiCAST system, a semantic analysis of text translation to Russian SL was carried out [16]. Fusion of feature technique was used in [17] for the recognition of ISL through K-nearest correlated neighbor classification, which was extended by the fusion of SIFT and HOG descriptors adding substantial accuracy for the recognition of the gestures. Dividing the text processing into front-end and back-end modules, a system for conversion of Thai language followed by speech synthesis module in [18] achieved an accuracy of 97.83% and a rate of 97.61% outside set. As both text and numerical data cannot be treated through a single system, two systems were combined in the form of data mining and text mining [19] to achieve data analysis support. A nonparametric unsupervised classifier algorithm, termed as the k-nearest neighbors' algorithm (KNN), performs classification or predictions using the proximity of elements about individual data points. It is mostly used for the classification of data. Moment techniques and KNN classifier are used to generate SL for ISL [20]. This resulted in 82% accuracy with good quality English being generated. SVMs are supervised learning algorithms that are used for regression analysis and classification. It is a binary, robust, and non-probabilistic linear classifier. To handle nonlinear data, kernel techniques were used in SVMs, and it is used in recognition of Bangladeshi hand SLs, the linear binary pattern (LBP) and SVM are used [21]. SVM was used for hand sign classification, which can recognize signals using one or two hands but not multiple hands. The use of Bluetooth for communication [22] gave a better performance than ZigBee or RF-link and other protocols. This model, a 3D inception, is one of the top-tier models in Action Recognition [23] to develop SL recognition. Modified CNN is a new Action Recognition model with good training accuracy. Semantic analysis as a concept and the associated components like water, keywords, semantic model, stop words, etc. were used in [24] in an attempt to fix the meanings of analyzing and outlining documents, and text documents.

12.3 SPEECH RECOGNITION

Speech recognition (SR) is an ability that empowers a program to convert an individual's speech to a written format. It centers on the interpretation of discourse from a verbal arrangement to a text one, though voice acknowledgment simply tries to distinguish a singular client's voice. Organizations can customize and use SR technology for their own requirements like brand recognition, language weighting, speaker labeling, acoustics training, and profanity filtering. Due to the constraints imposed by varieties of human speech, the development of SR has become a difficult task. The development of algorithms depends on statistics, mathematics, and linguistics, and it has become one of the most complex areas of research in computer science. The components of a SR involve input of speech, extraction of features, development of feature vectors, decoding them, and finally an output in the word format. The appropriate output is generated through the process of decoding with the use of acoustic models, a dictionary consisting of pronunciation, and models of languages to determine the appropriate output. The normal undertaking of our application involves utilizing a huge jargon, speaker free, and nonstop speech recognizer. This technology is assessed on the basis of its Word Error Rate (WER) and accuracy. WER is a metric used mostly to measure the performance of a SR algorithm and is generally used in the machine translation systems. The main hurdle in this measurement is in the different lengths of the sequence of words which is tested and the matching word sequence which is supposed to be accurate. The computation of word error is done using the formula $WER = (S + D + I)/(S + D + C)$, where S is the number of substitutions, D is the number of deletions, I is the number of insertions and C is the number of correct words.

Further developments in SR applications and gadgets use AI and ML techniques. They incorporate punctuation, grammar, design, and piece of sound and voice signs to comprehend and handle human speech. In this application, we use Google API, as mentioned in [25], which has the least WER compared to Sphinx and Microsoft API. With respect to % of WER, Google API is the most efficient one and is more convenient API to be used.

12.3.1 Google API

Google Speech-to-Text empowers the simple joining of Google speech recognition techniques advances into engineering applications. It sends voice and gets a textual record from the Speech-to-Text API administration. It allows developers to change voice to message in more than 125 languages and variations. It is easy and uses one of the most advanced DNN algorithms for automatic SR [26]. The features possessed are speech adaptation, domain-specific models, and streaming speech recognition.

12.3.2 Sign language

SLs are those languages that convey meanings through the visual-manual modality. The existence of deaf people has necessitated the existence of such languages for communication. These languages form the core of languages for deaf. Completely or partially deaf people use these languages in communication by using SL along with the normal common languages. As on record, there are 150 SLs according to the 2021 edition of Ethnologue. According to another survey, there are over 200 such languages as per the SIGN-HUB Atlas of SL structures. The latest estimation made in 2021 the most popular and widely used SL is the Indo SL having 151 SLs in their list.

SLs are expressed through manual articulations in combination with non-manual elements. SLs are full-fledged natural languages with their own grammar and lexicon, although they are not universal [10]. These are usually not mutually intelligible [11].

Parallel to spoken languages, the SLs have also been represented through some features like hand shape, orientation location, movement, and non-manual expression. Phenomes is the term used for these in SLs also because of the similarity in their functionalities. More generally, both sign and spoken languages share the characteristics that linguists have found in all natural human languages, such as semanticity, transitoriness, arbitrariness, and cultural transmission.

SL-based communication is a correspondence framework utilizing signals that are deciphered visually. Numerous individuals in hearing-impaired communities utilize gesture-based communications as their essential method for communication. These people include both hearing-impaired and normal individuals who communicate through SL. For some, hearing-impaired individuals, communication through SL fills in as their essential, or local, language. Dialects can be passed on in various manners known as modalities. These modalities incorporate discourse and sign. Essentially, American SL (ASL) and British SL (BSL) share a similar methodology and sign. However, they are entirely different dialects. This clarifies the way that communication via gestures is not the same in different countries and even from one local to another inside a single country. Etymologists believe both spoken and marked communication to be kinds of regular language, implying that both arose through a theoretical, extended maturing process and developed over the long run without careful planning.SL ought not to be mistaken for non-verbal communication.

12.3.3 Indian Sign Language (ISL) and alphabets

ISL is the predominant gesture-based communication in South Asia. As of now, roughly 7 million hearing-impaired individuals in the south Asian continent use ISL as their essential language. ISL signs follow a specific request, similarly as communicated in English. Nonetheless, in ISL one sign can communicate implying that would require the use of a few words in speech [27]. ISL ought not to be mistaken for non-verbal communication. ISL is utilized in the hearing-impaired communities all over India. However, ISL isn't utilized in deaf schools to teach hearing-impaired kids. Educator preparing programs don't place instructors toward instructing techniques that utilize ISL.

There is no instructing material that joins communication through SL. ISL mediators are a critical prerequisite at foundations and spots where communication among hearing-impaired and hearing individuals happens.

The grammar of ISL uses syntax and spatial grammar. For example:

PASS and FAIL – The handshape for the sign is something similar; however, they move in contrary ways.
MONEY, PAY, and RICH – They have the equivalent handshape yet a better place of enunciation and development design.
THINK, KNOW, and UNDERSTAND – The spot of articulation is the head, which is something very similar for all signs.

The SL letters are manual letters in order; that is an arrangement of addressing every one of the letters of a letter set, utilizing just the hands. Making words utilizing a manual letter set is called fingerspelling. Manual letters in order are a piece of gesture-based communication. For ISL, the one-gave manual letter set is utilized. Fingerspelling is utilized to compliment the vocabulary of ISL where spelling singular letters of a word is preferred or just alternative, The Indian manual alphabets are presented in Figure 12.2.

12.4 IMPLEMENTING ISL IN AISLI

The first language used by hearing-impaired individuals is ISL. Communication through sign which is independent of verbal communication and also correspondence through people. Hence, it is very smooth and conveys the thoughts of people efficiently.

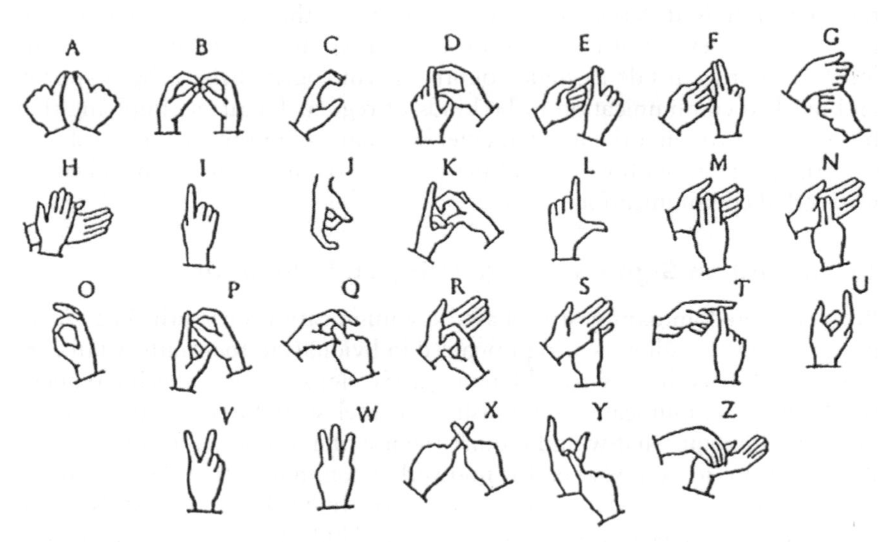

Figure 12.2 Indian manual alphabets.

The speech recognition engine of Google which generates text from speech is given as input to the ISL database in our approach.

The roles played by the ISL dataset are to include:

- Various around 1,000 ISL prerecorded GIFS where each single video clip compares to the essential words or phrases of the vocabulary in English.
- The manually generated independent alphabet.

The perceived contents represent the contents portrayed in Figure 12.3; e.g. individual alphabet and individual expressions. First, the word will be matched if it's a basic phrase or sentence and is available in the ISL GIFs database.

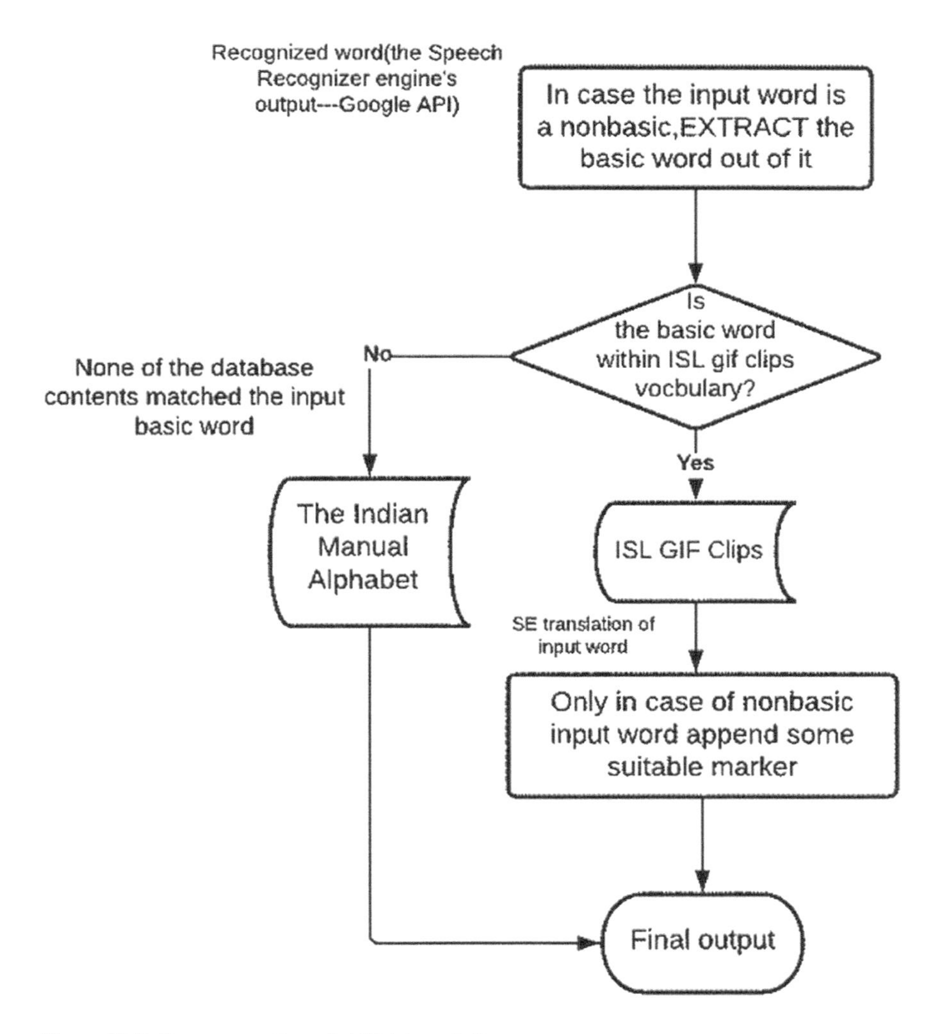

Figure 12.3 Demonstration of AISLI translation.

Figure 12.4 Interface.

If matched to the data set, then the following GIF will be displayed. Furthermore, assuming the word is not basic one, the fundamental word will be extricated out of it and matched with the ISL GIF dataset. In case the basic word was out of the ISL data set vocabulary, basically, the first word before extraction will be totally fingerspelled. If the word is not matched to the data set available the word or the phrase string will be counted, eventually each identified alphabet will be manually spelled through applying it to the Indian Manual Letters data set according to the order of the alphabets in the string counted. Note that the last yield portrayed in Figure 12.4 below can be a combination of GIFs and fingerspelling.

12.4.1 AISLI demonstration

This is the opening interface of the hearing impairment assistant, which gives us the option to start the application using two options:

1. Speak: This allows the user to start the interaction and get the resulted ISL Visual from the given Speech.
2. Exit: This option allows the user to exit the Program.
3. Detecting speech: In this step recognition of speech takes place (Figure 12.5).

   The application will ask to say something, and then it will turn on the microphone of the system to listen to the voice. The program will listen for a specific amount of time if no speech is detected then

```
(base) C:\Users\Hp\Desktop\tarp>python main.py
Speak...
```

Figure 12.5 Speech recognition.

```
(base) C:\Users\Hp\Desktop\tarp>python main.py
Speak...
good morning
```

Figure 12.6 Speech-to-text conversation.

it will ask again to say something. Here in this scenario, we spoke the word "good morning" and program recognized the word and thus converted it into text.

4. Speech-to-text conversation/translation: In this step, speech is converted/translated to text (Figure 12.6).

   The application in the previous scenario listened to speech and detected the word "good morning". Speech recognized is then converted to text for further manipulation.

Text Manipulation to Derive the Required ISL

In the previous scenario, we saw the program successfully detect the word "good morning" now the application matches the word initially with a database of GIFs that show the animation of the word or a phrase (Figures 12.7 and 12.8).

If the phrase or word is not present in the database, then the application will show the detected phrase in ISL visual letter by letter, it counts the letters of the string and then displays every letter, fingerspelling them in ISL manual alphabets.

Figure 12.7 Text manipulation to derive SL.

Figure 12.8 Resultant SL from text manipulation.

12.5 CONCLUSIONS

In this chapter, we have proposed AISLI, using Machine Translation and NLP Techniques for translating English speech to ISL in form of Graphics Interchange Format (GIF) and letters as output according to the phrase or word recognized. The Google Application Program Interface (API) is utilized as the audio acknowledgment engine for AISLI. For interpretation, the ISL structure is followed. It is expected that it will make the process for hearing-impaired individuals to comprehend simpler and speak with others, who do not know the language, using NLP and Machine Translation as the principal approach. We trust that this application stands out well and will be noticed well toward the above issue and will be helpful in eliminating the boundary between the hearing-impaired people and individuals who can hear. As of today, applications like AISLI are still not being used by the 0.24 million hearing impaired in India and are not readily available to them.

REFERENCES

1. Wikipedia Page. https://en.wikipedia.org/wiki/Indo-Pakistani_Sign_Language, last accessed 2021/09/24.
2. Tripathy, B.K., Sharma, A., Natarajan, P., Swarnalatha, P., Lakkshmanan, A., Krishnan, N. Image transformation using modified K-means clustering algorithm for parallel saliency map, *IJET*, 3(5), 347–353 (2013).

3. Tripathy, B.K., Ghosh, A., Panda, G. K. Adaptive K-means clustering to handle heterogeneous data using basic rough set theory. In: Meghanathan N., Chaki N., Nagamalai D. (eds.) *Advances in Computer Science and Information Technology. Networks and Communications. CCSIT 2012. Lecture Notes of the Institute for Computer Sciences, Social Informatics and Telecommunications Engineering*, vol. 84, pp. 193–201. Springer, Berlin, Heidelberg (2012).
4. Tripathy, B. K., Basu, A., Govel, S. Image segmentation using spatial intuitionistic fuzzy C means clustering. In: *2014 IEEE International Conference on Computational Intelligence and Computing Research*, pp. 1–5, Coimbatore (2014).
5. Tripathy, B.K., Deepthi P.H., Mittal, D. *Hadoop with Intuitionistic Fuzzy C-means for clustering in Big Data, Advances in Intelligent Systems and Computing*, vol. 438, pp.599–610. Springer, Singapore (2016).
6. Gupta, A., Tripathy B. K. Implementing GloVe for context based k-means++ clustering. In: *2017 International Conference on Intelligent Sustainable Systems (ICISS)*, pp. 1041–1046 (2017), DOI: 10.1109/ISS1.2017.8389339.
7. Tripathy, B. K., Mittal, D. Hadoop based uncertain possibilistic kernelized c-means algorithms for image segmentation and a comparative analysis. *Applied Soft Computing*, 46, 886–923 (2016).
8. Debgupta R., Saha A., Tripathy B.K. A faster fuzzy clustering approach for recommender systems. In: Bhateja V., Satapathy S., Zhang YD., Aradhya V. (eds) *Intelligent Computing and Communication, ICICC 2019, Advances in Intelligent Systems and Computing*, vol. 1034, pp. 315–324. Springer, Singapore (2019), DOI: 10.1007/978-981-15-1084-7_30.
9. Tripathy, B.K. and Deepthi, P. H. *Fuzzy set and rough set based Data Clustering Algorithms in Decision Making, IGI Edited volume: Fuzzy and Rough Set Theory in Organizational Decision Making.* Arun Kumar, S., Gao, X.-Z., Abraham, A. (eds.), pp. 116–136, Chapter-6. IGI Global, Hershey, PA (2016).
10. Mariappan, H. M., Gomathi, V. Real-time recognition of Indian sign language. In: *IEEE International Conference on Computational Intelligence in Data Science (ICCIDS)*, pp. 1–6 (2019). DOI: 10.1109/ICCIDS.2019.8862125.
11. Ahmed, M., Idrees, M., Ul Abideen, Z., Mumtaz, R., Khalique, S. Deaf talk using 3D animated sign language: A sign language interpreter using Microsoft's kinect v2. *SAI Computing Conference (SAI)*, pp. 330–335 (2016). DOI: 10.1109/SAI.2016.7556002.
12. Tripathy, B. K., Baktha, K. Investigation of recurrent neural networks in the field of sentiment analysis. In: *The Proceedings of IEEE International Conference on Communication and Signal Processing (ICCSP17)*, pp. 2047–2050, IEEE, Melmaruvathur (2017).
13. Tripathy, B. K. and Anuradha, J. *Soft Computing Advances and Applications*, Cengage Learning Publishers, New Delhi (2015).
14. Chowdhury, A. R., Biswas, A., Hasan, S. M. F., Rahman, T. M., Uddin, J. Bengali sign language to text conversion using artificial neural network and support vector machine. In: *2017 3rd International Conference on Electrical Information and Communication Technology (EICT)*, pp. 1–4 (2017). IEEE. DOI: 10.1109/EICT.2017.8275248.
15. Dutta, K. K., Raju K. S. K., Kumar, G. S., Swamy, B.S.A. Double handed Indian Sign Language to speech and text. In: *Third International Conference on Image Information Processing (ICIIP)*, pp. 374–377 (2015). DOI: 10.1109/ICIIP.2015.7414799.

16. Dutta, K. K., Bellary, S. A. S. Machine learning techniques for Indian sign language recognition. In: *International Conference on Current Trends in Computer, Electrical, Electronics and Communication (CTCEEC)*, pp. 333–336 (2017). DOI: 10.1109/CTCEEC.2017.8454988.
17. Grif, M., Manueva, Y. Semantic analyses of text to translate to Russian sign language. In: *11th International Forum on Strategic Technology (IFOST)*, pp. 286–289 (2016). DOI: 10.1109/IFOST.2016.7884107.
18. Gupta, B., Shukla, P., Mittal, A. K-nearest correlated neighbor classification for Indian sign language gesture recognition using feature fusion. In: *International Conference on Computer Communication and Informatics (ICCCI)*, pp. 1–5 (2016). DOI: 10.1109/ICCCI.2016.7479951.
19. Lin, X., Yang, J., Zhao, J. The text analysis and processing of Thai language text to speech conversion system. In: *The 9th International Symposium on Chinese Spoken Language Processing*, pp. 436–436 (2014). DOI: 10.1109/ISCSLP.2014.6936630.
20. Sai, V. S. V., Champawat, Y. S., Tripathy, B. K. Recommendation system based on text analysis. *International Journal of Innovative Technology and Exploring Engineering (IJITEE)*, 8(9), 2351–2354 (2019).
21. Patel, U., Ambekar, A. G. Moment based sign language recognition for Indian languages. In: *International Conference on Computing, Communication, Control and Automation (ICCUBEA)*, pp. 1–6 (2017). DOI: 10.1109/ICCUBEA.2017.8463901.
22. Santa, U., Tazreen, F., Chowdhury, S. A. Bangladeshi hand sign language recognition from video. In: *20th International Conference of Computer and Information Technology (ICCIT)*, pp. 1–4 (2017). DOI: 10.1109/ICCITECHN.2017.8281818.
23. Sengupta, A., Mallick, T., Das, A. A cost effective design and implementation of arduino based sign language interpreter. In: *IEEE 2019 Devices for Integrated Circuit (DevIC)*, pp. 12–15 (2019). DOI: 10.1109/DEVIC.2019.8783574.
24. Gunawan, H., Thiracitta, N., Nugroho, A. Sign language recognition using modified convolutional neural network model. In: *Indonesian Association for Pattern Recognition International Conference (INAPR)*, pp. 1–5 (2018). DOI: 10.1109/INAPR.2018.8627014.
25. Golani, N., Khandelwal, I., Tripathy, B.K. Hybrid intelligent techniques in text mining and analysis of social networks and media data, Chapter-1. In: Banati, H., Bhattacharyya, S., Mani, A., Köppen, M. (eds.) *Hybrid Intelligence for Social Networks*, pp. 1–24. Springer, Cham (2017).
26. Këpuska, V., Bohouta, G. Comparing speech recognition systems (Microsoft API, Google API and CMU Sphinx). *International Journal of Engineering Research and Application*, 7(3), 20–24 (2017).
27. Google Cloud Speech to Text Website. https://cloud.google.com/speech-to-text, last accessed 2021/09/24.

Chapter 13

Fragile medical image watermarking using auto-generated adaptive key-based encryption

Subhrajit Sinha Roy and Abhishek Basu
RCC Institute of Information Technology

Avik Chattopadhyay
University of Calcutta

CONTENTS

13.1 INTRODUCTION

Telemedicine is the modern trend of smart healthcare systems that aims to provide healthcare services at distance. Patients and doctors may interact with each other on certain issues through telemedicine. Here, the electronic patient reports (EPRs), along with the corresponding pathological test images, play a central role in diagnosis [1]. Any small alteration in the medical images may lead to false diagnosis, and thus, tamper detection is essential for medical images. Fragile image watermarking could be effective in this purpose. However, visual distortion that occurred due to the insertion of the watermark is a matter of concern. Besides, it is also important for the EPRs to be sent to the receiver in a secured way as it carries patients' personal information. Thus, to maintain privacy of the EPRs and offering enough data capacities for these EPRs are another two challenges for the researchers and developers.

Several image watermarking techniques are already inexistence to serve different services like security, tamper detection, robustness of the embedded information, privacy to the hidden information, etc. Spatial domain techniques [2, 3] have been developed for easy execution with high imperceptibility, whereas frequency domain-based methods provide higher

DOI: 10.1201/9781003381167-13

security and robustness [4–7]. Often some complementary tools are utilized to get better performance. For instance, human visual system-based techniques [8] are involved to improve visual transparency and hiding capacity, error control codes and encryption methods [9–12] are used to enhance security; and most recently, the hybrid mode-based and intelligent technique-based approaches [13–16] are also developed in search of some satisfactory outcomes.

This way, different methodologies have been developed at different times to solve different specific issues. However, due to the contradictory behaviors among the important features [17] of any image watermarking scheme, i.e. robustness, imperceptibility, and payload or data hiding capacity, an ultimate solution has not been achieved yet.

Here for tamper detection, a new fragile color image watermarking scheme has been proposed in the spatial domain that also intends to provide immense privacy and data capacity for the EPRs. In this approach, the medical cover image is segmented into two parts – regions of interest (ROI) and regions of non-interest (RONI). A fragile watermark is embedded into the ROI and an encrypted version of EPRs is embedded into the RONI. The fragile watermarking is performed in a reversible way to ROI portions are unaffected. Although it is a fragile watermarking technique, EPRs are made comparatively robust. It is because, for any slightest modifications or noises, the watermark may be destroyed. But till, diagnosis may be continued if the concerned doctor found high similarities between the EPRs and the image. Acute reliability is provided to the EPRs to prevent any type of misappropriation. The proposed embedding and extracting processes for the watermark and EPRs have been discussed in Section 13.2. In Section 13.3, the performance of this scheme is discussed with some test images, and a comparative study with some other existing frameworks is also done. Finally in Section 13.4, this chapter is concluded.

13.2 PROPOSED METHODOLOGY

13.2.1 Watermark embedding

This proposed method embeds a fragile binary watermark into the medical images for tamper detection, i.e. to identify whether any kind of modifications or misappropriations has performed on the image or not. Along with that, it inserts the EPRs into RONI, i.e. the regions, which are not considered for diagnosis purpose. As it is already mentioned earlier that the EPRs carry personal information about the patients, and for the same, EPRs are required to be transmitted confidentially. Thus, an encrypted form of EPR is embedded instead of the original version. The EPR can be decrypted only through the original watermark, along with an auto-generated key. Thus, high reliability also sustains for the EPR. The overall embedding process, shown in Figure13.1, is carried out through the following steps.

i. The color medical image (I) is taken and decomposed into three color planes. Let the red, green, and blue color components are R, G, and B, respectively. Naturally, these R, G, and B can be considered as three separate gray-scale images, of which pixel values vary within 0–255.

ii. Based on the green plane pixel intensity values, ROI and RONI segmentation is done with respect to a particular threshold value (t). Thus, the segmentation map (S) is converted basically a binary image, pixels, having values less than t, are set to '0' and indicate RONI, and pixels, having values minimum t or greater than that, are set to 1 to indicate ROI. Small RONI bubbles in ROI are adjusted to ROI before generating the ultimate segmentation map. The total number of pixels in RONI is calculated as p.

iii. The watermark of size $M \times M$ is taken and based on the value of p, it is resized to $N \times N$, such that $N2 \leq p < (N+1)^2$. Along with resizing the watermark, a key (K) of size N×N is also generated in such a manner that for any element $k(i,j) \in K$,

$$k(i,j) = \left(i^{j\text{mod}7}\right)\text{mod } 2$$

The EPR image is first resized to $N \times N$ and then encrypted through K to form the encoded EPR (E).

$$E = EI \oplus K$$

Where, E_I is the resized EPR image having $N \times N$ number of pixels. The most interesting feature of this scheme is that the key and the size of the watermark are adaptive, i.e. these may not be the same for every image.

iv. Considering the segmentation map as a mask, the least significant bits (LSBs) from the pixels in ROI regions of blue plane are taken to form a binary sequence (L) of length N^2.

v. Fragile watermarking is done by performing XOR operation between the resized watermark (W) bits and the ROI LSBs of the blue plane.

vi. The sequence L is XORed with E, and embedded into the RONI pixels of blue plane through bit substitution with (5,1) repetition code in higher bit-planes. This provides robustness to the ($L \oplus E$) data. Parallel to that, ($W \oplus E$) data is embedded into the red-plane RONI pixels in the same manner. Being XORed with watermark, EPR information becomes reliable, and for applying the repetition code, robustness for EPR is improved. The green plane is kept unchanged because the segmentation map, the key, and the size of watermark and EPR are estimated from it.

vii. Finally, from the modified red and blue planes, along with the green plane form the ultimate watermarked image.

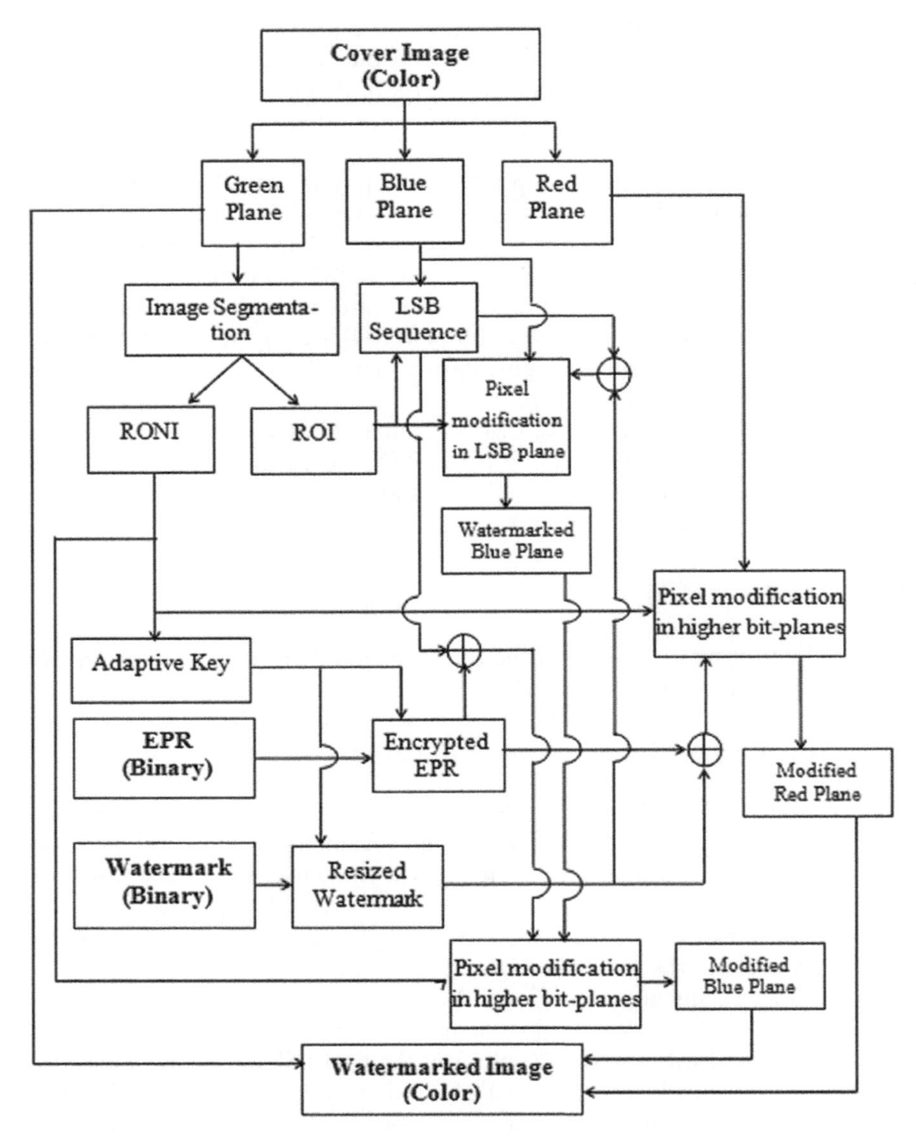

Figure 13.1 Block diagram for the proposed watermarking scheme.

13.2.2 Extraction of watermark and EPR

Only a proper extraction of the embedded information can validate the authenticity of the received images. The following steps are involved in the proposed method for extracting the watermark and EPR (Figure 13.2).

i. After receiving the watermarked color image, it is decomposed into red, green, and blue planes. Then as described in the embedding procedure, the segmentation map and the key (K) are generated, and the watermark, present at the receiving end, is resized to the size of K. Let this resized watermark be denoted by W'.

ii. The bits are extracted from the higher bit-planes of the RONI pixels of the red plane of the received image, and a similarity estimation is performed for every set of extracted bits from each pixel because (5,1) repetition code was used to embed these information bits. The similarity estimation for five extracted bits (b_1, b_2, ... b_5) is carried out through the following equation,

$$\text{Estimated bit} = [b_1b_2(b_3+b_4+b_5) + b_3b_5(b_4+b_2+b_1)] + [b_4b_5(b_2+b_1) + b_2b_3b_4]$$

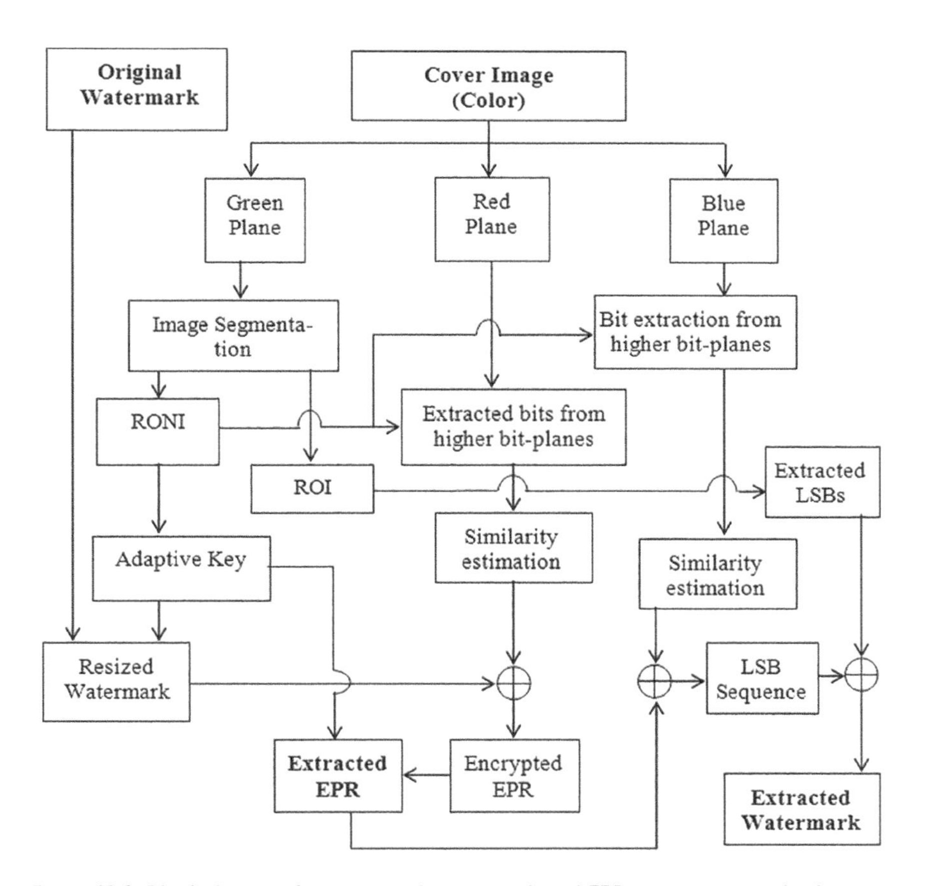

Figure 13.2 Block diagram for proposed watermark and EPR extraction method.

Consider E' as the bit array extracted from the red plane. Another bit array (L') will be obtained from the higher bit-plane RONI pixels of the blue plane.

iii. The encrypted EPR will be obtained from E', when it is XORed with W'. Another XOR operation, applied between this encrypted EPR and L', will provide the LSB sequence, which was used to embed the fragile watermark.

iv. Then the extracted LSB sequence is XORed with the LSBs of the ROI pixels of the blue plane to provide the watermark (W_X).

v. This recovered watermark W_X is then compared to for validation and tamper detection.

This way, the watermark is extracted along with the EPR. It is to be noted that for any slight modification to the image, watermark will be distorted to indicate the presence of tamper; however, the EPR may sustain yet, so that the diagnosis may be continued based on that if the concerned doctor seems to do the same.

13.3 RESULTS AND DISCUSSION

The efficiency of any digital image watermarking scheme is quantified through three major parameters imperceptibility, robustness, and hiding capacity. In the purpose of analyzing these qualities, first the outputs of the proposed methodology have been obtained using a binary watermark, a typical EPR image, and a set of medical color images, which are shown in Figures13.3a, band 13.4a, respectively. The size of the cover images is chosen as 512 × 512. As it is already mentioned that the size of the watermark and the EPR image, to be embedded into the cover, are adaptive in nature, initially it is set to 384×384.

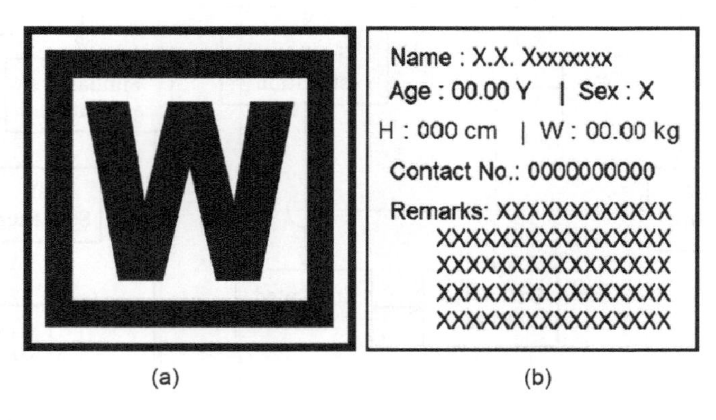

Figure 13.3 (a) Binary watermark. (b) EPR Image.

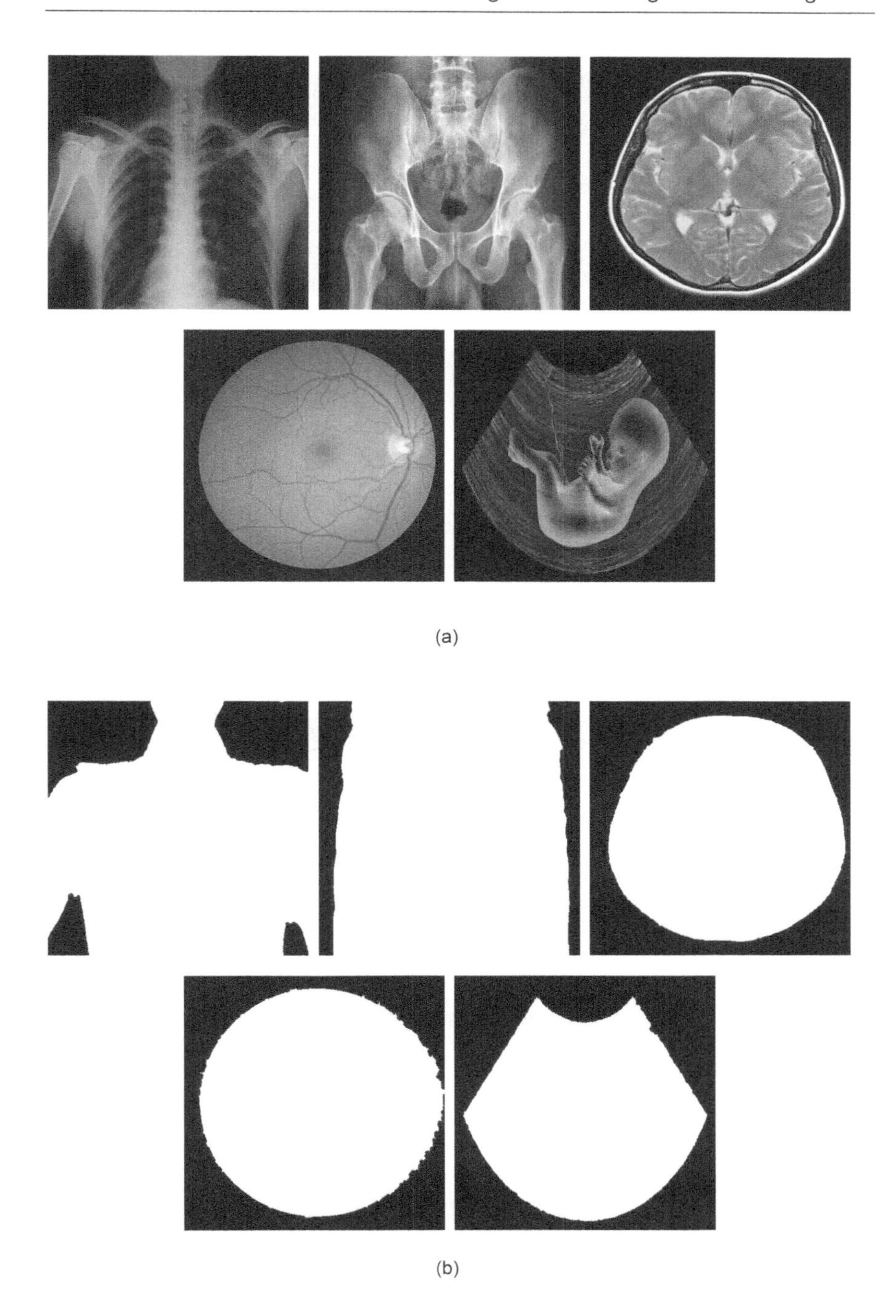

Figure 13.4 (a) Color medical images, used as cover. (b) Corresponding ROI and RONI segmentation maps.

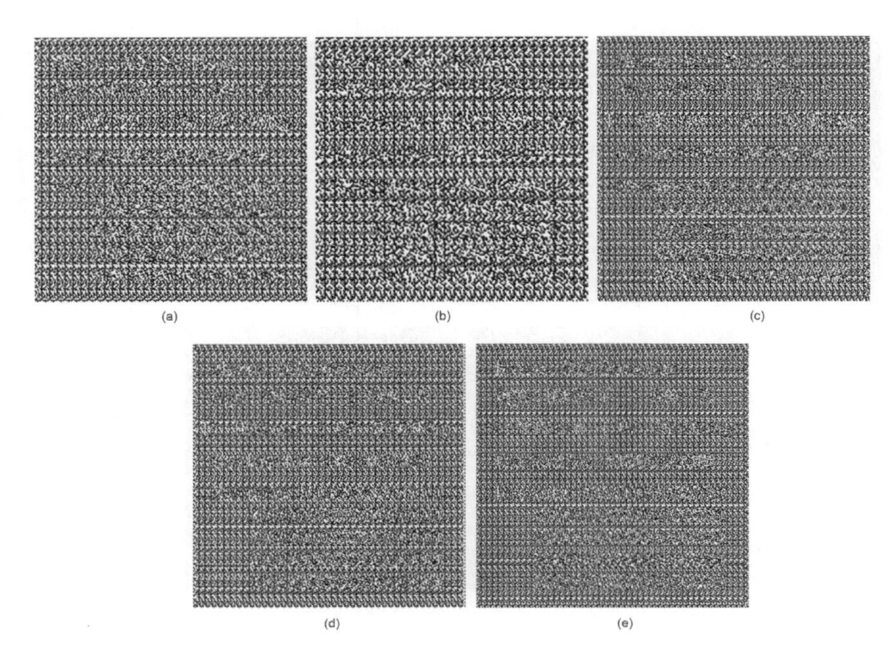

Figure 13.5 Encrypted EPRs generated, respectively, for the cover images of Figure13.4a.

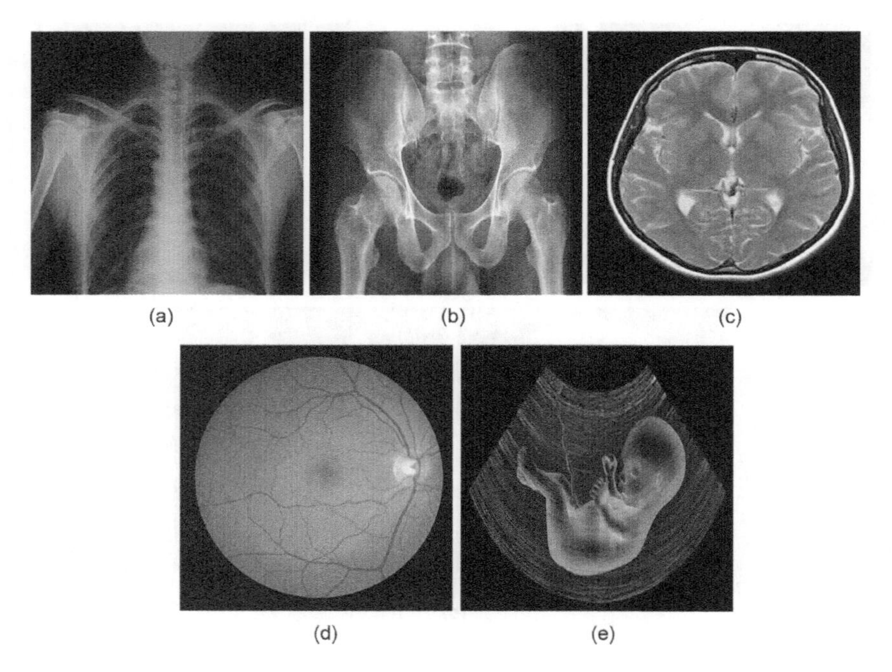

Figure 13.6 Watermarked images obtained after fragile watermarking for the cover images of Figure13.4a, respectively.

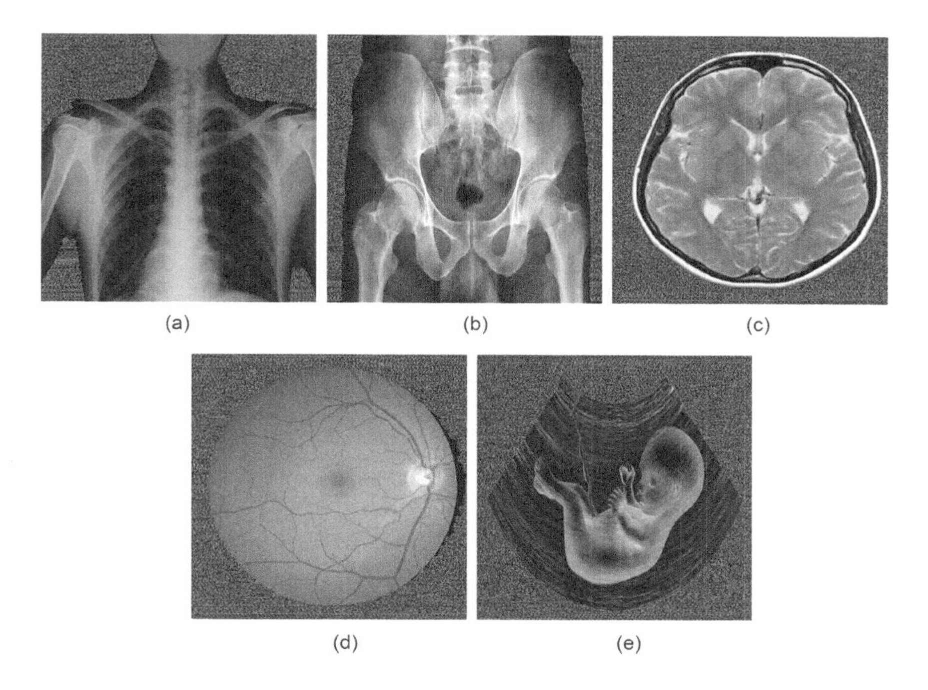

Figure 13.7 Watermarked images after embedding the robust information into images of Figure 13.6, respectively.

After applying the proposed embedding process on different medical images, the average number of embedded information bits is obtained as 79,356 for each of the watermark and EPR image. Thus, the average hiding capacity for the proposed scheme is approximated as 0.61 bits per pixel.

Quality metrics for images [18] are often used to estimate the aesthetic distortion present in a modified or affected image with respect to its corresponding original image. Although there are many quality metrics are available in this field of research, a few of the most conventional and standard metrics are employed here to assess the imperceptibility and robustness of this technique. Considering the medical images as the originals, imperceptibility has been analyzing through some universally used metrics after embedding the fragile watermark only, and then the same analysis is performed after embedding the robust information (i.e. encrypted $EPR^{\oplus}$ resized watermark and encrypted $EPR^{\oplus}$ LSB sequence). These imperceptibility assessments are given in Tables13.1 and 13.2, respectively. The necessary input, output, and intermediate images, required to realize the system performance, are shown in Figures 13.3–13.7. The high PSNR values and the corresponding images after fragile watermarking depict that the ROI of the images is affected in extremely low scale; moreover, these affected

Table 13.1 Imperceptibility after fragile watermarking

Images	*PSNR (dB)*	*NC*	*SSIM*	*UIQI*
X-ray image 1	63.2237	1.0000	0.9999	0.9987
X-ray image 2	65.1400	1.0000	1.0000	0.9993
MRI image	61.0232	1.0000	0.9998	0.9987
Retinal fundus image	61.2009	1.0000	1.0000	0.9917
USG image	60.1636	1.0000	0.9999	0.9994

Table 13.2 Imperceptibility after embedding EPR and LSB sequence into the watermarked image

Images	*PSNR (dB)*	*NC*	*SSIM*	*UIQI*
X-ray image 1	11.5815	0.9967	0.7621	0.8435
X-ray image 2	13.6903	0.9919	0.8460	0.9016
MRI image	9.5392	0.9970	0.6333	0.7619
Retinal fundus image	9.6511	0.9895	0.6440	0.7554
USG image	8.6178	0.9906	0.5435	0.6989

bits can be corrected after a lossless transmission as the fragile watermark is extracted in a reversible way. Visual quality degrades after robust data insertion, but for being embedded into RONI, it does not hamper the diagnosis process.

The watermark is made fragile to detect any slightest modifications to the image. To verify this, a set of attacks have been applied to the watermarked images, and it is shown in Figure13.8 that the extracted watermarks are completely destroyed for the attacks. However, in contrast, it is reflected in Figure13.9 that the EPR can sustain most of the attacks. Geometrical and cropping attacks are not acceptable. For medical images, as these affect the images in a large scale, and thus, the EPRs also get affected more for these attacks. The robustness for the EPR is quantified in terms of some quality metrics, given in Table 13.3 by considering the resized EPR as the original image and the extracted EPR as the noisy image.

Finally, this scheme is compared with a few existing frame works by means of imperceptibility and payload. The robustness of the scheme is not included in comparison because the watermark does not provide any robustness at all. However, the robustness offered by the EPR confirms that the goal of embedding this information is fully achieved. The PSNR value, obtained after fragile watermarking, is considered as the parameter for comparing the imperceptibility; and it is evident that it is too good for this proposed scheme. The average hiding capacity (Table 13.4), offered by this proposed scheme, is relatively higher than the others also.

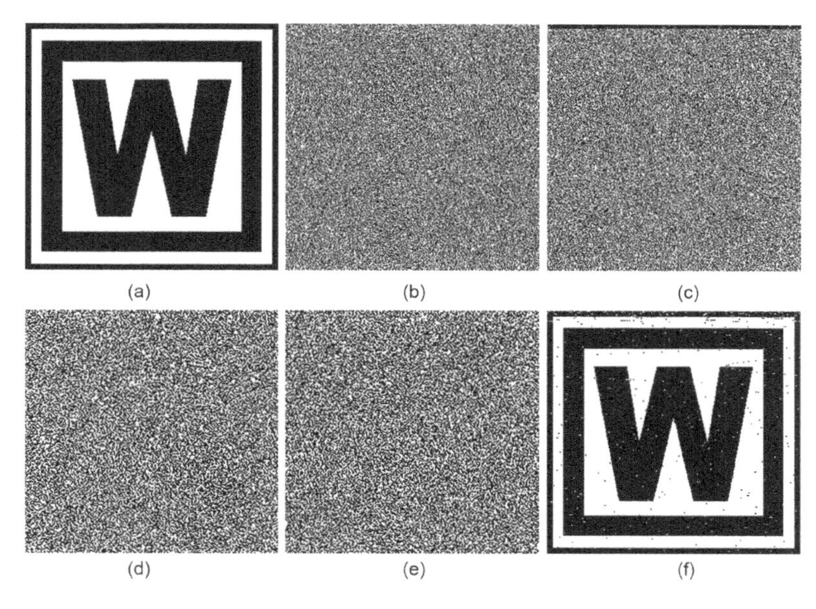

Figure 13.8 Extracted watermark after applying attacks. (a) Without attack. (b) JPEG compression (50%). (c) Salt and pepper. (d) Gaussian noise. (e) Scaling. and (f) Low-pass filtering.

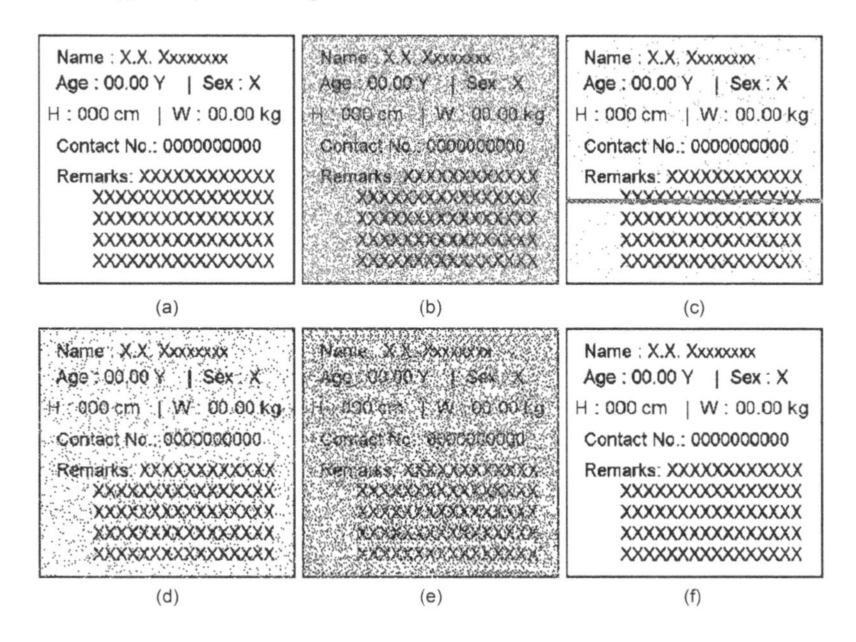

Figure 13.9 Extracted EPR after applying attacks. (a) Without attack. (b) JPEG compression (50%). (c) Salt and pepper. (d) Gaussian noise. (e) Scaling. and (f) Low-pass filtering.

Table 13.3 Quantitative analysis of robustness for the EPR data

Attacks	*BER (%)*	*NC*	*SSIM*	*UIQI*
No attack	0	1	1	1
JPEG compression (50%)	13.2145	0.8675	0.9960	0.8676
Salt and pepper	1.2203	0.9896	0.9996	0.9878
Gaussian noise	3.3965	0.9663	0.9992	0.9661
Scaling	16.5904	0.8317	0.9947	0.8336
Low-pass filtering	0	1	1	1

Table 13.4 Comparison of system proficiency with existing state-of-the-art frameworks

Attacks	*PSNR (dB)*	*Payload (bpp)*
Proposed scheme	62.15	0.61
Nagm and Elwan [13]	55.41	1
Alshanbari [19]	48.67	0.088
Kahlessenane et al.[5]	74.16	0.125
Gull et al. [20]	51.26	1
Singh [21]	30.42	0.018

13.4 CONCLUSION

A spatial domain-based image watermarking scheme has been developed in this chapter for providing authenticity and security to medical images as well as EPR data. An auto-generated encryption key is utilized to encode EPR with watermark after being resized adaptively. The efficiency of the proposed method is estimated in terms of imperceptibility, robustness, and hiding capacity. A relative study with some recent techniques has established the fact that this proposed scheme is proficient enough to reach the goal of this work.

REFERENCES

1. Singh, A. K., Kumar, B., Singh, G., and Mohan, A. (2017) Medical image watermarking techniques and applications. *Multimedia Systems and Applications*. Springer. DOI: 10.1007/978-3-319-57699-2.
2. Singh, A. K., Sharma, N., Dave, M., and Mohan, A. (2012) A novel technique for digital image watermarking in spatial domain. *2012 2nd IEEE International Conference on Parallel, Distributed and Grid Computing*, Waknaghat, pp. 497–501.
3. Sinha Roy, S., Basu, A., and Chattopadhyay, A. (2021) Secured diabetic retinopathy detection through hard exudates. In: Pan I., Mukherjee A., Piuri V. (eds.) *Proceedings of Research and Applications in Artificial Intelligence. Advances in Intelligent Systems and Computing*, vol. 1355, pp. 213–221. Springer, Singapore.

4. Liu, J., Li, J., Ma, J., Naveed Sadiq, U.A.B, and Yang, A. (2019) A robust multi-watermarking algorithm for medical images based on DTCWT-DCT and Henon map. *Appl. Sci.* 9(4): 1–23.
5. Kahlessenane, F., Khaldi, A., Kafi, R., et al. (2021) A DWT based watermarking approach for medical image protection. *J. Ambient Intell. Human Comput.* 12: 2931–2938.
6. Kahlessenane, F., Khaldi, A., Kafi, R., et al. (2021) A robust blind medical image watermarking approach for telemedicine applications. *Cluster Comput.* 24: 2069–2082.
7. Elbasi, E. and Kaya, V. (2018) Robust medical image watermarking using frequency domain and least significant bits algorithms. *2018 International Conference on Computing Sciences and Engineering (ICCSE)*, Kuwit, pp. 1–5.
8. Kavitha, V., Palanisamy, C., and Sureshkumar, T. (2020) perceptual masking based medical image watermarking using DTCWT and HVS. *J. Crit. Rev.* 7(2): 561–567.
9. Thakur, S., Singh, A.K., Kumar, B., and Ghrera, S.P. (2020) Improved DWT-SVD-based medical image watermarking through hamming code and chaotic encryption. In: Dutta D., Kar H., Kumar C., Bhadauria V. (eds.) *Advances in VLSI, Communication, and Signal Processing. Lecture Notes in Electrical Engineering*, vol. 587, pp. 897–906. Springer, Singapore.
10. Terzija, N., Repges, M., Luck, K., and Geisselhardt, W. (2002) Digital image watermarking using discrete wavelet transform: Performance comparison of error correcting codes. In: J.J. Villanueva (ed.) *Visualization, Imaging, and Image Processing.* Acta Press, Marbella.
11. Priya, S. and Santhi, B. (2019) A novel visual medical image encryption for secure transmission of authenticated watermarked medical images. *Mobile Networks Appl.* 26: 2501–2508
12. Nayak, M.R., Tudu, B., Basu, A., and Sarkar, S.K. (2015) On the implementation of a secured digital watermarking framework. *Inf. Secur. J. A Glob. Perspect.* 24(4–6): 118–126.
13. Nagm, A. and Safy Elwan, M. (2021) Protection of the patient data against intentional attacks using a hybrid robust watermarking code. *Peer J Comput. Sci.* 7: e400.
14. Madhu, B. and Holi, G (2021) CNN approach for medical image authentication. *Indian J. Sci. Technol.* 14(4): 351–360.
15. Phan, T.H.D., Bui, V.H., Nguyen, T.A., and Hoang, T.M. (2020) A novel watermarking scheme based on the curvelet transformation method for medical images. *2020 7th NAFOSTED Conference on Information and Computer Science (NICS)*, Hochiminh, pp. 379–383.
16. Nawaz, S.A., Li, J., Liu, J., Bhatti, U.A., Zhou, J., and Ahmad, R.M. (2020) A feature-based hybrid medical image watermarking algorithm based on SURF-DCT. In: Liu, Y., Wang, L., Zhao, L., Yu, Z. (eds.) *Advances in Natural Computation, Fuzzy Systems and Knowledge Discovery. ICNC-FSKD 2019. Advances in Intelligent Systems and Computing*, vol. 1075, pp. 1080–1090. Springer, Cham.
17. Sinha Roy, S., Basu, A., and Chattopadhyay, A. (2019) *Intelligent Copyright Protection for Images*, 1st ed., CRC, Taylor and Francis, New York, NY.

18. Kutter, M. and Petitcolas, F.A.P. (1999) A fair benchmark for image watermarking systems. In: P. W. Wong and E. J. Delp III (eds.) *Electronic Imaging '99, Security and Watermarking of Multimedia Contents*, vol. 3657, pp. 226–239. SPIE, San Jose, CA.
19. Alshanbari, H.S. (2021) Medical image watermarking for ownership & tamper detection. *Multimed. Tools Appl.* 80: 16549–16564.
20. Gull, S., Loan, N.A., Parah, S.A. et al. (2020) An efficient watermarking technique for tamper detection and localization of medical images. *J. Ambient Intell. Human Comput.* 11: 1799–1808.
21. Singh, A.K. (2019) Robust and distortion control dual watermarking in LWT domain using DCT and error correction code for color medical image. *Multimed. Tools Appl.* 78: 30523–30533.

Chapter 14

Designing of a solution model for global warming and climate change using machine learning and data engineering techniques

Soumit Roy
Analytics Presales and Solution Practice Lead, Jade Global

Rik Das
Xavier Institute of Social Service

Rupal Bhargava
upGrad Education Pvt. Ltd.

Ayush Gupta and Sourav De
Cooch Behar Govt. Engineering College

CONTENTS

14.1 INTRODUCTION

In 20th century, the planet is facing multiple major environmental challenges including global warming, melting of glaciers, and multiple natural disasters [1,2]. Climate change is influenced by global warming [3–5]. Pollution is the source of all issues, which is invoked by the usage of nonrenewable energy resources. Pollution has become uncontrollable in spite of human beings getting equipped with modern technologies and systems. Recent literature [6] has claimed that one of the major factors in greenhouse gas emission is CO_2 apart from other greenhouse gases, which also contribute to climate change prediction. Greenhouse gas emission is controlled by multiple features [7].

DOI: 10.1201/9781003381167-14

One of the important features is waste material recycle [8]. Recycle of waste material and environment management for areas where the temperature is increasing are also leading issues in society [9]. Environmental informatics can help create data engineering platforms to integrate all parameters to provide a 360° view on the global warming issue [10]. Research work on environmental protection [11] has highlighted the application of data engineering in facilitating the process. The work in [12] proposes a unique plug and play data model to fit and absorb the data from any source system and provide KPI. However, tracking and relating with temperature increase is the key unique area which were not covered by fellow researcher [13,14].

Although it has improved a lot in a few of the countries, there is no formal system that exists which can track the generation vs. recycle items, and there are no key performance indicators (KPI) mentioned to improve this. We have observed a positive correlation of temperature with CO_2 emission and population and have also found a wide usage of data engineering to control global warming. As a part of this work, we came up with IPA (Integrate, Predict, and Act) methodology, which will contribute to climate hazard reconciliation process as well. This chapter suggests a design of such data engineering system focused on global warming and has proposed Green Machine Automation Model (GMAD) architecture in pursuit of controlling global warming using data engineering and machine learning.

14.2 RELATED WORK

Global warming is one of the most critical trending problems in the world. Identifying the most common contributing factors or features that are increasing global warming is one of the key research areas for decades. Latest models resulted by near past researchers conclude that earth surface warming is accelerated. Current model prediction technique reveals that global warming will take a rapid rise [15]. The proposed modeling from previous research advocates

1.5°C temperature surge is likely to take place by 2030. The work of [6] has described the relation between temperature change and CO_2 relationship using few country data. He found that the linear model used for the analysis is revealing high prediction error.

Primarily, researchers are only focused on predicting temperature [16,17] or identifying the CO_2 emission trend [18]. It is observed that detailed solution of application of data modeling to predict climate change is rarely seen in recent literature [7,19,20].

Data science and machine learning are being used as key accelerators to take proactive action for controlling global warming. Trends and features that impact global warming are identified to help the affected region using data analytics. Conversion of urban area from countryside is one of the key reasons for global warming. In the recent past, research on climatology of urban area has been done both in a structured or unstructured way. To control global warming, environmental informatics can be used with a wide application of cutting-edge technologies to control the factors like global forest change, planning green and smart urban architecture, planning on conversion of energy usage (nonrenewable to renewable) [9].

The solution approach provided in the current work is in contrast with most of the existing propositions, which focus on normal data modeling rather than data model for environmental issues like global warming.

14.3 SOLUTION TECHNIQUE

This study has proposed the IPA methodology (Integrate, Predict, and Act) which will contribute to climate hazard reconciliation process as an integrated solution platform. Figure 14.1 shows the iterative IPA cycle.

14.3.1 Prediction technique selection

Our research work mainly deals with a numerical variable, which varies with time variable. Therefore, the point of interest is to model time series for numbers. Robustness of model needs to be the priority to ensure generalization with unseen data. By using scatter plot, the characteristics of target variables are identified. As both of our target variables are continuous in nature, we have selected time series -ARIMAX, SARIMAX to predict the target variable. The presence of an exogenous variable that affects the business directly can improve accuracy. SARIMAX is used to improve accuracy

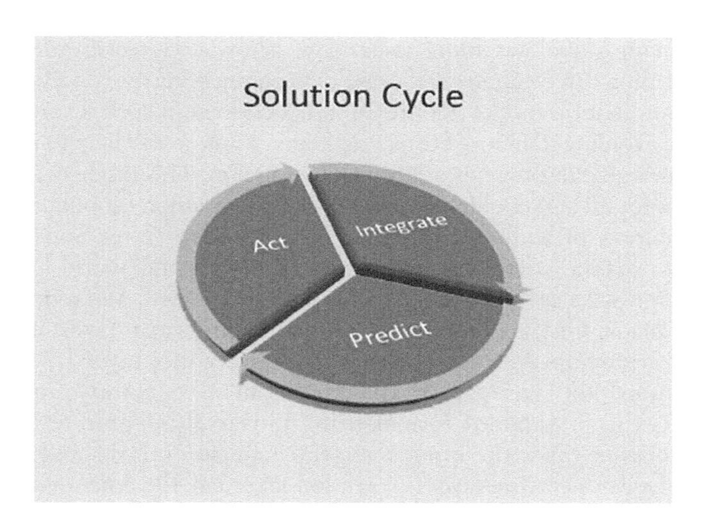

Figure 14.1 IPA solution cycle.

with an exogenous variable [21]. Equation (14.1) shows the dependent variable evaluation equation with the independent variable and exogenous factor.

$$y_t = \beta_0 + \beta_1 X_{1,t} + \beta_2 X_{2,t} + \ldots. + \beta_k X_{k,t} + \left(\frac{\theta_q(B)\theta_Q(B^s)}{\phi_p(B)^{\phi}{}_p\left(B^s\right)(1-B)(11-B)^d(1-B^s)^D} \in_t \right) \tag{14.1}$$

Root mean square error gives the standard deviation of the model prediction error. It is the difference between the values predicted by a model and the values actually observed. A smaller value of RMSE indicates better model performance. Equation (14.2) shows the RMSE calculation using observed temperature and forecast temperature.

Every model has a prediction error. Standard deviation of prediction error is denoted by Root mean squarer. RMSE also defines the variance between the predicted value and actual value. A smaller value of RMSE indicates better model performance. Equation (14.2) shows the RMSE calculation using observed temperature and forecast temperature.

$$\text{RMSE} = \sqrt{\frac{\sum_{t=1}^{n} (\text{observed temp} - \text{forecast temp})^2}{n}} \tag{14.2}$$

14.4 SOLUTION ARCHITECTURE

One of the key goals of this study is to get a reliable control system, which can deal with global warming issue. The below architecture defines how global warming can be controlled using data engineering and analytics. Green machine automation model (Figure 14.2) is designed in such a way so that it can ensure avoiding data loss for recycle items. Thus, it will help to reduce and control renewable resource usage and CO_2 emission. This model will open the data access for all stakeholders to take prescriptive action on polluting zone.

This solution provides tight integration between states and countries. This helps to take corrective action for countries and world governance body. Region-wise carbon tax system can be reconciled, and a new process can be in place. GMAD has five layers: data acquisition layer, integration layer, operational layer, tracking layer, and governance layer.

Data acquisition layer gathers the data from state, county, and country data sources in a standard format using universal adaptor or API. Data integration layer takes the input from the acquisition layer and integrates them and feeds operation layer. Operation layer has the data engine, which prepares and cleans the data to feed it into model. Model learned from repetitive data and feed the output to integration layer. Integration layer

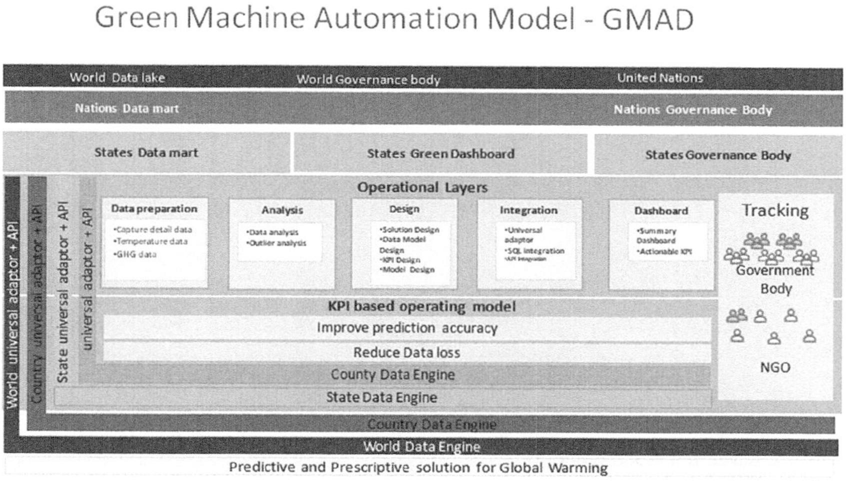

Figure 14.2 GMAD a sustainable control system.

again gives the input to the dashboard. Dashboards have specific KPI to calculate the performance. KPI data are fed to governance body to take corrective action. The total process is iterative for continuous improvement of the control system and increase efficacy of predictive models.

14.5 RESULT AND ANALYSIS

Climate data is increasing day by day with the increase of population. Temperature is getting increase with the increase of greenhouse gas emission. Climate scientists are warning about global warming and pollution. However, world leaders and a lot of countries have ignored like they have ignored the alert about the COVID-19 pandemic. By 2050 world can be in the red zone if we will not take corrective action. Reduction of global fossil fuel consumption is very much needed to protect human beings from mass extinction due to global warming impacts. Global temperature trend depends on land temperature trend and ocean temperature trend. Global temperature in the pre-industrial years was trending on both sides, as it was rising in some years and it was as also decreasing in some years. But after 1950, the global temperature is rising very fast, as increment in 1° effects very much.

Figure 14.3 shows the trend of monthly global average temperature distribution from 2006 to 2015. It is observed that January has the lowest average temperature and December has second lowest. May has the highest average temperature and June has second highest average temperature.

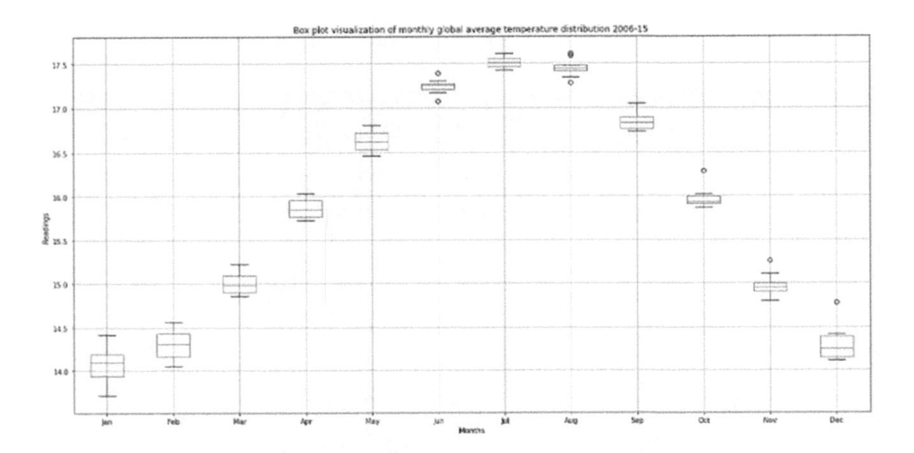

Figure 14.3 Temperature trend.

We have used times series model to get the predicted result. At first, we have applied seasonal decompose on the base temperature data and checked the components. Figure 14.4 shows the details of the decomposed component. There are three major components –trend, seasonal and residual. Trend is rapidly increasing. After 1960, it increased a lot, and after 1980, it has increased rapidly. Therefore, it's a factor of concern until we take strong corrective measures. Distribution of seasonal and residual parameter is uniform.

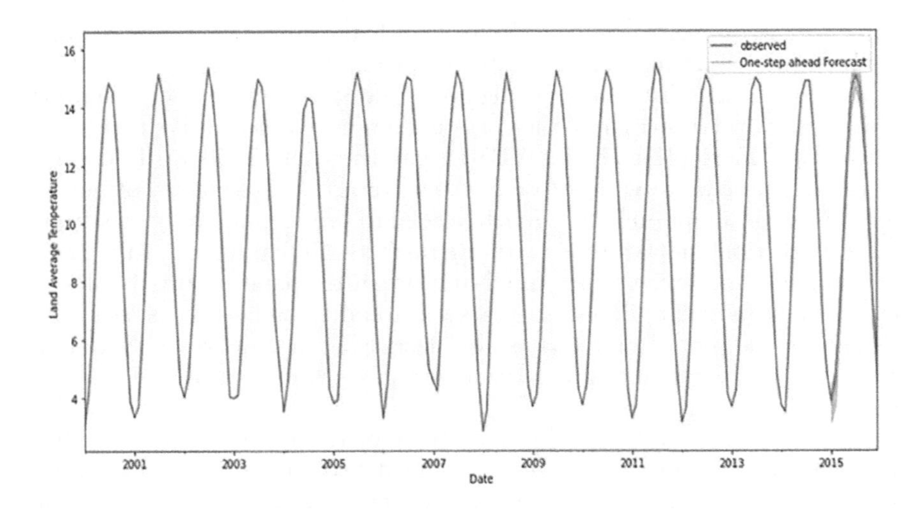

Figure 14.4 Stationary time series distribution.

SARIMAX model is used with .05 alpha. Figure 14.4 shows good result in observed and one-step-ahead plot. The mean squared error of this forecast came around .09, which is encouraging.

A few other algorithms like Logistic Regression, Linear Discriminate Analysis, KNearest Neighbors, Regression Trees, Gaussian Naive Bayes, and Support Vector Machines are also implemented to observe the outcomes in comparison to SARIMAX. As 15 year of temperature data is used (Global temperature datatest of Berkely earth), five-fold stratified cross-validation is used to resample the data for the evaluation of machine learning models. This will shuffle the data and split it into multiple folds. Stratified cross-validation is very effective in reshuffling the data. Figure 14.5 shows the result of the comparison for various algorithms.

Although the LDA, KNN, CART, and SVM values are very close, the SVM algorithm gives the best results with accuracy of 77%. It's a decent value, but not perfect. If the temperature accuracy is nearby 98% then only it will be possible to take corrective actions on time. Creating sustainable environment is not an instant process. It can only be done by continual improvement of all processes. Therefore, the SVM model is not a feasible option. As temperature data has exogenous and seasonal factor, time series algorithm is the best fit for predicting the output variable. Moreover, data is not linear in nature and it has some seasonality. Therefore, time series is the best option for forecasting climate change. SARIMAX gives 99% of accuracy. Thus, SARIMAX is the suggested algorithm.

Second quadrant is for formalizing the recycle ecosystem. Solid waste is not managed properly in most of the developing countries. Ten percent of

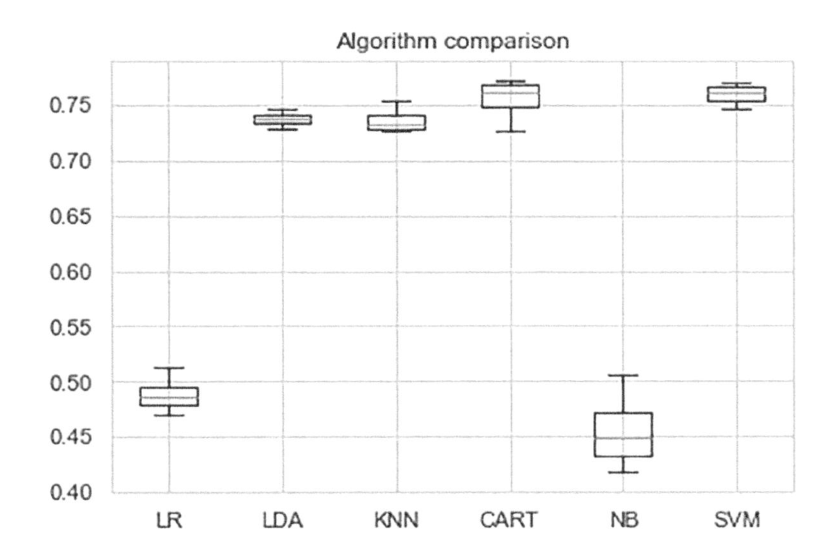

Figure 14.5 Model comparison.

the waste is only processed through formal waste management system and rest and burned. Therefore, these create a huge environmental issue. Oceans are also being contaminated by waste, which is not managed properly. Ten percent of the emission comes from the management of solid waste. If corrective action is not taken then emission for solid waste management is set to reach 2.6 B tons by 2050. To get a sustainable climate we should concentrate on formalizing recycle ecosystem.

14.6 CONCLUSION

World population is increasing everyday with time. In the last 10 years, the population has increased by 7%. Fossils fuel utilization has also increased due to industry and other human needs. In addition, 84% of the world's energy is still supplied by fossil fuels. Global electricity demand is predicted to be increased by 4.6% in 2021. Global CO_2 emission is forecasted to jump by 1.5 B tons in 2021. Unless government bodies take corrective action to cut emissions, we will see a worse situation in 2022. Despite the usage of renewable energy coal usage has also increased due to the demand. More than 80% of coal demands are coming from Asia, majorly from China. Renewable energy usage has increased by 8% in 2021. Energy production by windmill is set to grow by 17% by 2021. World leader's climate summit organized by Joe Biden (US president) in April 2021 has put some clear direction to world leaders. There is no system to track and gather data of this nature from different countries. This study gives a holistic and robust methodology to get a sustainable climate. This study gives a 360° view of global warming scenario. An integrated solution architecture which can be used by any source state or country as it is powered by universal adaptor. Assisting global leaders and environment activists is the key outcome of this research. Global warming is a wide range of issue across the nations. In this work, we have shown the result focused on temperature and emission prediction. After predicting the emission, we have established a solution model for a control system. This control system has the capability to reduce the global warming if we can place it and utilize properly to facilitate the governance body, which has a major role to control this climate issue.

REFERENCES

1. Cheshmehzangi, A. (2020) 'The analysis of global warming patterns from 1970s to 2010s', *Atmospheric and Climate Sciences*. doi: 10.4236/acs.2020.103022.
2. Hansen, J., et al. (2006) 'Global temperature change', *Proceedings of the National Academy of Sciences of the United States of America*. doi: 10.1073/pnas.0606291103.

3. Intergovernmental Panel on Climate Change. (2007) *Climate Change 2007 - The Physical Science Basis: Working Group I Contribution to the Fourth Assessment Report of the IPCC. Science.*
4. Cheshmehzangi, A. and Dawodu, A. (2018) *Sustainable Urban Development in the Age of Climate Change: People: The Cure or Curse.* doi: 10.1007/978-981-13-1388-2.
5. Biskaborn, B. K., et al. (2019) 'Permafrost is warming at a global scale', *Nature Communications.* doi: 10.1038/s41467-018-08240-4.
6. Kadam, M., Kanoo, N. and Zheng, Y. (2017) 'Climate change: Relationships to CO_2 emission and locations', in *RIIT 2017- Proceedings of the 6th Annual Conference on Research in Information Technology.* doi: 10.1145/3125649.3125-654.
7. Christensen, T. H., et al. (2009) 'C balance, carbon dioxide emissions and global warming potentials in LCA-modelling of waste management systems', *Waste Management and Research.* doi: 10.1177/0734242X08096304.
8. Gentil, E., Christensen, T. H. and Aoustin, E. (2009) 'Greenhouse gas accounting and waste management', *Waste Management and Research.* doi: 10.1177/0734242X09346702.
9. Paul, P., et al. (2020) 'Environmental informatics vis-à-vis big data analytics: The geo-spatial & sustainable solutions', *SSRN Electronic Journal.* doi: 10.2139/ssrn.3670083.
10. Yang, C., et al. (2019) 'Big earth data analytics: A survey', *Big Earth Data.* doi: 10.1080/20964471.2019.1611175.
11. Letrache, K., El Beggar, O. and Ramdani, M. (2018) 'Green data warehouse design and exploitation', in *ACM International Conference Proceeding Series.* doi: 10.1145/3289402.3289529.
12. Kimball, R., et al. (2008) *The Data Warehouse Lifecycle Toolkit*, Wiley.
13. Yener, F. and Yazgan, H. R. (2019) 'Optimal warehouse design: Literature review and case study application', *Computers and Industrial Engineering.* doi: 10.1016/j.cie.2019.01.006.
14. Herden, O. (2015) 'Data Warehouse', in *Taschenbuch Datenbanken.* doi: 10.3139/9783446440265.014.
15. Xu, Y., Ramanathan, V. and Victor, D. G. (2018) 'Global warming will happen faster than we think', *Nature.* doi: 10.1038/d41586-018-07586-5.
16. Wang, C., Zhang, Q. and Zhang, W. (2020) 'Corporate social responsibility, green supply chain management and firm performance: The moderating role of big-data analytics capability', *Research in Transportation Business and Management.* doi: 10.1016/j.rtbm.2020.100557.
17. Skytt, T., Nielsen, S. N. and Jonsson, B. G. (2020) 'Global warming potential and absolute global temperature change potential from carbon dioxide and methane fluxes as indicators of regional sustainability – A case study of Jämtland, Sweden', *Ecological Indicators.* doi: 10.1016/j.ecolind.2019.105831.
18. Lotfalipour, M. R., Falahi, M. A. and Bastam, M. (2013) 'Prediction of CO_2 emissions in Iran using grey and ARIMA models', *International Journal of Energy Economics and Policy,* 3(3): 229–237.
19. Doreswamy, G. I. and Manjunatha, B. R. (2017) 'Hybrid data warehouse model for climate big data analysis', in *Proceedings of IEEE International Conference on Circuit, Power and Computing Technologies*, ICCPCT, 2017. doi: 10.1109/ICCPCT.2017.8074229.

20. Hassani, H., Huang, X. and Silva, A. E. (2019) 'Big data and climate change', *Big Data and Cognitive Computing*. doi: 10.3390/bdcc3010012.
21. Hyndman, R. J. and Fan, S. (2010) 'Density forecasting for long-term peak electricity demand', *IEEE Transactions on Power Systems*. doi: 10.1109/TPWRS.2009.2036017.

Chapter 15

Human age estimation using sit-to-stand exercise data-driven decision-making by neural network

Susmita Das
Narula Institute of Technology

Dalia Nandi
Indian Institute of Information Technology Kalyani

Biswarup Neogi
JIS College of Engineering

Debashis De
Maulana Abul Kalam Azad University of Technology, West Bengal

CONTENTS

15.1 INTRODUCTION

We discussed biological age (A_{bi}) and chronological age (A_{ch}) in years concerning the life expectancy prediction of human beings. Many studies have focused on this issue where only featuring facial expressions are considered to detect age. The experimental data collection for human beings is hard to follow up human subjects throughout the aging process due to their long life span. The present work aims to determine biological age in years and

DOI: 10.1201/9781003381167-15

therefore as predictors of longevity. Muscular strength evaluation during sitting and rising activity is one of the most critical factors to determine the biological age of the human body [1]. A particular age group is involved in experimenting in this chapter. This chapter included a wide range of ages to achieve the predicted biological age. Incorporating artificial intelligence in longevity detection is a different approach than the other traditional techniques [2]. Utilizing the neural network-based method to detect the lifespan is a challenging aspect where many data are involved. The discussion of recent research gives the idea of the efficient method for life expectancy prediction using a neural network [3]. Deep learning is the justified method to meet the need for biological age extraction [4]. The big data analytics platform gives almost 98% accurate predictions regarding the healthcare topic of human beings [5]. EMG (Electromyography)-based output during the movement of human body parts is the dependent factor for the life expectancy prediction [6]. EMG signal processing is the necessary parameter to achieve proper outcomes during the movements of the body parts [7,8]. The investigatory research work is followed regarding physical activities for the aged and physically challenged people [9,10].

15.2 LITERATURE REVIEW

The comparison related to the mortality, aging, longevity, biological age, health condition, and prediction algorithm between the proposed work and the state of the art is presented in Table 15.1.

15.2.1 Motivation

Previous research works on the mortality investigation of human beings show the training of physical exercises and cardiological activities. EMG signal has not been utilized for the age estimation, while this is the most reliable method to achieve information about the bioelectric phenomena [11]. The process of lifespan prediction is the most relevant and technically innovative field in recent healthcare technology advancements. We achieve royal aging using preventive healthcare technologies for senior citizens.

15.2.1.1 Specific contribution

Based on the comparative analysis shown in Table 15.1, the utilization of the EMG signal during sit-to-stand exercise without any support and the other health parameters is the contribution to the current research work. Therefore, the overall activities of body parts and their functionality are the critical factors for the age prediction model developed using artificial neural network (ANN) with approx. 22.41% error output efficiency [12].

Table 15.1 Table of comparison on the features of the health status prediction

Features of the works	*Previous works*	*Current work*	*Status of the current work*
Mortality	Musculoskeletal fitness assessed by Sit-to-Rise Test (SRT), 2014 [1]	Human Age estimation using body-health parameters and SRT	Better estimation of age with more parameters
Aging and longevity	Synthetic age prediction using body parameters, 2019 [2]	Biological and chronological age comparison	The new approach to age prediction
Biological age	Biological age prediction using locomotor activities, 2018 [3]	Biological age prediction using body parameters and exercise	Utilized neural network with sufficient body-activity parameters
Health condition	Body health status checking, 2016 [4]	Body fitness checking with the decision of attention needed as per estimated age	Body fitness information with more parameters
Age prediction algorithm	Neural network-based algorithm for EMG signal of human body movement, 2019 [5]	Neural network-based algorithm for different health parameters of the human body during sit-to-stand test	The better decision-making process for human body fitness test

15.3 METHODOLOGY

Chronological age (A_{ch}) refers to the actual time duration of the living period. Biological age (A_{bi}) refers to the physiological age. The process of age prediction can be arranged in a proper sequence. We have considered five main age groups for the experimental procedure. Those are children, teenagers, adults, middle-aged, and senior citizens who are between 2 and 10 years; 11 and 20 years; 21 and 35 years; 36 and 59 years; and 60 and 85 years, respectively. We have selected two groups such as adults and middle-aged volunteers, for further tests among these age groups. Moreover, we verified the system's reliability when different accurate results are achieved simultaneously concerning other age groups and relative health parameters.

We present the chronological and biological age concept in Figure 15.1. The middle line in Figure 15.1 represents the reference age where the biological and the chronological ages are the same. The line below the reference level shows where A_{bi} is less than the $A_{ch,}$ and the line above the reference level shows where A_{bi} is greater than the A_{ch}. We performed some basic tests for different age groups to acquire input data. The measurements include body temperature, weight, blood pressure, and height [13,14]. Noninvasive

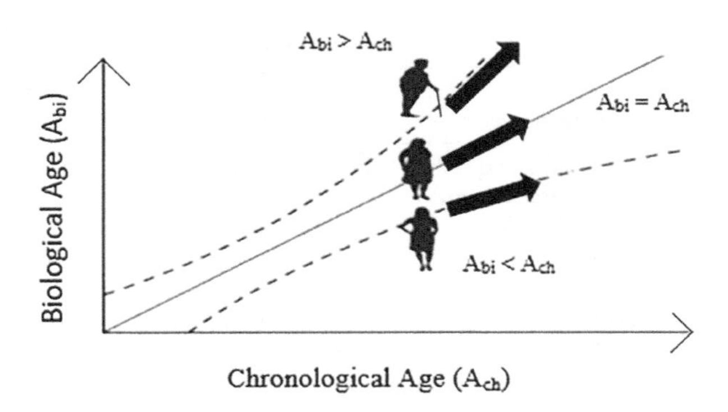

Figure 15.1 Pictorial representation of chronological age (A_{ch}) vs. biological age (A_{bi}) of human.

EMG sensors are placed on the calf muscles in the lower limbs because only through this muscle the output of the lower limb movements can be detected. We completed these data acquisitions, and they are arranged in the required format for further use. We collected the output observations of muscular activities from muscle groups of the human lower limb during sitting and standing exercises. We have taken these readings from the EMG signals for age prediction. Furthermore, we collected these readings repeatedly for different age groups of people in the postures such as sitting and standing so that the efficiency of muscles can be observed and analyzed due to the aging effect [15,16].

We have considered two ways of measurement in this research work. In the standing to sitting posture, at first, the subject is in the standing posture on his bare feet and would be asked to gently sit down on the firm ground without using any form of support. In the sitting to a standing posture, the subject sitting on firm ground would be asked to gently stand up on their bare feet without using any form of support. We recorded the electrical signals produced by the muscles during these courses of motion. In both cases, if the subject needs additional help to do the task, the type of support should be mentioned, i.e., with hands or a stick, and data should be recorded accordingly.

EMG is a diagnostic test that is performed to record the electrical activity of muscles at rest and during contraction. After collecting data, we arranged it according to the age division. Finally, a data training algorithm is run with the various EMG signals and obtained body parameters. This algorithm helps to predict the biological age of the subject. Furthermore, after being modified, this algorithm is the key to finding out the life expectancy of a subject.

We present the sit-to-stand exercise shown in Figure 15.2. In the context of aging, the physical capacity becomes very low toward the end of the active life of the workers doing laborious jobs physically. Therefore, only

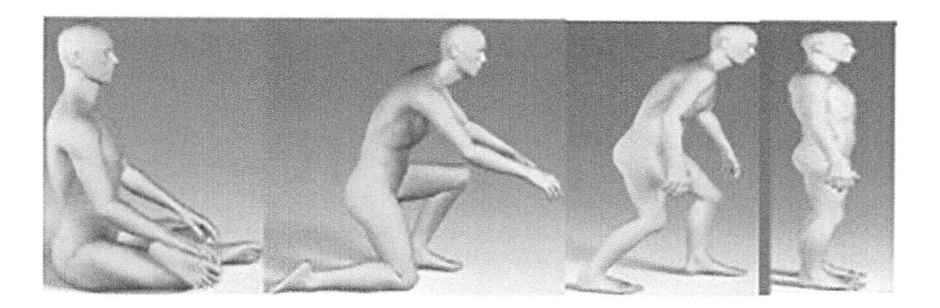

Figure 15.2 Schematic diagram of the sit-to-stand exercise.

work can describe the physical changes connected with age resulting in the demand for the production system for older workers.

In the beginning of the flowchart in Figure 15.3, the proposed system takes four primary parameters as input along with EMG. The parameters are height, weight, body temperature, and blood pressure as health parameters. We set these acquired parameters as the input and the person's chronological age is given as the output response to the model for training. The reference values are the healthy persons' data that person's age group needs to possess. We enter these data into the system, and it defines the input parameters to the program logic. As a result, it provides the output as per the trained model through training. After getting the patient's approximate biological age, the system checks for the patient's health condition. It compares the biological age with the patients' chronological age. The biological

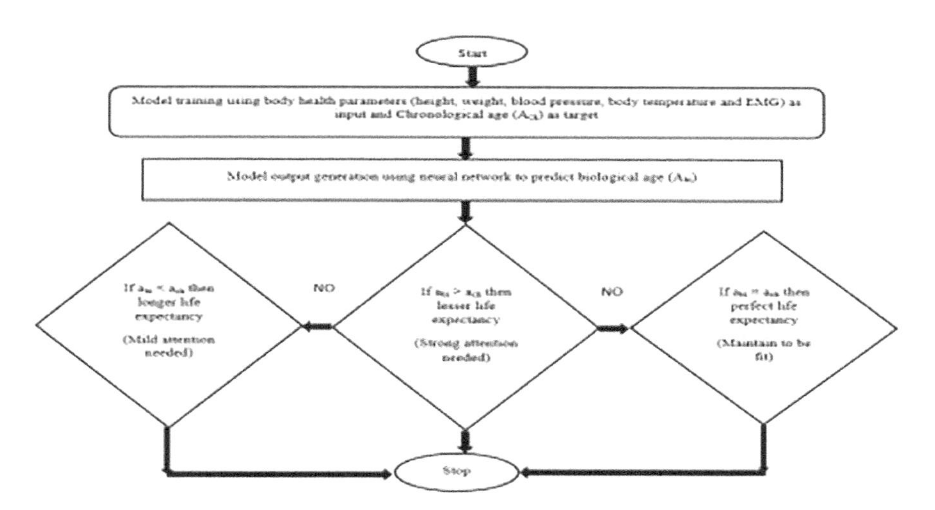

Figure 15.3 Flowchart of the age prediction process and required action.

ages are compared with the subjects' chronological ages, and it is observed that the values are matched with little error percentage. The training process is started by utilizing the known values of the health condition and the sit-to-stand EMG output [17,18].

Then the trained model shows the validation of the input data concerning the known chronological age [19,20]. The first condition is that the biological age (A_{bi}) is greater than the chronological age (A_{ch}). This scenario is common nowadays due to increased pollution and anxiety among people. These persons become sexagenarians. It signifies that the aging process of the individual is speedy, due to which they look much older than they usually are [21–23]. The second scenario comes when the biological age is equal to the chronological age. This shows that the patients' age increases proportionally with their health and the number of days on this planet. The third scenario is the rarest scenario when the biological age is less than the chronological age. In this scenario, it is noticed that these people don't age a lot. After some primary years of their existence, their age will become relatively constant, as if their age clock has stopped. As a result, sometimes, a septuagenarian looks like a quinquagenarian. When any cases arise, the system triggers individual alarms for individuals. For the first case, when ($A_{bi}>A_{ch}$), strong attention is needed to be healthy. For the second case, when ($A_{bi}=A_{ch}$), the proper maintenance to be fit is expected. For the third case, when ($A_{bi}<A_{ch}$), mild attention is needed to be healthy. Some case studies for this practical experiment regarding age calculation are given below in Tables 15.2–15.5 for ages such as 21, 30, 42, and 50 years:

Table 15.2 Body parameters of the persons with the chronological age (A_{ch}) 21 years

Height (cm)	*Weight (kg)*	*Blood pressure (mmHg)*	*Body temperature (° F)*	*EMG during exercise (V)*
158.1	61	130/80	98.3	0.97
158.5	62	130/80	97.3	0.92
158.9	60.5	130/80	97.8	0.99
157.9	61	130/80	97.9	0.94
158.3	61	130/80	98.3	0.93

Table 15.3 Body parameters of the persons with the chronological age (A_{ch}) 30 years

Height (cm)	*Weight (kg)*	*Blood pressure (mmHg)*	*Body temperature (° F)*	*EMG during exercise (V)*
190.1	90.5	130/95	98	0.91
190.2	91	130/95	98	0.86
190.3	90.8	130/95	98	0.85
158.3	80.5	130/90	97.8	0.9
158.5	79.9	130/90	97.8	1.08

Table 15.4 Body parameters of the persons with the chronological age (A_{ch}) 42 years

Height (cm)	*Weight (kg)*	*Blood pressure (mmHg)*	*Body temperature (° F)*	*EMG during exercise (V)*
153.3	66	130/90	97	0.88
153.1	65	130/90	97	0.85
153.4	66.2	130/90	97	0.84
170	75	120/80	97.2	0.84
170	75.5	120/80	97.2	0.85

Table 15.5 Body parameters of the persons with the chronological age (A_{ch}) 50 years

Height (cm)	*Weight (kg)*	*Blood pressure (mmHg)*	*Body temperature (° F)*	*EMG during exercise (V)*
154.2	69.5	131/92	97.3	0.91
154.2	70	131/94	97.3	0.88
154.1	69.9	130/95	97.4	0.87
154	69.8	131/92	97.6	0.82
154.4	69.7	130/92	97.3	0.90

The mean error values of predicted biological age compared to chronological age regarding the data presented in Tables 15.2–15.5, respectively are 0.08, 0.32, 0.69, and 1.40.

15.4 RESULTS AND DISCUSSION

The feedforward backpropagation algorithm has been implemented in the neural network training process. Here the specific and mostly age-related body parameters have been considered to estimate the biological age (A_{bi}) of the human being. Typically, this supervised algorithm is used in natural language and image processing, but this has been used in a different field following the human nervous system. In this method, the errors are reduced with repeated validation checking. After training, the input data with various iterations up to 32 epochs, this performance curve has been observed in Figure 15.4, where the test curve (red line) is best fitted with the validation curve (green line) with validation performance 6.3428 at epoch 26. The best-fitted curve shows that the biological age estimation is near the chronological age. Finally, the number of epochs considered for the test and the mean square error values are presented to show the detailed behavioral analysis of the validation curve for age calculation.

Figure 15.5 shows the accuracy of the training curve appears with the value of gradient=0.70105 at epoch 32, with mu=0.01 at epoch 32, and

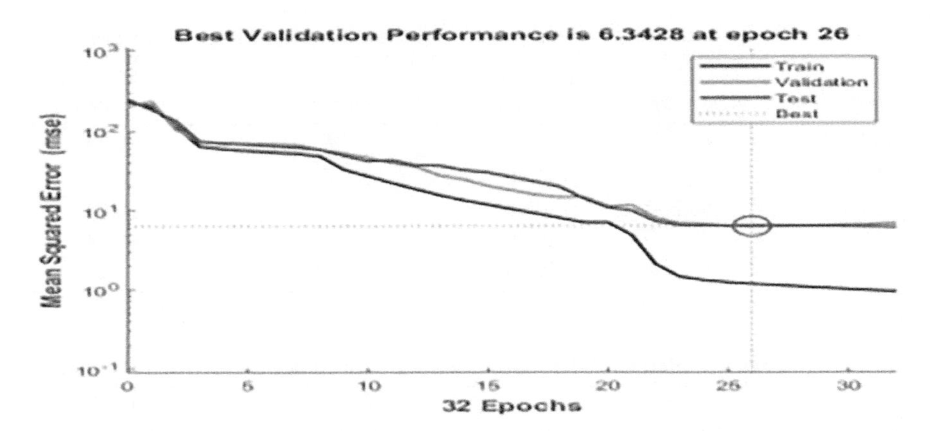

Figure 15.4 The performance curve of the trained model with the best-fitted curve for the biological age estimation.

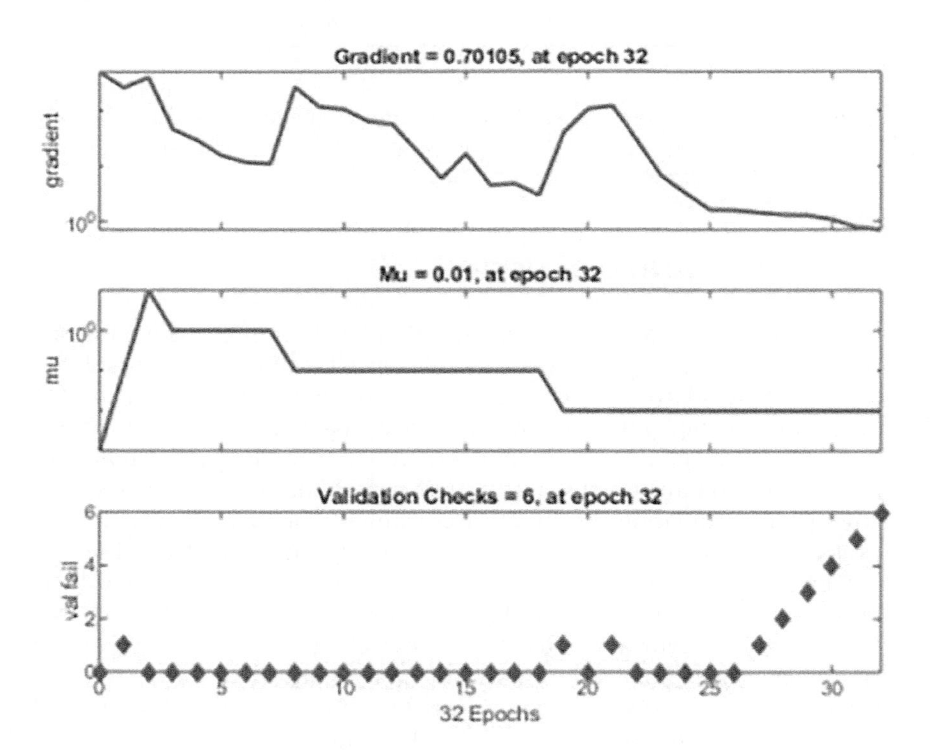

Figure 15.5 The performance curve of the trained model with validation for the biological age estimation.

validation checks=6 at epoch 32. Furthermore, the validation curve concerning the test data has been presented for output estimation satisfaction. Therefore, at the mentioned points of the given curves, the estimated output of the training shows the best performance.

Here the regression curve has the most data proximity with fit among 'training' and 'all' iterations, especially in the first (blue line) and last (black line) graphs with 0.99355 and 0.98469 regression values, respectively shown above in Figure 15.6. In this graphical presentation, the training, validation, test, and data are indicated with the best fit curve for the biological age estimation using the subjects' health parameters and exercise data. As per the other references, the used methods show the model training process less efficient because of the lack of compatibility. But in this training method, the proper validation has been observed concerning the reference values where the selected parameters and the trained model are properly defined for the specific application.

In Table 15.6, a comparative analysis has been presented regarding human beings' chronological and biological age using a neural network prediction algorithm. The decisions are made according to the designed logic using fitness measures. The age estimation error percentage ranges from approx. 0.09% to 22.41%. When the error percentage value is higher than 10%, strong attention is recommended; in the case of lower than 10%, mild attention is recommended; otherwise, it is decided as a fit person with a negligible error percentage value.

The prediction of the biological age of human beings can be performed using the classification of the observed data from the experimented subjects also. This process is another efficient way to reach the goal to predict age.

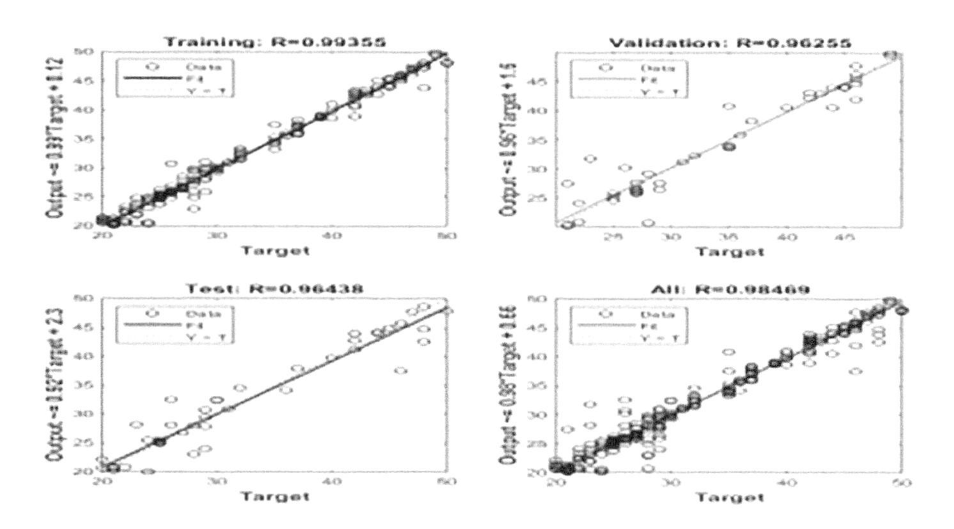

Figure 15.6 Regression curve of the trained model for the age prediction.

Table 15.6 Output prediction result of biological age of human body using neural network algorithm

Chronological age (A_{ch})	*Biological age (A_{bi})*	*Errors (A_{ch} − A_{bi})*	*Error (%)*	*Decision on fitness*
21	20.57	0.43	2.05	Fit
22	26.93	−4.93	22.41	Strong attention needed
23	26.81	−3.81	16.56	Strong attention needed
24	27.68	−1.76	7.33	Mild attention needed
25	22.44	2.55	10.2	Mild attention needed
26	24.13	1.89	7.27	Mild attention needed
27	25.20	1.80	6.66	Mild attention needed
28	22.95	5.05	18.03	Strong attention needed
29	29.17	−0.17	0.59	Fit
30	30.07	−0.07	0.23	Fit
31	30.85	0.15	0.48	Fit
32	34.14	−2.14	6.69	Mild attention needed
35	30.83	4.16	11.88	Strong attention needed
36	34.80	1.20	3.33	Mild attention needed
37	37.09	−0.09	0.24	Fit
39	39.52	−0.52	1.33	Fit
40	39.45	0.55	1.37	Fit
42	43.82	−1.82	4.33	Mild attention needed
43	42.96	0.04	0.09	Fit
44	43.38	0.62	1.41	Fit
45	44.49	0.50	1.11	Fit
46	46.88	−0.88	1.91	Fit
47	46.40	0.60	1.28	Fit
48	49.80	−1.80	3.75	Mild attention needed
49	48.11	0.89	1.82	Fit
50	48.77	1.23	2.46	Fit

15.5 CONCLUSIONS AND FUTURE SCOPE

The current focus is to predict biological age and provide health care alerts. However, as a challenging aspect, detecting age by analyzing various health parameters, this technology is considered as an emerging technique.

A new attempt toward the approach of the device gives the ultimate framework to estimate the ages of human beings in different ranges of age. Body mass index (BMI) is another vital piece of information to state health status. The techno-commercial aspect of healthcare facilities such as body fitness measurement and life expectancy prediction can be generated with advanced information on mortality.

REFERENCES

1. Brito L.B., Ricardo D.R., Araújo D.S., Ramos P.S., Myers J., Araújo C.G., Ability to sit and rise from the floor as a predictor of all-cause mortality. *Eur J Prev Cardiol* 21(7): 892–898 (2014).
2. Zhavoronkov A., Mamoshina P., Vanhaelen Q., Scheibye-Knudsen M., Moskalev A., Aliper A., Artificial intelligence for aging and longevity research: Recent advances and perspectives. *Ageing Res Rev* 49: 49–66 (2019). ISSN 1568-1637.
3. Pyrkov T.V., Slipensky K., Barg M., et al., Extracting biological age from biomedical data via deep learning: too much of a good thing? *Sci Rep* 8: 5210 (2018).
4. Sahoo P.K., Mohapatra S.K., Wu S. Analyzing healthcare big data with prediction for future health condition. *IEEE Access* 4: 9786–9799 (2016).
5. Luo J., Liu C., Yang C. Estimation of EMG-based force using a neural-network-based approach. *IEEE Access* 7: 64856–64865 (2019).
6. Moore J.H., Raghavachari N. Artificial intelligence based approaches to identify molecular determinants of exceptional health and life span-an interdisciplinary workshop at the national institute on aging. *Front Artif Intell* 2: 12 (2019).
7. Thomas R., Michael P., Michael B., Barnes P. Life expectancy for people with disabilities. *Neurorehabilitation* 27(2): 201–209 (2010).
8. Pollock R.D., Carter S., Velloso C.P., Duggal N.A., Lord J.M., Lazarus N.R., Harridge S.D. An investigation into the relationship between age and physiological function in highly active older adults. *J Physiol* 593(3): 657–680 (2015).
9. Jin J., Jin K.J., Adibi K.S., Adibi S. *Systematic Predictive Analysis of Personalized Life Expectancy Using Smart Devices, Technologies*. MDPI, Switzerland (2018).
10. Zhavoronko A., Li R., Ma C., Mamoshina P. Deep biomarkers of aging and longevity: From research to applications. *Aging* 11(22): 10771–10780 (2019).
11. Rezaeian A., Rezaeian M., Khatami S.F. et al. Prediction of mortality of premature neonates using neural network and logistic regression. *J Ambient Intell Human Comput* 21: 1–14 (2020).
12. Kim J.C., Chung K. Neural-network based adaptive context prediction model for ambient intelligence. *J Ambient Intell Human Comput* 11: 1451–1458 (2020).
13. De D., Mukherjee A., Sau A., Bhakta I. Design of smart neonatal health monitoring system using SMCC. *Healthcare Technol Lett* 4(1): 13–19 (2017).
14. Banerjee A., Chakraborty C., Kumar A., Biswas D. Chapter 5: Emerging trends in IoT and big data analytics for biomedical and health care technologies. In: Valentina Emilia Balas, Vijender Kumar Solanki, Raghvendra Kumar, Manju Khari (eds.), *Handbook of Data Science Approaches for Biomedical Engineering*. Academic Press, Cambridge, MA, 121–152 (2020). ISBN 9780128183182.
15. Lai D., Chen Y., Luo X., et al. Age estimation with dynamic age range. *Multimedia Tools Appl* 76: 6551–6573 (2017).

16. Li X., Makihara Y., Xu C., et al. Gait-based human age estimation using age group-dependent manifold learning and regression. *Multimedia Tools Appl* 77: 28333–28354 (2018).
17. Xia L., Yang J., Han T., et al. A mobilized automatic human body measure system using neural network. *Multimedia Tools Appl* 78: 11291–11311 (2019).
18. Yao L., Yang W., Huang W. A fall detection method based on a joint motion map using double convolutional neural networks. *Multimedia Tools Appl* 81(4): 4551–4568 (2022). https://doi.org/10.1007/s11042-020-09181-1
19. Guo G., Guowang M., Fu Y., Huang T.S. Human age estimation using bio-inspired features. *2009 IEEE Conference on Computer Vision and Pattern Recognition*, Miami, FL, pp. 112–119 (2009).
20. Ashiqur R.S. Quantifying human biological age: A machine learning approach. Graduate Theses, Dissertations, and Problem Reports. 7376, West Virginia University (2019).
21. Nguyen D.T., Cho S.R., Shin K.Y., Bang J.W., Park K.R. Comparative study of human age estimation with or without preclassification of gender and facial expression. *Sci World J* 2014: 15 (2014).
22. Li X., Makihara Y., Xu C., et al. Gait-based human age estimation using age group-dependent manifold learning and regression. *Multimed Tools Appl* 77: 28333–28354 (2018).
23. Cao D., Lei Z., Zhang Z., Feng J., Li S.Z. *Human Age Estimation Using Ranking SVM*. HoHai University, Center for Biometrics and Security Research & National Laboratory of Pattern Recognition, Institute of Automation, Chinese Academy of Sciences, China Research and Development Centre for Internet of Thing. 7th Chinese Conference, CCBR 2012, Guangzhou (2012).

Chapter 16

Feature-based suicide-ideation detection from Twitter data using machine learning techniques

Dhrubasish Sarkar, Poulomi Samanta, and Piyush Kumar
Amity University Kolkata

Moumita Chatterjee
Aliah University

CONTENTS

16.1 INTRODUCTION

Mental health is an important and highly debated topic for any generation, and it is becoming increasingly prominent globally. This condition can be influenced by a number of factors such as stress, trauma or traumatic events, anxiety, hopelessness, physical or mental disease, dealing with challenges, and dealing with personal troubles, among other things. Suicidal ideation is more common in those suffering from major depression. Suicidal death is comparable to one person dying every 40 seconds, according to World Health Organization (WHO) mortality figures. As a result, WHO has issued new guidelines to help countries improve suicide prevention and care [1].

DOI: 10.1201/9781003381167-16

In today's world, social media is one of the venues where people actively participate in their everyday lives. There are some sites, in particular, where the majority of users post their daily status. Twitter, Facebook, WhatsApp, YouTube, and Instagram are among these platforms. People discovered that these are the most convenient and cost-effective ways to connect with individuals, groups, and to use other useful entities such as daily updates, status, posts, and so on.

People post status updates on social media platforms, allowing us to identify mental health problems and suicidal attempts. Posts and status updates allow them to express their emotions and find relief from the stress they are feeling. When a person is in a difficult situation, their word choices change to reflect the situation, and in some cases, emojis help us understand whether the individual's post is positive or negative. Examining social media posts, according to researchers, can help with the identification of depression and other mental health issues. These online activities have encouraged them to create cutting-edge potential medical treatments and early suicide detection techniques. This is performed by employing machine learning and Natural Language Processing (NLP) methodologies to detect suicidal thoughts in user posts.

To better understand and recognize suicidal thoughts from user postings, the goal of this study is to examine the identification of suicidal thoughts through online user-generated material. This work covers a detailed assessment of emoticon usage, language preferences, time of posting analysis, and subject descriptors for understanding suicidal thoughts from user discussions. A variety of features were retrieved from the data, and four learning algorithms were utilized to predict suicidal ideation. Statistical studies of the postings in connection to the retrieved features were also undertaken, and several relevant facts such as tweet length, emoticon usage, temporal behavior, and language usage were revealed. Suicidal and non-suicidal writings usually discuss a wide range of issues that aid in our understanding of the two types. Latent Dirichlet Allocation is used to extract a collection of latent topics from normal and suicidal texts. Finally, suicidal ideation in user postings is detected using these feature sets in conjunction with classification algorithms. Each feature's performance is investigated both independently and in various combinations.

This chapter is organized as follows; the related works are discussed in Section 16.2. Section 16.3 describes the proposed methodology. Experimental analysis is detailed in Section 16.4. Section 16.5 concludes the chapter.

16.2 RELATED WORK

We looked at the gender differences in suicidal inclination among teenagers [2]. The report also states which gender has a higher likelihood of attempting suicide. We discovered several characteristics based on scenarios that link to suicidal tendencies among students based on worldwide data among students in another study [3]. Anxiety, insecurity, and loneliness were identified by the authors as important elements that can impair students' mental health. The authors in reference [4] describe various sorts of suicidal conduct. The authors looked at the outbreak as well as the risk of suicide ideation and protective variables. Another work [5] categorizes various risk factors for this topic among youth. Suicidal behavior is one of the world's most significant issues. Another study [6] gives some data concerning suicide, suicidal thoughts, and suicide attempts. Using a psychiatric healthcare dataset and a hybrid machine learning model, the authors of reference [7] provided a rule-based method for categorizing the occurrence of suicidal thoughts. Another study uses machine learning and NLP of electronic health records to explore suicidal behavior among psychiatrically hospitalized teenagers [8]. The authors of reference [9] looked into how machine learning could be used to identify suicidal ideation. Suicide-ideation applications are examined using a variety of data sources, including questionnaires, suicide notes, and online user content. The authors of reference [10] examined the use of NLP and machine learning approaches to predict suicide symptoms and ideation in persons released from psychiatric facilities. The authors of reference [11] overcome this issue by using Twitter to detect suicide profiles. For recognizing users with suicidal ideation, they gathered information from user posts as well as their profiles. The authors of reference [12] described a method for evaluating user topic descriptions, a language preference, and online social media such as Twitter and Reddit. By analyzing behavioral, relational, and multimodal data from several social platforms and developing algorithms to identify users with suicidal ideation, the authors of reference [13] outlined a strategy for assessing the suicide risk of Spanish-speaking users on social media. The authors of reference [14] used a machine learning classifier to see if the amount of uneasiness for a suicide-ideation tweet could be exclusively identified from its text The authors of reference [15] constructed a multimodal depression dictionary learning model that may be utilized to diagnose depressed Twitter users by mining numerous feature groups comprised of data with depression symptoms from both online social behaviors and clinical data. The authors of reference [16] related anxiety depression to agitation, aberrant cognitive processes, and insomnia. They proposed a model for predicting anxious depression using real-time tweets and linguistic clues. To categorize the positive and negative sentiments of Twitter users, the authors of reference [17] used a variety of machine learning methods. They then compared the findings to deep learning techniques. The authors of reference [18] construct a training and test dataset on depression using Twitter messages, and the Naive Bayes classifier is used to categorize the data. The proposed model in this chapter builds on the work done by the authors in references [18,19].

16.3 PROPOSED METHODOLOGY

16.3.1 Data set exploration and preparation

Using Twitter data, we train our classification algorithm for identifying suicidal thoughts. Using Twitter APIs, we acquired a data collection of 1,88,704 tweets and 1,169 individuals. The training data set is manually annotated to create two data sets: one for suicide-indicative posts and one for non-suicidal posts.

Data from online social media cannot be utilized directly for feature extraction due to the existence of numerous noises in the raw data. Language and spelling problems, emoticons, and other undesired features may be included in the data, further complicating the situation. Therefore, data preparation is necessary to get a high-quality and trustworthy dataset for use in our investigation. The numerous data-preprocessing procedures employed in our model are depicted in Figure 16.1.

16.3.2 Feature extraction and analysis

To give high accuracy in suicide-ideation detection, feature extraction is a key stage in acquiring extensive information about users. This study gathered several features from tweets, including TF-IDF, sentiment analysis, emoticons, temporal, statistics, and topic-based features.

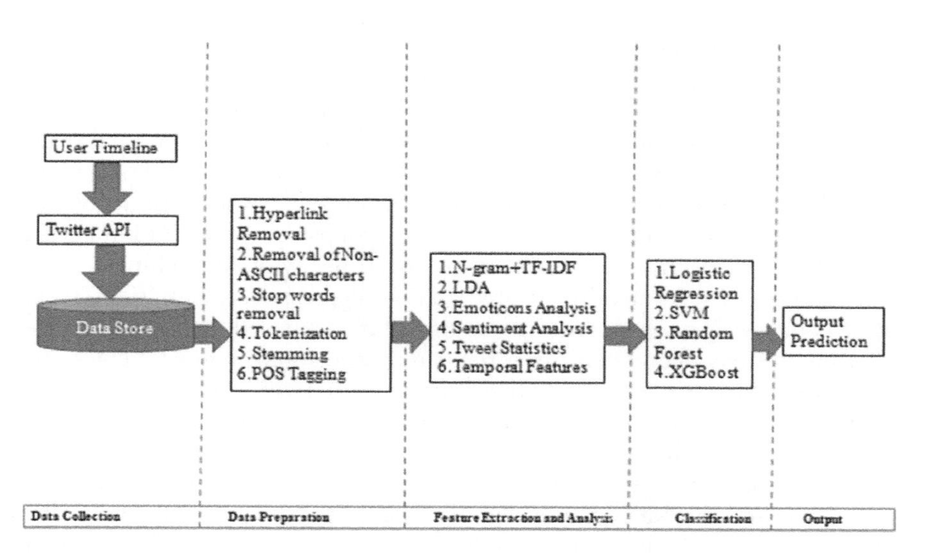

Figure 16.1 Proposed methodological framework for suicide-ideation detection of Twitter data.

Emojis/emoticons sentiment: Users utilize emojis to express their feelings using nonverbal aspects and simple symbols. Emojis may help us understand the emotion of any text or tweet, and knowing how to differentiate between good and negative sentiment content is critical. Users' tweets typically include a plethora of emoticons that may be classified as positive, negative, or neutral.

In this work, various statistical measures are used to calculate a user's emoji usage. Several data concerning the employment of emoticons were discovered during the analysis of suicidal ideation texts. Suicidal persons use less emojis on average, and the number of emojis used in suicidal messages was shown to be substantially lower than in normal posts. The most often used emojis among suicidal-tendency users are folded hands, loudly crying face, expressionless face, screaming in fear face, pleading face, disappointed face, perplexed face, and face without mouth. The most often used emojis among non-suicidal inclined users are: face with tears of pleasure, rolling on the floor laughing, loudly weeping face, crimson heart, clapping hands, winking face, and smiling face with heart eyes. Based on such data, we may conclude that suicidal-tendency users use negative emoticons more frequently than normal users.

Tweet statistics: This group contains statistical measures derived from user tweets. Simple, brief sentences are used in certain communications, whereas complicated sentences and large paragraphs are used in others. The number of tweets and their durations, the number of tweets per user connected to suicide, and the proportion of suicide-related postings to total posts per user are all considered. Figure 16.2

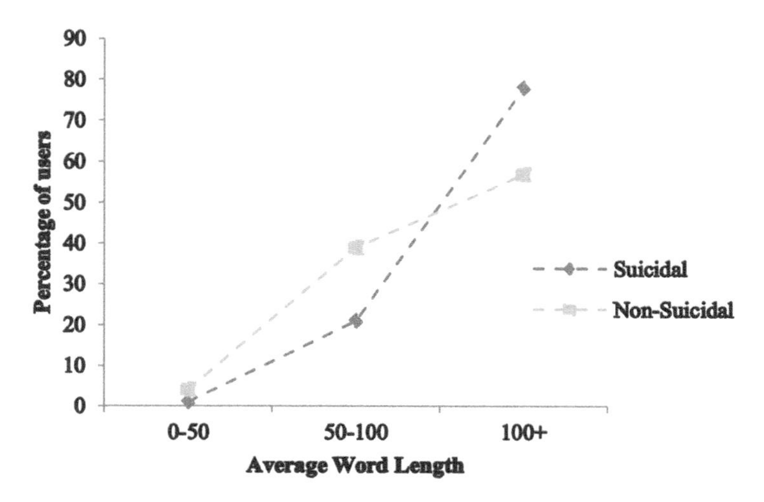

Figure 16.2 Mean length of tweets for suicidal and normal users.

depicts the average length of tweets for suicidal and non-suicidal users. The mean text length of users with suicide intents is much larger than that of non-suicidal users, according to the data. This is because suicidal users are more likely to have a mental condition or social issues, which are reflected in their posts.

Sentiment analysis: Sentiment analysis is an effective method for identifying emotions in user texts. This approach works well for writings with a subjective context, such as suicide or depression. In general, sentiment analysis classifies emotions as good, negative, or neutral. By examining a user's tweets, it is possible to anticipate if he is expressing positive, negative, or neutral feelings. Sentiment analysis of user tweets from both suicide and non-suicidal datasets reveal that most users in the suicidal dataset have a sentiment score between −0.5 and −1, but most users in the non-suicidal dataset have a sentiment score above −0.5. Analysis of the datasets reveals that suicidal ideation users are more prone to write tweets with negative sentiment compared to non-suicidal users. The graph in Figure 16.3 depicts the frequency of negative tweets used by suicidal and non-suicidal users.

Topic features: Non-suicidal postings differ from suicide-ideation posts in that they include a wide variety of topics, which can assist us discern between the two. We developed a topic model for deriving latent themes from the suicidal dataset using Latent Dirichlet Allocation (LDA). LDA determines the likelihood of each tweet belonging to a distinct subject given a sequence of tweets and a number of subjects

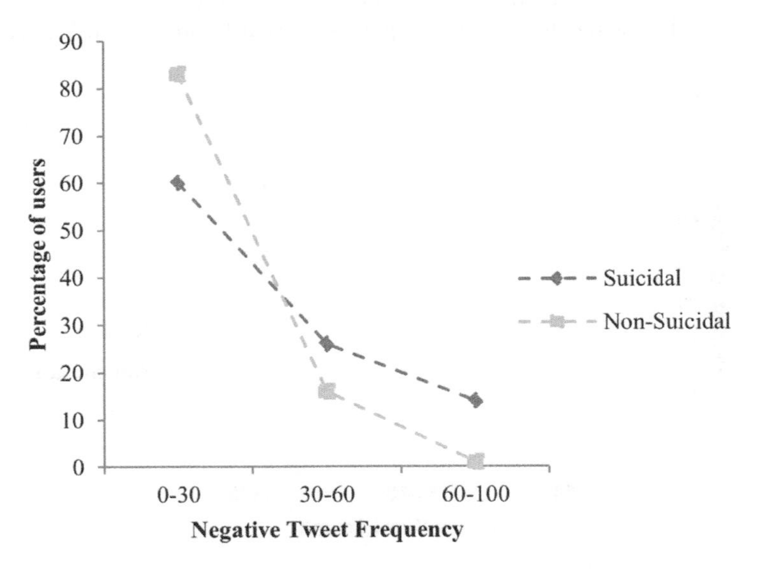

Figure 16.3 Frequency of negative tweets for suicidal and normal users.

k, where *k* is determined empirically. We tested with several numbers of topics for this study and decided that 20 was a reasonable amount based on the accuracy ratings given by the classifiers. By picking 20 themes, integrating them with additional variables, and using Logistic Regression as a classifier, we acquire an accuracy score of 87%.

N-gram and TF-IDF: With this approach, the impact of frequently occurring, less informative tokens is lessened by accounting for less frequent, more informative tokens. Tokens with a greater tf -idf value can be discovered only in one post and not in others. The stop words were initially deleted from the dataset for the investigation, and the term –document matrix was reduced to the most prevalent unigrams and bigrams. For each post category, only the list of 100 frequently unigrams and bigrams is picked. Negative emotions, moods, self-obsession, fury, hopelessness, hostility, interpersonal interactions, and the usage of the present tense are all associated with suicide ideation, according to the research. Lexicons with terminology such as prior events, social interactions, and family-oriented terms are included in the regular articles. Figure 16.4 depicts a word cloud of suicidal and non-suicidal data.

Temporal features: This feature helps us comprehend how people on social media post their content at various times of the day, for example, AM or PM as shown in Figure 16.5. The suicidal data collection has more instances of the value of AM than the non-suicidal data set, according to an analysis of the datasets. Suicidal and depressing thoughts are higher during the AM because of loneliness, break from work, absence of energy and differences in communications between light/darkness and the nervous system. Specifically, in the case of users with suicidal tendencies, 73% were found to be more active between 6 pm and 6 am, while 27% was shown to be more involved during

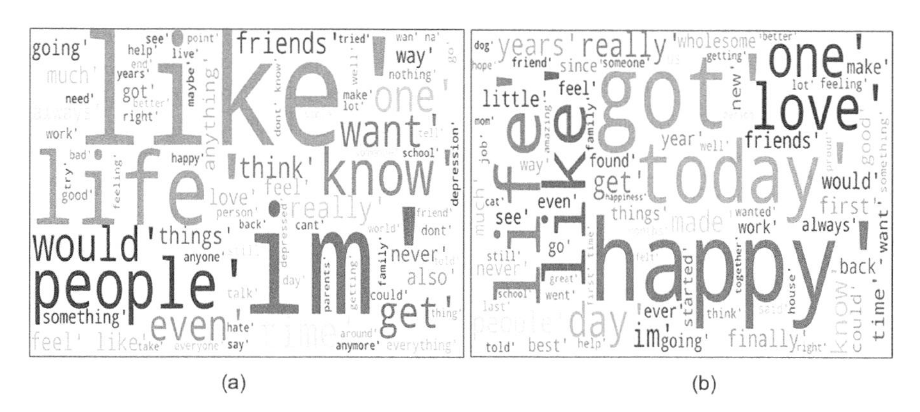

Figure 16.4 Word cloud. (a) Suicidal data. (b) Non-suicidal data.

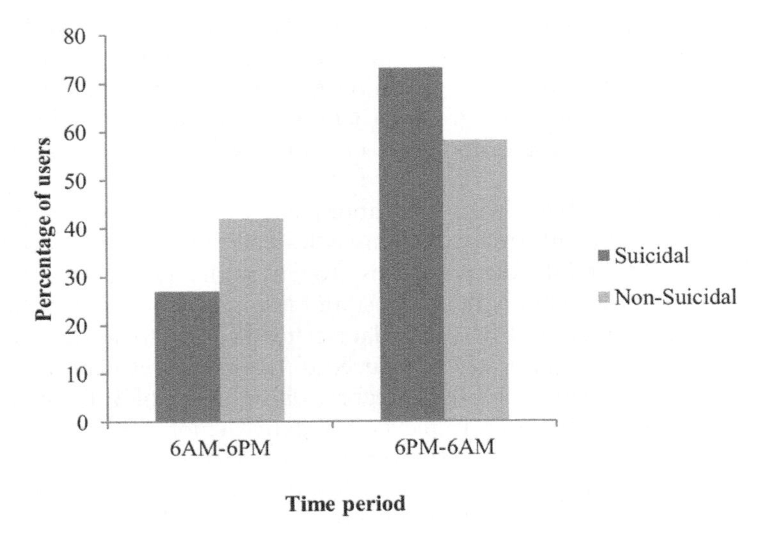

Figure 16.5 Temporal analysis for suicidal and non-suicidal posts.

the day. In the case of users who did not have suicidal tendencies, 58% were active during the night, and 42% during the day. According to another research, just 6% of users had suicidal thoughts between 12 and 6 pm, compared to 21% of users who didn't have suicidal thoughts.

16.4 EMPIRICAL EVALUATION

16.4.1 Classification model

Suicidal ideation detection on social media is a classification issue requiring supervised learning. A supervised classification model is trained to learn the function $y_i=F(x_i)$, where $y_i=1$ implies suicidal content and $y_i=0$ suggests otherwise, given a data set of tweets $x_i=1....n$ and their labels $y_i=1...n$. For the suicidal ideation classification problem, we used four classical learning algorithms. Logistic Regression, Random Forest, Support Vector Machine, and XGBoost are among the classifiers used in the model's development.

16.4.2 Classification results

The purpose of this study is to determine if a user is depressed and has suicidal thoughts by analyzing their comments. The whole dimension feature space acquired from the dataset is first subjected to text classification algorithms. We used four major classifiers to understand the significance of the different features retrieved from the dataset. Each classifier used all of the feature categories. Evaluation metrics are used to assess the performance of the aforementioned strategies. It can be computed in four different ways: True positive (TP): the cases of suicide ideation that are positive and predicted as positive. True negative (TN): the cases of suicide ideation that are negative and predicted as negative. False positive (FP): the suicide-ideation cases that are actually negative, but predicted as positive. False negative (FN): the suicide-ideation cases that are actually positive but predicted as negative. This study considers the following assessment metrics: (i) accuracy, (ii) recall, (iii) precision, and (iv) F-measure. The information in the confusion matrix, which contains the predicted outcome for each test sample, is used to generate these measures. The evaluation metrics are defined as follows [20]:

Accuracy is the proportion of the outcomes that are correctly predicted.

$$\text{Accuracy} = \frac{\text{TP} + \text{TN}}{\text{TP} + \text{TN} + \text{FP} + \text{FN}} \tag{16.1}$$

Precision is the proportion of true positive to the cases that are predicted as positive.

$$\text{Precision} = \frac{\text{TP}}{\text{TP} + \text{FP}} \tag{16.2}$$

Recall is the percentage of true positives to the cases that are actually positive.

$$\text{Recall} = \frac{\text{TP}}{\text{TP} + \text{FN}} \tag{16.3}$$

F-Measure is the average of precision and recall. F-Measure is calculated as follows:

$$F\text{-Measure} = 2.\frac{\text{Precision} * \text{Recall}}{\text{Precision} + \text{Recall}} \tag{16.4}$$

The studies are evaluated using 10-fold cross validation on all levels of test data. We examined the efficiency of the linguistic characteristics using classifiers and observed a boost in prediction. LR classifier gives an accuracy of 76% and an F1-score of 0.75 using the LDA feature. The accuracy of Trigram+TF-IDF with SVM classifier is 74%, and the F1-score is 0.73.

When using SVM, temporal analysis as a single feature offers a 74% accuracy rate. As a single feature, Sentiment analysis works well. Using the XGBoost model, it has a 78% accuracy (78, 0.78). The SVM model obtains 88% accuracy (88, 0.87) when used in conjunction with other features. We aimed to understand which combination of linguistic, statistical, topic, temporal, and emotional characteristics worked best in terms of suicide-ideation tweet classification accuracy. The linguistic feature was combined with the temporal features and sentiment analysis. In our study, integrating all of the features yielded the greatest performance for detecting suicidal thoughts. It beats other feature combinations, including the LDA+ Trigram+TF-IDF+Tweet Statistics (87%, 0.87), and LDA+ Trigram+TF-IDF+ SA+EA (87%, 0.87). The SVM is the best classifier in most circumstances, followed by LR and XGBoost. Table 16.1 contains a full examination of the findings for combined features.

Table 16.1 Machine learning classifier performance metrics based on statistical, linguistics, emoticons, sentiment and temporal features

	Logistic regression	*Random forest*	*Support vector machine*	*XG boost*
Trigram+TF-IDF				
Accuracy	.74	.63	**.74**	.71
Precision	.73	.80	**.73**	.70
Recall	.74	.53	**.73**	.70
F1	.74	.44	**.73**	.70
Cohen Kappa	.72	.71	**.72**	.70
LDA				
Accuracy	**.76**	.61	.70	.71
Precision	**.75**	.30	.70	.69
Recall	**.75**	.50	.70	.69
F1	**.75**	.38	.70	.71
Cohen Kappa	**.81**	.47	.69	.68
Tweet Statistics				
Accuracy	.70	.65	.70	**.71**
Precision	.69	.57	.70	**.71**
Recall	.71	.65	.70	**.71**
F1	.70	.59	.70	**.71**
Cohen Kappa	.72	.52	.74	**.69**
Temporal Features				
Accuracy	.72	.66	**.74**	.71
Precision	.71	.80	**.73**	.70
Recall	.72	.56	**.74**	.71
F1	.72	.51	**.74**	.71
Cohen Kappa	.76	.57	**.67**	.71
Sentiment Analysis (SA)				
Accuracy	.75	.69	.74	**.78**
Precision	.72	.68	.75	**.77**

(Continued)

Table 16.1 (Continued) Machine learning classifier performance metrics based on statistical, linguistics, emoticons, sentiment and temporal features

	Logistic regression	*Random forest*	*Support vector machine*	*XG boost*
Recall	.71	.69	.77	**.72**
F1	.73	.68	.75	**.78**
Cohen Kappa	.73	.476	.67	**.68**
Emoticons Analysis (EA)				
Accuracy	**.77**	.65	.75	.73
Precision	**.75**	.72	.73	.76
Recall	**.76**	.56	.74	.73
F1	**.77**	.52	.75	.77
Cohen Kappa	**.82**	.817	.74	.68
LDA+Trigram+TF-IDF+SA+EA				
Accuracy	**.86**	.61	.70	.71
Precision	**.85**	.30	.70	.69
Recall	**.85**	.50	.70	.69
F1	**.85**	.38	.70	.71
Cohen Kappa	**.64**	.53	.70	.65
LDA+Trigram+TF-IDF+Tweet Statistics				
Accuracy	.85	.61	**.87**	.81
Precision	.84	.81	**.86**	.80
Recall	.84	.50	**.87**	.79
F1	.85	.39	**.87**	.80
Cohen Kappa	.76	.74	**.69**	.78
LDA+ TF-IDF +Trigram+ +SA+EA+ Temporal features +Tweet Statistics				
Accuracy	.83	.63	**.88**	.80
Precision	.84	.81	**.86**	.80
Recall	.84	.50	**.87**	.79
F1	.84	.69	**.87**	.81
Cohen Kappa	.69	.43	**.58**	.71

The bold values signify the chosen (best fit) combinations

16.5 CONCLUSION AND FUTURE SCOPE

The purpose of this research is to gain a better knowledge of the public's perspective of suicide by examining suicide-ideation materials published by people online. We collected and analyzed data from Twitter to better understand suicidal intentions and behavior. Several features from the data set are retrieved, including linguistic, topic, sentiment analysis, TF-IDF, and temporal, and it is demonstrated experimentally that all of the features are effective in distinguishing suicidal posters from regular ones. It may be concluded from this research that correct feature selection can lead to high prediction performance.

Despite our proposed method's good success, there is a lot of room for more advancement and research in this area. Although this study analyzes temporal data, we have simply divided the day into two slots (6 AM-PM, 6 PM-AM). We intend to divide the day into several time slots to understand more about the frequency of time of the user posting and how it links to suicidal ideation. Furthermore, additional features such as word embedding may increase the accuracy of detecting suicidal thoughts. We hope that our findings will serve as a basis for future research in this area.

REFERENCES

1. World Health Organization. One in 100 deaths is by suicide, https://www.who.int/news/item/17-06-2021-one-in-100-deaths-is-by-suicide, last accessed 2021/10/18.
2. Dholariya, P. Suicidal tendency in youth in relation to their gender (2017). doi: 10.13140/RG.2.2.23058.35523.
3. Pandey, A.R., Bista, B., Dhungana, R.R., Aryal, K.K., Chalise, B., Dhimal, M. Factors associated with suicidal ideation and suicidal attempts among adolescent students in Nepal: Findings from global school-based students health survey. *PLoS ONE* 14(4): e0210383 (2019). doi: 10.1371/journal.pone.0210383.
4. Nock, M.K., Borges, G., Bromet, E.J., Cha, C.B., Kessler, R.C, Lee, S. Suicide and suicidal behavior. *Epidemiologic Reviews* 30(10): 133–54 (2008). doi: 10.1093/epirev/mxn002.
5. Bilsen, J. Suicide and youth: Risk factors, 9: 540 (2018). doi: 10.3389/fpsyt.2018.00540. PMID: 30425663; PMCID: PMC6218408.
6. Klonsky, E.D., May, A.M., Saffer, B.Y. Suicide attempts, and suicidal ideation. *Annu Rev Clin Psychol* 12: 307–330 (2016). doi: 10.1146/annurev-clinpsy-021815-093204. Epub. PMID: 26772209.
7. Fernandes, A., Dutta, R., Velupillai, S. Identifying suicide ideation and suicidal attempts in a psychiatric clinical research database using natural language processing. *Sci Rep* 8: 7426 (2018). doi: 10.1038/s41598-018-25773-2.
8. Nock, M.K., Borges, G., Bromet, E.J., Cha, C.B., Kessler, R.C., Sing, L. Suicide and suicidal behavior. *Epidemiologic Reviews* 30: 133–154 (2018). doi: 10.1093/epirev/mxn002.
9. Ji, S., Pan, S., Li, X., Cambria, E., Long, G., Huang, Z. Suicidal ideation detection: A review of machine learning methods and applications. *IEEE Transactions on Computational Social Systems* 8(1): 214–226 (2020).
10. Cook, B.L., Progovac, A.M., Chen, P., Mullin, B., Hou, S., Baca-Garcia, E. Novel use of natural language processing (NLP) to predict suicidal ideation and psychiatric symptoms in a text-based mental health intervention in Madrid. *Computational and Mathematical Methods in Medicine* 2016: 8708434 (2016). PMID: 27752278; PMCID: PMC5056245.

11. Mbarek, A., Jamoussi, S., Charfi, A., Hamadou, A.B. Suicidal profiles detection in twitter. 289–296 (2019). doi: 10.5220/0008167602890296.
12. Ji, S., Yu, C.P., Fung, S., Pan, S., Long, G. Supervised learning for suicidal ideation detection in online user content. *Complexity* 1–10. doi: 10.1155/2018/6157249 (2018).
13. Ramírez-Cifuentes, D., Freire, A., Baeza-Yates, R., Puntí, J., Medina-Bravo P., Velazquez D.A., Gonfaus, J.M., Gonzàlez, J. Detection of suicidal ideation on social media: Multimodal, relational, and behavioral analysis. *Journal of Medical Internet Research* 22(7): e17758 (2020). doi: 10.2196/17758.
14. O'Dea, B., Batterham, P.J., Calear, A.L., Paris, C., Chsistensen, H. Detecting suicidality on twitter. *Internet Interventions* 2(2): 183–188 (2015).
15. Shen, G., Jia, J., Nie, L., Feng, F., Zhang, C., Hu, T., Chua, T., S., Zhu, W. Depression detection via harvesting social media: A multimodal dictionary learning solution. *Proceedings of the Twenty-Sixth International Joint Conference on Artificial Intelligence (IJCAI-17)*, Melbourne, (2017).
16. Kumar, A., Sharma, A., Arora, A. Anxious depression prediction in real-time social data. *Proceeding of International Conference on Advanced Engineering, Science, Management and Technology – 2019 (ICAESMT19)*, Dehradun (2019).
17. Lora, S.K., Sakib, N., Antora, S.A., Jahan, N. A comparative study to detect emotions from tweets analyzing machine learning and deep learning techniques. *International Journal of Applied Information Systems* 12(30): 6–12 (2020).
18. Rao, T.S.M., Kompalli, V.S., kompalli, U. A model to detect social network mental disorders using AI techniques. *Journal of Critical Reviews* 7(15): 2582–2587 (2020). ISSN-2394-5125.
19. Chatterjee, M., Samanta, P., Kumar, P., Sarkar, D. Suicide ideation detection using multiple feature analysis from twitter data. *2022 IEEE Delhi Section Conference (DELCON)* (pp. 1–6). New Delhi, IEEE (2022).
20. Kumar, P., Samanta, P., Dutta, S., Chatterjee, M., Sarkar, D. Feature based depression detection from twitter data using machine learning techniques. *Journal of Scientific Research* 66(2), 220–228 (2022).

Chapter 17

Analyzing the role of Indian media during the second wave of COVID using topic modeling

Anchal N G, Akshay Sriram, Jitin Jain Mathew, Lakshmi Shankar Iyer, and Tripti Mahara
Christ University

CONTENTS

17.1 INTRODUCTION

COVID-19, one of the most severe pandemics the world has seen in a century, was first reported in an unexpected outbreak in Wuhan, China. The disease that started as a local epidemic took no time to evolve into a pandemic. Most of the countries responded swiftly with nationwide lockdowns to prevent the outbreak of disease. The period from March 2020 to December 2020 was termed as the first wave of Coronavirus outbreak. In India, the first wave peaked in September 2020 with approximately 95,000 active cases per day (Ranjan et al., 2021). By the end of December 2020, India saw a steady decline in the number of cases and by mid of January 2021, the reported cases were less than 20,000 per day (NetDesk, 2021). With numbers declining rapidly while many countries were going through the second wave, some felt that India had averted the second wave. With this, a fake sense of normalcy started prevailing throughout the country. All stakeholders including the government started violating COVID appropriate behavior protocols. The effects were seen soon as India started observing the increase in the number of cases after February 11, 2021. This was nearly 6 months after the first peak was observed in September 2020. This was the signal of the second wave in the country. On April 19, 2021, the number of

DOI: 10.1201/9781003381167-17

reported new cases was approximately 0.3 million which was already three times the values in the first peak. Unlike the first wave, the current spread of the virus mutants reached remote locations and entire families got infected. Moreover, more people in younger age groups of 30–50 years, who go out to work, were found to be contracting the virus. The second wave also saw a spate of reinfections and much more side effects as compared to the first wave. Everywhere there was uncertainty and panic.

In such a situation the role of the media, which is considered as the fourth pillar of democracy (Kumar & Singh, 2019) along with The executive, The Legislature and The Judiciary as the three main pillars of Indian constitution (Aladia, 2017) cannot be undermined. Media presents accurate, relevant, and latest information about the important events taking place within and across the country covering a wide range of perspectives. It also helps in forming opinions on various issues along with the capability to make a judgment. Thus, any country's social, economic, political, and international affairs are largely reported and influenced by the media.

Media can be classified into print media, electronic media, and social media. Print media is the oldest media comprising newspapers, magazines, and journals. It is considered as the media for the masses because of its wide reach. Newspapers are one of the cheapest forms of print media with a wide distribution network. A newspaper plays a vital role in disseminating information about events to the public. Traditional newspapers have a new makeover in terms of electronic or e-newspaper, which is a self-contained and reusable version of traditional newspaper (Panda & Swain, 2011). The audience for e-newspaper is seeing rapid growth because of internet penetration and portability (Ahlers, 2006). This portability factor is regarded as one of the prominent reasons for the shift in news consumption habits (Mudgal & Rana, 2020).

In the COVID pandemic, the role and contribution of e-newspapers cannot be undermined. As everyone was stuck at home due to either lockdown or safety reasons, the e-newspapers became one of the sources to connect with the outside world. This is also supported by the fact that the audience for online news sources is increasing at a rapid pace because of huge internet penetration (Mudgal & Rana, 2020). The role of mass media in addressing public health issues cannot be underestimated, and the researchers wanted to find the role those Indian newspapers played during the second wave of COVID. This became the motivation of the study, and to find the answers, the following objectives were identified:

Q1: What are the major topics of discussion in the media during the second wave of COVID in India related to the subject?

Q2: What are the trends in topic identification during the period of the study?

The analysis was done using the LDA (Latent Dirichlet Allocation) algorithm to identify topics that were discussed in the newspaper. LDA is an unsupervised technique used in Text Analytics that clusters the document based on the contents to discover relevant topics. The structure of this chapter is as follows. Section 17.2 consists of a Literature Survey followed by Methodology in Section 17.3. Results and discussion are presented in Section 17.4 followed by Conclusion in Section 17.5.

17.2 LITERATURE SURVEY

Considerable work in different domains is going on to address various issues associated with COVID-19 pandemic. Application of Text Analytics to understand the sentiments of various stakeholders and Topic Modeling to understand the topics related to a particular situation are the areas that are extensively being researched during these tough times. For instance, Liu et al. (2020), Patel et al. (2021), and Wan et al. (2021) understood the role of media by analyzing the news articles in China, Canada, Brazil, and America, respectively during the initial outbreak of COVID. Song et al. (2021) developed a model to categorize disinformation during COVID. The main aim of this model was to segregate disinformation to prevent confusion among citizens and reduce distrust in policymakers due to false information.

Similarly, Ghasiya and Okamura (2021) analyzed a database of approximately 100,000 COVID-19 news articles published across four countries: the UK, Japan, South Korea, and India to understand the most common published themes. Ng et al. (2021) analyzed news media narratives across 20 countries related to pandemic to reveal that the narratives were on the themes: Pre-pandemic, Early, Peak, and Recovery. Amara et al. (2021) and Jang et al. (2021) analyzed the content posted on social media platforms like Facebook and Twitter to demonstrate that extracted topics depict the chronological sequence of events during the pandemic and that trends of topic are related to public health promotions and interventions.

LDA is one of the most utilized techniques for topic modeling and is used in this research. It is an unsupervised machine learning algorithm and therefore does not require any prior information about the input data. It is a generative probabilistic model, which is often used to find relevant topics in a set of documents. Each document usually comprises a mix of latent topics. Each word in the document is assigned to a topic. Every topic is a bag of words. The number of topics needs to be defined in the beginning by the researcher. These topics depict the common occurrence of words in several documents that were provided as the input.

The LDA model, as shown in Figure 17.1, has three levels of representation namely corpus, document, and word levels (Blei et al., 2003).

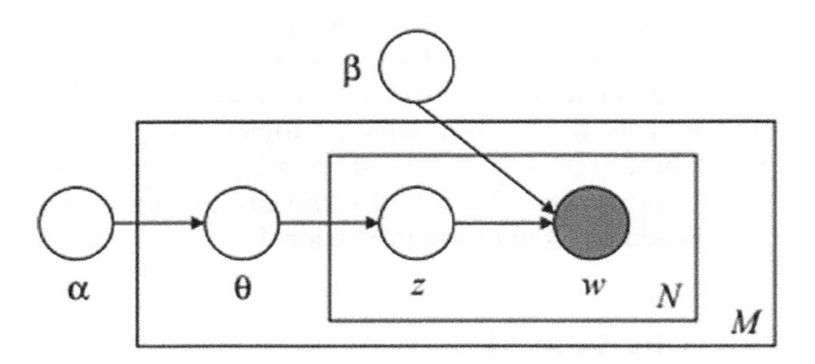

Figure 17.1 LDA model.

The parameters are defined at the corpus level, document level, and word level. α and β are corpus-level parameters. They are assumed to be sampled once while generating a corpus. The variable θ is a document-level variable. It is sampled once per document. Z and w are word-level variables. They are sampled once in each document for each word.

Using the value of corpus-level parameters (α, β) that are defined in the beginning, θ and z are probabilistically estimated. Finally, the word 'w' is estimated by selecting a topic using topic distribution in a document and a word through a word distribution within a topic. In brief, LDA works as follows:

1. The parameters, as mentioned above, are initialized along with the number of topics 'k'.
2. Each word is assigned to a particular topic randomly using Dirichlet distribution.
3. The above steps are repeated for all the words in the corpus.

17.3 METHODOLOGY

This section discusses the details of the methodology used. It consists of data collection, data pre-processing, and topic modeling.

17.3.1 Data collection

News articles are one of the important sources of secondary data. A total of 4902 related to COVID-19 were extracted from prominent Indian e-newspapers such as *The Times of India*, *Hindustan Times*, *The Indian Express*, *The Hindu*, and *The Deccan Chronicle* collected from March 15, 2021 to May 31, 2021 as the second wave emerged during March–May 2021 in India. Web scraping code written in Python was used to scrap the data. Keywords like "COVID", "COVID second wave in India", "Pandemic second wave in India" were used to extract the articles. The articles were collated in the excel sheet. A cell in an excel sheet consisted of a newspaper article. Along with content, the date of publication was also captured in the excel sheet to facilitate trend analysis.

17.3.2 Data pre-processing

Data pre-processing was done before applying LDA in Python. The data processing consisted of three steps. In the first step, the stop words are removed, and the next section includes removing punctuations, white spaces, and numbers from the articles. The main reason for removing the numbers is that it does not add any value to the text. In the next step, the document is converted to lowercase. Lemmatization is performed to convert a word into its root form, preserving the meaning of the root word. For example, *caring* will become *care*. Lemmatization is preferred over stemming as stemming could result in the root word with no proper meaning. For example, *caring* would become *car* which is not correct. After completing the pre-processing, tokens are generated. The document term "matrix" is generated using tf-idf technique. The document term "matrix" is used as the input to the LDA algorithm. Figure 17.2 depicts the steps of data pre-processing.

17.3.3 Topic modeling

Gensim package of Python is used to perform topic modeling, and pyLDAvis is used for visualization purposes. There are two hyperparameters: alpha and theta. For each document, alpha controls the prior distribution over topic weights, whereas eta is the parameter of prior distribution over word weights in each topic. Default values of these parameters are selected while modeling. LDA is used to extract keywords for the number of topics defined. For the newspapers, each article is taken as a document for analysis, and 5–14 topics were extracted and analyzed. Coherence value was used to select the number of topics. The highest coherence score of 0.645 was obtained for nine taking into account the interpretability and meaning.

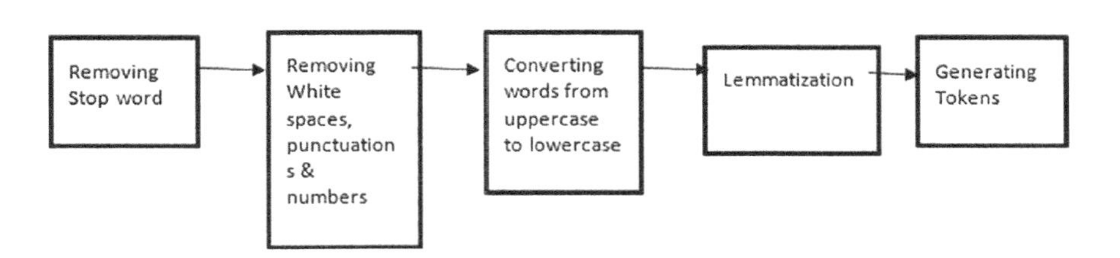

Figure 17.2 Data pre-processing.

After setting the number of topics to nine, the top ten most frequent words were extracted for each topic using the collapsed Gibbs sampling method for LDA.

17.4 RESULTS AND DISCUSSION

Table 17.1 depicts the top ten keywords and the associated topics that emerged when LDA was applied. These topics range from the role of decision-making authorities to vaccine production, drive and trials along with discussion on various issues related to resources. For example, the topic bed scam neither confirms nor rejects the situation's reality. However, it depicts how Tejasvi Surya, a BJP Member of Parliament, exposed a bed allotment scam in Bangalore, Karnataka. The bed allotment fraud, on the other hand, was highly relevant, but the masterminds behind it remain a mystery, according to the topic extracted. This topic mainly focuses on the bed scam. Similarly, the topic COVID crisis dimensions deal with various crises that were relevant during the second wave.

Figure 17.3 depicts the count of articles over a period of the data collection, which indicates the most discussed topics. Figure 17.4 depicts the number of articles in a bubble format. Figure 17.5 depicts the trend of the articles published topic-wise.

The results show that for the topic "Covid Crisis Dimension", the number of articles published week-wise shows a rising trend from Mid-March to the end of April. This was the time when the crisis was at its peak, and there were various reasons that led to it, and hence the trend is rising in this period. There is not much variation in terms of the articles published on the topic "Decision making Authorities" as they were always involved in the process and were discussed. The period from the last week of April saw a tremendous increase in demand for oxygen especially in two states such as Delhi and Maharashtra. In addition, oxygen was not available to meet the demand. Thus, this duration saw a rising trend for the articles published for the topic "Medical Oxygen Logistics". The vaccine trials started earlier, but as the articles are taken from March 1, 2021, we still observe news about the topic "Vaccine trials", but there is no pattern observed. World across after successful trials of the various vaccines were conducted, and in parallel, the pharma industry started the production. Also, in the second wave, there was vaccine shortage, and hence we see a trend in the number of articles published on the topic "Vaccine Production". On the other hand, the vaccine drive for health line workers began on January 16, 2021, whereas for the common mass, senior citizens began in April 2021, and hence we observe a trend for the articles related to "Vaccine Drive" during the complete period. The number of cases started increasing sharply in the first week of April resulting in shortage of various resources. This led to a rise

Table 17.1 Keywords and associated topics

COVID crisis dimension	*Decision-making authorities*	*Medical oxygen logistics*	*Vaccine drive*	*Vaccine trials*	*Vaccine production*	*Resource constraints*	*Recovery of resources*	*Bed scam*
Assistance	PM	Production	Health workers	Immune	Serum Institute	Delhi	vial	court
People	CM	Tanker	Drive	Clinical	Raw material	cylinder	fake	bed
Bed	Central Government	Railway	Beneficiary	Ffficacy	Capacity	bed	accuse	BBMP
Cylinder	Ministry	Demand	Private	Phase	Bharat Biotech	center	Remdesivir	War room
Family	Commissioner	Lmo (Liquid medical oxygen)	Hospital	Variant	technology	health	find	Surya
Doctor	Police	Express	Arogya Setu	Testing	Patent	shortage	medicine	Scam
Death	ICMR	Transport	Age group	Safety	manufacturing	Oxygen concentrators	cylinder	Bench
Ambulance	Officer	Capacity	Covaxin	Develop	Vials	private	recover	South Bangalore
ICU	Hospital	Steel Plant	Stock	Antibody	Trade	vaccine	seize	Raid
Life	High court	Gas	Covishield	participant	Poonawala	need	crime	Navneet Kalra

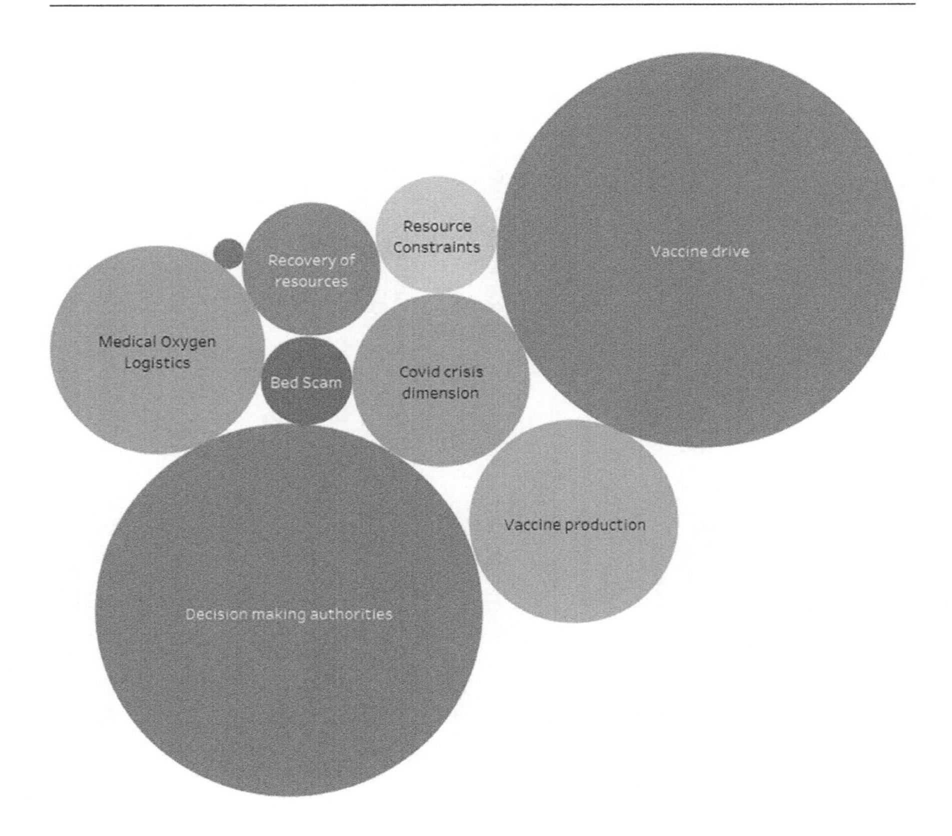

Figure 17.3 Topic-wise count of articles.

in the number of articles related to the topic "Resource constraints". Also, hoarding and black marketing of oxygen, drugs, and injections started at the same time resulting in "Bed Scam" related articles and "Recovery of Resources".

The analysis of the newspaper articles published in the prominent Indian newspapers during the second wave of COVID in India depicts that most of the articles focused on reporting the ground reality news. It includes news on topics like shortages of various resources, reporting about the scam, and the role of the decision-making authorities. The role of the mass media is not only limited to reporting but also to create awareness and educate the masses about various current happenings. As the pandemic was one of its type and the number of cases increased drastically from the first wave and there were large casualties as compared to the first wave, the newspapers should have demonstrated empathy by also publishing articles about well-being, mental health, and managing the crisis in a better way. The people went through a long period of lockdowns and social isolation. They relied

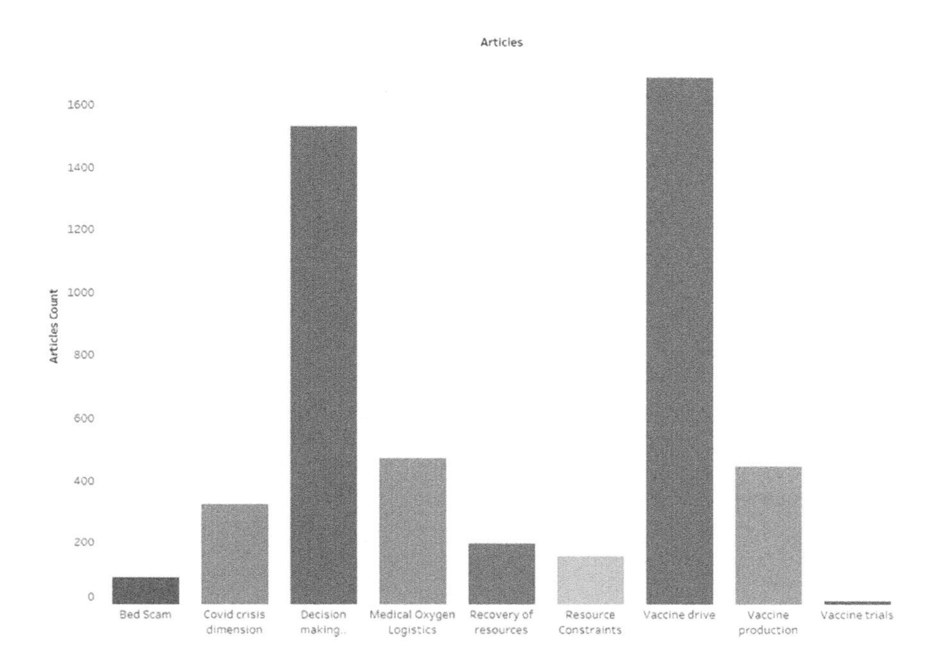

Figure 17.4 Topic-wise articles.

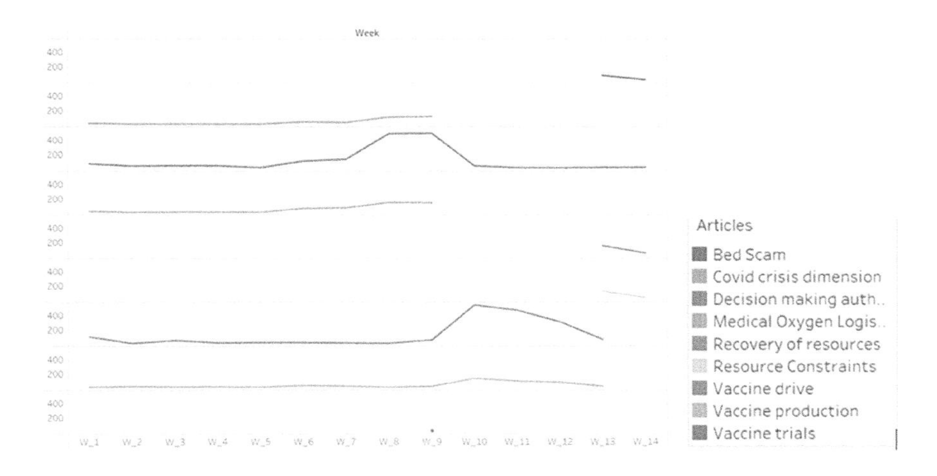

Figure 17.5 Weekly trend.

on news sources available from mass media like newspapers and TV channels. This reliance could have been channelized to make the life of people more sorted and understand various facets of the pandemic.

17.5 CONCLUSION

The research used Topic Modeling to extract important topics from the news published during the second wave of COVID in India. The results suggest nine topics with major focus on vaccination, resources, decision-making authorities, and crisis. Other than these results, this study also helps in understanding the focus of the newspapers during such a crucial period. As observed, most of the news were on the current situation with very negligible importance to other issues like improving mental health, coping with loss and stress, expert opinions, etc. that would have helped in easing the situation. Even though the study has analyzed 4,902 articles published across various Indian newspapers, it is not possible to cover all the aspects of such a wide topic. Also, the process of labeling the topics is manual and highly depends on the researcher's understanding and knowledge. The future scope of work also includes performing sentiment analysis on the news articles to understand the orientation of the articles and comparing the articles from different countries during the second wave to understand the role of media.

REFERENCES

Ahlers, D., (2006), News consumption and the new electronic media, *The International Journal of Press/Politics*, 11(1), 29–52, doi: 10.1177/1081180X05284317.

Amara, A., Hadj Taieb, M.A. & Ben Aouicha, M., (2021), Multilingual topic modeling for tracking COVID-19 trends based on Facebook data analysis. *Applied Intelligence*, 51, 3052–3073, doi: 10.1007/s10489-020-02033-3.

Aladia, L.L.M., (2017). Role of legislature and judiciary in present Indian scenario, *Journal of Emerging Technologies and Innovative Research*, 4(1), 681–686.

Blei, D.M., Ng, A., & Jordan, M., (2003), Latent dirichlet allocation, *Journal of Machine Learning Research*, 3, 993–1022.

Ghasiya, P., & Okamura, K., (2021), *Investigating COVID-19 News Across Four Nations A Topic Modeling and Sentiment Analysis Approach*, IEEE Access, Australia

Jang, H., Rempel, E., Roth, D., Carenini, G., & Janjua, N.Z., (2021), Tracking COVID-19 discourse on twitter in North America infodemiology study using topic modeling and aspect-based sentiment analysis, *Journal of Medical Internet Research*, 23(2), 1–12.

Kumar, P., & Singh, K., (2019), Media, the fourth pillar of democracy a critical analysis, *International Journal of Research and Analytical Reviews*, 6(1).

Liu, Q., Zheng, Z., Zheng, J., Chen, Q., Liu, G., Chen, S., Chu, B., Zhu, H., Akinwunmi, B., Huang, J., Zhang, C.J.P., & Ming, W., (2020), Health communication through news media during the early stage of the covid-19 outbreak in china digital topic modeling approach, *Journal of Medical Internet Research*, 22(4), doi: 10.2196/19118.

Mudgal, R., & Rana, P., (2020), Future of print and e-newspaper in India a critique, *International Journal of Multidisciplinary Research*, 6(5).
NetDesk., (2021), *Coronavirus Updates*. Hindu, May2021. Accessed in July 2021.
Ng, R., Chow, T.Y.J., & Yang, W., (2021), News media narratives of COvid-19 across 20 countries early global convergence and later regional divergence, *PLOS ONE*, doi: 10.1371/journal.pone.0256358.
Panda, K.C., & Swain, D., (2011), E-newspapers and e-news services in the electronic age an appraisal, *Annals of Library and Information Studies*, 58, 55–62.
Patel, J., Desai, H., & Okhowat, A., (2021), The role of the Canadian media during the initial response to the COVID-19 pandemic a topic modelling approach using Canadian broadcasting corporation news articles, *JMIR Infodemiology*, 1(1).
Ranjan, R., Sharma, A., & Verma, M.K., (2021), Characterization of the second wave of COVID-19 in India, *medRxiv*, doi: 10.1101/2021.04.17.21255665.
Song, X., Petrak, J., Jiang Y., Singh, I., Maynard, D., & Bontcheva, K., (2021), Classification aware neural topic model for COVID-19 disinformation categorisation, *PLOS ONE*, doi: 10.1371/journal.pone.0247086.
Wan, X., Lucic, M.C., Ghazzai, H., & Massoud, Y., (2021), Topic modeling and progression of American digital news media during the onset of the COVID-19 pandemic, *IEEE Transactions on Technology and Society*, doi: 10.1109/TTS.2021.3088800.

Chapter 18

Hardware-efficient FIR filter design using fast converging flower pollination algorithm – a case study of denoising PCG Signal

Poulami Das
Department of Computer Engineering
K.C. College of Engineering and Management Studies and Research

Avishek Ray
Department of Electronics and Telecommunication Engineering
K.C. College of Engineering and Management Studies and Research

CONTENTS

18.1 INTRODUCTION

In the present digital age of electronic appliances, dealing with signals has become a part of everyday modern life. During transmission via any medium, signals get affected by unwanted components, which is adverse but inevitable. Elimination of such unwanted components, termed as *noise*, from transmitted signals proved to be an important as well as puzzling task for the researchers from the initial days of signal processing. Among a significant number of

DOI: 10.1201/9781003381167-18

techniques proposed for the removal of noise from signals, the use of digital filters (Sharma, 2009; Mitra, 2013) has proven effective in several ways. The design of digital filters involves obtaining a perfect set of coefficients using programmable optimization algorithms.

Digital filters can be classified into two types: finite impulse response (FIR) filter (Sharma, 2009; Mitra, 2013) and infinite impulse response (IIR) filter (Sharma, 2009; Mitra, 2013). Parallel to the windowing method (Kaiser, 1966; Harris, 1978; Sharma, 2009) and frequency sampling method (Gold & Jordan, 1969), FIR filters can also be implemented using optimized filter coefficients (McClellan & Parks, 1972; Reddy & Sahoo, 2015). Filter coefficient optimization is considered as a problem for minimizing errors in passband and stopband. Error in the frequency bands states deviancy between designed and ideal filter responses.

Heuristic algorithms have found its application in filter coefficient optimization. However, this class of algorithms has the lacunae in determining the local optimal solutions due to convergence speed; to overcome this, "metaheuristic algorithms" (Yang, 2011) were proposed. Genetic algorithm (GA) (Thapar et al., 2012; Yang, 2014), ant colony optimization (ACO) (Tsutsumi & Suyama, 2014; Sasahara & Suyama, 2015), particle swarm optimization (PSO) (Yang, 2014), BAT algorithm (Yang, 2010, 2014), cuckoo search algorithm (CSA) (Yang, 2008; Yang & Deb, 2009, 2010), and flower pollination algorithm (FPA) (Yang 2013, 2014) are some nature inspired metaheuristic algorithms.

In the proposed technique, advancement on a metaheuristic algorithm FPA (Yang, 2014) is performed by updating the replacement approach of the worse nests, resulting in an innovative algorithm referred to as fast converging flower pollination algorithm (FFPA). This algorithm outpaces basic GA, PSO, CSA, and FPA based on convergence time. In the proposed technique, stated in this chapter, both FPA and FFPA have been used for obtaining optimized filter coefficients for implementing FIR filters. Responses of the designed filters are now close to the ideal filter response with cost of the adders kept at minimum. Lessening the number of adders denotes to the decrement in SPT terms.

18.2 PROBLEM FORMULATIONS

FIR filters stated in Equation (18.1),

$$H(z) = \Sigma h(n) z^{-n},\ n = 0, 1, \ldots, N \tag{18.1}$$

N is filter order and $h(n)$ denotes set of $(N+1)$ filter coefficients (Sharma, 2009). For FIR filters with symmetricity property, $h(n)$ stated by Equation (18.2).

$$h(n) = h(N-1-n), \quad n = 0,1,\dots,N-1 \tag{18.2}$$

For optimization of coefficients of symmetric filters, $(N/2)+1$ coefficients must be taken into consideration in case asymmetric filters all the $N+1$ coefficients get optimized. Based on filter order and symmetricity of filter coefficients categorizes the filters into following four types. Both Type I and Type II filters have symmetric coefficients but even and odd order, respectively.

18.2.1 Specification of desired filter characteristics

Filter design initiates with the specification of filter characteristics such as passband frequency (ω_p), stopband frequency (ω_s), passband ripple (δ_p), and stopband ripple (δ_s). The magnitude response has a peak deviation of δ_p in passband and maximum deviation of δ_s in stopband. Minimum stopband attenuation (A_s), maximum and minimum passband attenuation A_{p_1} and A_{p_2} respectively can be expressed in dB using the following equations:

$$A_s = -20\log_{10}\delta_s \tag{18.3}$$

$$A_{p_1} = 20\log_{10}(1+\delta_p) \tag{18.4}$$

$$A_{p_2} = 20\log_{10}(1-\delta_p) \tag{18.5}$$

18.2.2 Filter coefficients computation

Conventional equripple method of filter coefficient optimization uses a technique to minimize errors in passband and stopband (Aggarwal et al., 2013). The error function is given as (Singh & Josan, 2014),

$$E(\omega) = W(\omega)\left[H_d\left(e^{j\omega}\right) - H_a(e^{j\omega})\right] \tag{18.6}$$

$H_d\left(e^{j\omega}\right)$ and $H_a(e^{j\omega})$ are the ideal and estimated frequency response of the filter. Ideal frequency response of a lowpass filter is stated as in Equation (18.7) (Aggarwal et al., 2013; Singh & Josan, 2014).

$$\begin{aligned} H_d(e^{j\omega}) &= 1 \qquad \text{for } 0 \le \omega \le \omega_p \\ &= 0 \qquad \text{for } \omega_s \le \omega \le \pi \end{aligned} \tag{18.7}$$

$W(\omega)$ has control over error minimization in both the frequency bands. Using the key concept of the equiripple methodology, McClellan and Parks (1972) proposed an efficient algorithm for optimal filter design. Limitation of this strategy is the fixed value of $\frac{\delta_p}{\delta_s}$. To overcome this constraint, a mean square error (Dhabal & Sengupta, 2015)-based objective function has been adopted (shown in Equation 18.8.)

$$\varphi = \mu E_p + (1-\mu) E_s;\ 0 < \mu < 1 \tag{18.8}$$

For lowpass filter, E_p and Es are defined in Equations (18.9 and 18.10), respectively.

$$E_p = \frac{1}{\pi} \int_0^{\omega_p} \left(1 - H(\omega) - \delta_p\right)^2 d\omega \tag{18.9}$$

$$E_s = \frac{1}{\pi} \int_{\omega_s}^{\pi} \left(0 - H(\omega) - \delta_s\right)^2 d\omega \tag{18.10}$$

H denotes magnitude response of approximated filter. φ is stated as weighted sum of mean square errors E_p and E_s. Minimization of φ enhances performance of approximated filters.

18.2.3 Design filter architecture

Basic steps of FIR filter architecture implementation are multiplication and accumulation of filter coefficients with the input discrete time signal. Multiplier less filter circuits can be realized by representing coefficients as sums or differences of signed-power-two (SPT) terms (Solank, 2012). Transposed direct form of FIR filter stated in Equation (18.11),

$$Y(n) = \sum_{k=0}^{M-1} C_k X(n-k) \tag{18.11}$$

where $X(n)$ stands for input to the filter of order N, $C_0, C_1, \ldots, C_{M-1}$ represent filter coefficients, length of the filter is M (M=N+1) and $Y(n)$ states filter output.

Among several techniques for FIR filter design direct form and transposed direct form structures are shown below (Reddy, 2015) (Figures 18.1 and 18.2):

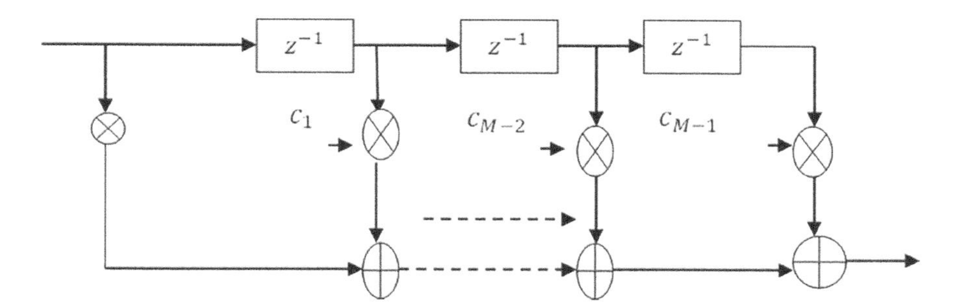

Figure 18.1 Direct form structure.

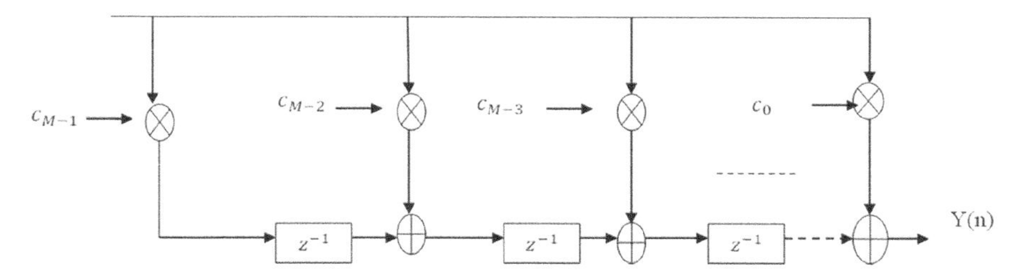

Figure 18.2 Transposed direct form structure.

18.3 HARDWARE-EFFICIENT FIR FILTER DESIGN USING FAST CONVERGING FLOWER POLLINATION ALGORITHM

18.3.1 Flower pollination algorithm

The FPA have proven its superiority in many aspects (Yang, 2013, 2014). Four basic rules are needed to be followed to implement the algorithm: (i) Biotic and cross-pollination are considered as global pollination, Pollen-carrying pollinator moves obeying Lévy flights. (ii) Abiotic pollination and self-pollination are used for local pollination. (iii) Pollinators are mainly insects, and they are responsible for development of flower constancy, which is equivalent to a reproduction probability that is proportional to the similarity of two flowers involved. (iv) Switching between global and local pollination can be controlled by a switch parameter $p \in [0,1]$, slightly biased toward local pollination.

Flower constancy can be described by the following equation.

$$x_i^{t+1} = x_i^t + \gamma L(\lambda)\left(c_{best} - x_i^t\right) \tag{18.12}$$

x_i^t stands for solution vector x_i at iteration t, c_{best} states current best solution. Scaling factor γ controls step size. $L(\lambda)$ is step size based on Lévy flights. L is generated by Lévy distribution. Local pollination is performed using Equation (18.13):

$$x_i^{t+1} = x_i^t + \epsilon\left(x_j^t - x_k^t\right) \tag{18.13}$$

If x_j^t and x_k^t come from same spices that means are selected from the same population, it is equivalent to a local random walk if ϵ is drawn from a uniform distribution in $[0,1]$.

18.3.2 Fast converging flower pollination algorithm

In FPA local random walk is performed using Eqn. 18.14, whereas in FFPA local random walk is performed using the following Eqn.

$$x_i^{t+1} = x_i^t + \epsilon\left(C_{best} - x_k^t\right) \tag{18.14}$$

In case of FFPA instead of using permute1 function, *CBEST* i.e. best nest till the latest iteration is used. This specialty of FFPA keeps the selection pressure toward the better solutions, thereby ensuring better results in fewer iterations. Moreover, this advancement of the algorithm does not flood the population on high fitness solutions.

The flowchart in Figure 18.4 briefs the implementation of a FIR filter using FFPA.

Following filter parameters order of the filter (N), word length of filter coefficients (B), passband edge frequency (ω_p), stopband edge frequency (ω_s), passband ripple (δ_p), stopband ripple (δ_s) and algorithm parameters objective function, Total number of flowers (n), Size of each flower (candidate solution), Total no. of iteration (MaxIteration), Upper bound (U_B) and Lower bound (L_B) are used as the input to the algorithm. In the flowchart, t refers to the current generation. The algorithm starts with the step of generating the initial population where each candidate solution represents a set of filter coefficients. As we are concerned about the symmetric filters, size of each candidate solution of the population is $K = \frac{N+1}{2}$. Initial population is generated randomly but within a specific range demarcated by lower and upper bounds, UL and UB respectively.

$$x_{j,i} = \text{rand}_j(0,1).\left(UB_j - UL_j\right) + UL_j \tag{18.15}$$

$rand_j(0,1)$ returns a uniformly distributed random number within range $(0,1)$ i.e. $0 < rand_j(0,1) < 1$. Generated number is multiplied with $\left(UB_j - UL_j\right)$

and then added to UL_j to obtain a number between UL_j & UB_j. The subscript j signifies that a new random value is generated for each element of a single candidate solution. Quality of each solution in the initial population is evaluated using a mean square error based objective function defined in Equation 18.8. For each solution, scaling factor (sf) is computed. Scaling factor is the ratio of 2^B to the maximum valued coefficient. Quantization of the coefficients is performed by multiplying them by 2^B and sf. Adder cost is estimated to implement filter using each solution in the population after performing common sub expression (CSE) elimination (Reddy & Sahoo, 2015). Implementation of filters requires structural adders and multiplier adders (Reddy & Sahoo, 2015; Das et al., 2018). Number of structural adders is estimated by computing the total number of additions required to obtain filter coefficients but after CSE elimination (Reddy & Sahoo, 2015; Das et al., 2018). Number of multiplier adders is same as the total number of SPT terms performed to represent the filter coefficients. Solution with the least value of φ is then stored in BEST. In the next step, a switching parameter $p \in [0,1]$ is chosen. For values of t less than MaxIteration new solutions are generated by local distribution or global distribution and solutions of the last population are replaced by the new better solutions. For each new solution, required number of adders is estimated to implement a filter. Solution with minimum φ value is stored in CURRENT_BEST. BEST is updated after each of the iterations if local best solution obtains better filter than the global best solution in other words it can be said that CURRENT_BEST is less than BEST. Algorithm terminates after maximum number of iterations occurred. Finally, an optimized set of filter coefficients is received from the algorithm. Optimized filter coefficients are capable of implementing a symmetric FIR filter with minimum adder cost (Figure 18.3).

18.4 SIMULATION RESULTS

In our research work 0.25π and 0.39π are used as the normalized pass band and stop band edge frequencies, respectively. Stop band attenuation is taken as 30. Pass band and stop band both are having ripple of 0.4. To obtain a benchmark filter again Parks-McClellan algorithm (PMA) (McClellan & Parks, 1972) is used. As we are concerned about the symmetric filters and the filter order is specified at 20, size of each candidate solution of the initial population of FFPA must be 11. After receiving an optimized solution as the output of the algorithm, using symmetricity property of the coefficients a complete set of filter coefficients can be obtained. Figure 18.4a contains magnitude response of different filters implemented using the coefficients optimized by flower pollination algorithm (FPA) but with different population size and specific number of total iterations of 500.

Figure 18.4a–h

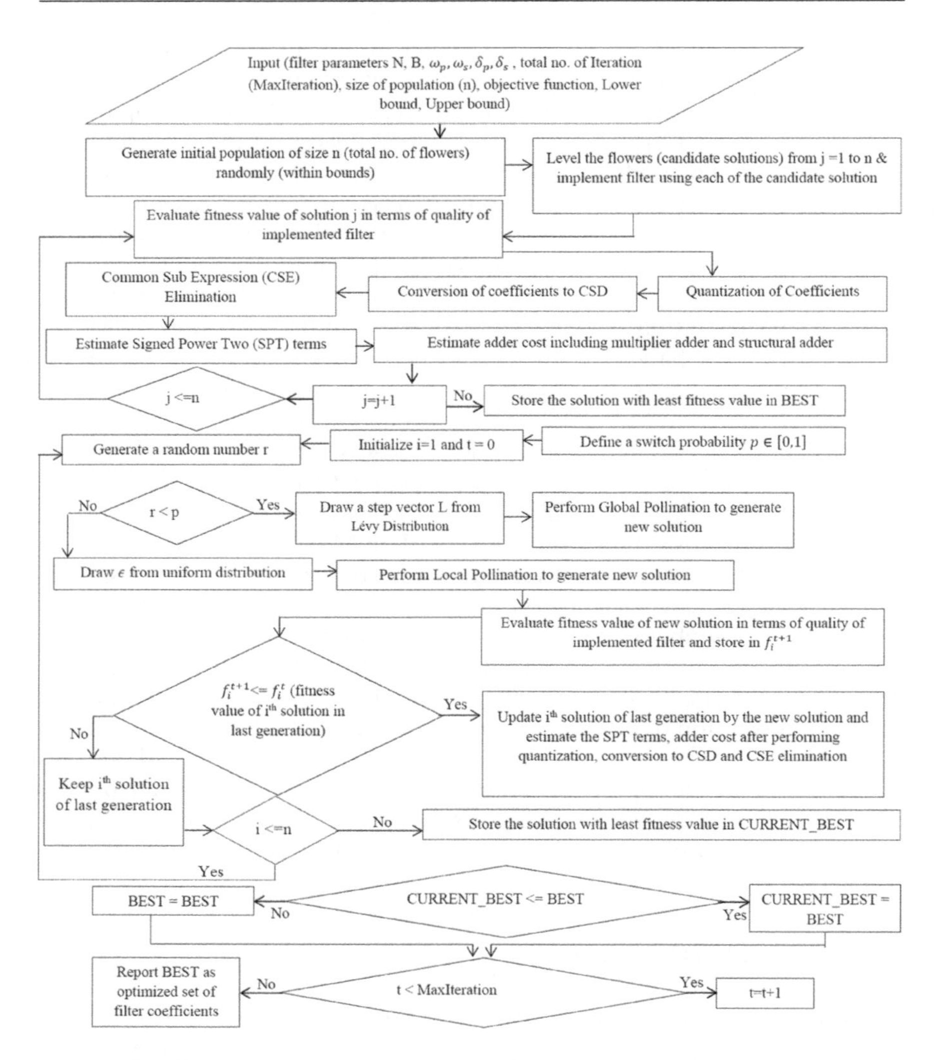

Figure 18.3 Filter design using fast converging flower pollination algorithm.

Requirement of adders in high quantitates adds to the filter cost. Adder cost of a filter can be reduced either by decreasing order of filter or by reducing number of nonzero bits present in the filter coefficients. Magnitude response of four different filters FFPA1, FFPA2, FFPA3, and FFPA4 are shown in Figure 18.5. FFPA1, FFPA2, and FFPA3 filters are implemented using the optimized coefficients obtained by FFPA with iterations 3000, population 20 but with different word length of coefficients. From Table 18.1 it can be observed that FFPA1 uses coefficients of word length 10 and obtains stop band attenuation of 37.61 but requires 16 adders and 23 SPT terms to

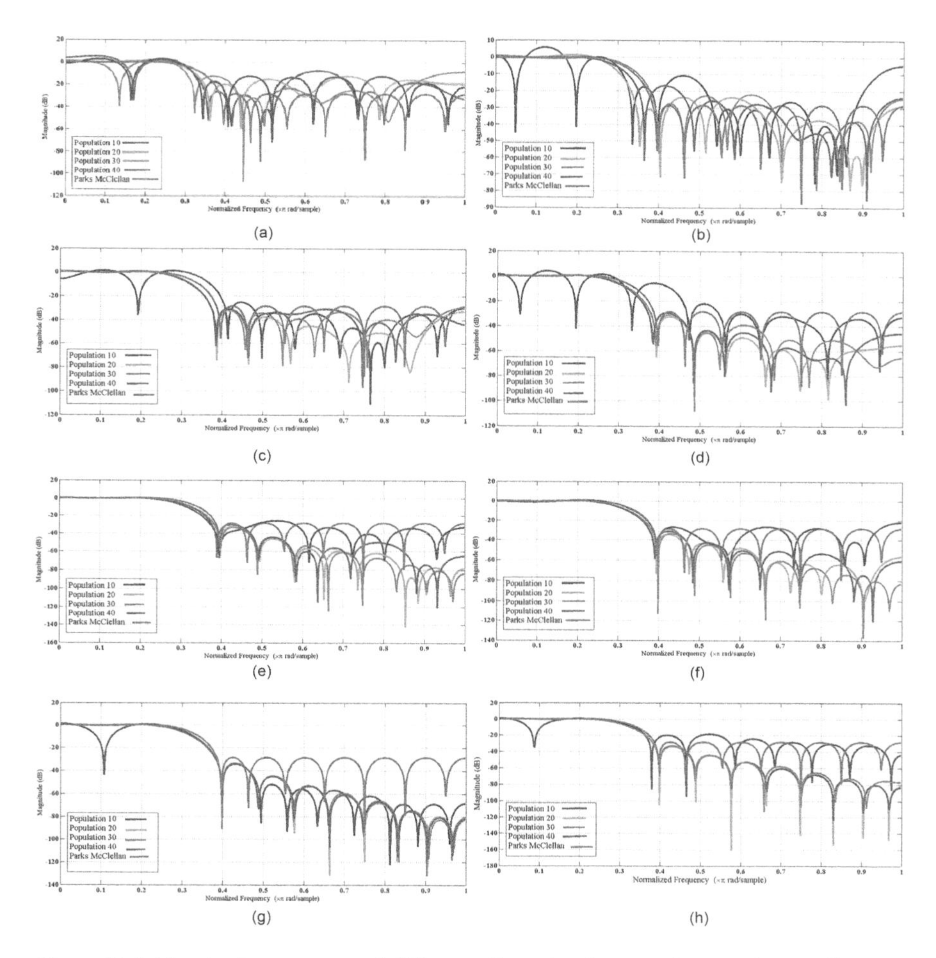

Figure 18.4 Magnitude response of different filters implemented using the coefficients optimized by flower pollination algorithm but with different population size and total number of iterations. (a) 500. (b) 1000. (c) 1500. (d) 2000. (e) 2500. (f) 3000. (g) 3500. (h) 4000.

be designed. Whereas by reducing the word length of the coefficients to 9 FFPA2 obtains stop band attenuation of 34.23 with only 12 adders and 18 SPT terms.

A comparative study of attenuation in stopband and adder costs of the filters implemented using optimized coefficients obtained by FFPA as well as the filters designed using the optimized coefficients obtained by different traditional optimization algorithms like GA, PSO, Differential Evolution (DE) (Karaboga & Cetinkaya, 2006; Chandra & Chattopadhyay, 2012) and CSA is performed in Table 18.1. Magnitude Response and Frequency Response of these filters are shown in Figure 18.6.

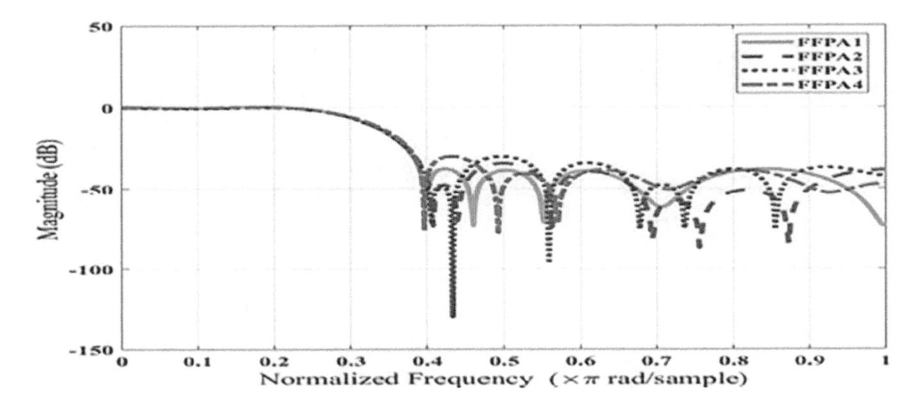

Figure 18.5 Magnitude response of filters implemented by the optimized coefficients obtained using FFPA1, FFPA2, FFPA3, and FFPA4.

Table 18.1 Comparative study of different properties of the designed filters

Filters	*N*	*WL*	*Asb (dB)*	*SPT*	*SPT gain (%)*	*MA*	*SA*	*TA*	*TA gain (%)*
PM	20	10	28.58	37		8	13	21	
GA	20	10	32.34	30	21.875	9	12	21	0
DE	20	10	32.86	32	15.625	7	10	17	19.0476
PSO	20	10	35.03	28	28.125	7	10	17	19.0476
CSA	20	10	35.70	28	28.125	7	10	17	19.0476
FPA1	20	10	37.61	23	37.8378	6	10	16	23.8095
FPA2	20	9	34.23	18	51.3513	6	6	12	42.8571
FPA3	20	8	30.10	14	62.1621	6	6	12	42.8571
FPA4	18	10	30.15	28	24.3243	7	10	15	28.5714

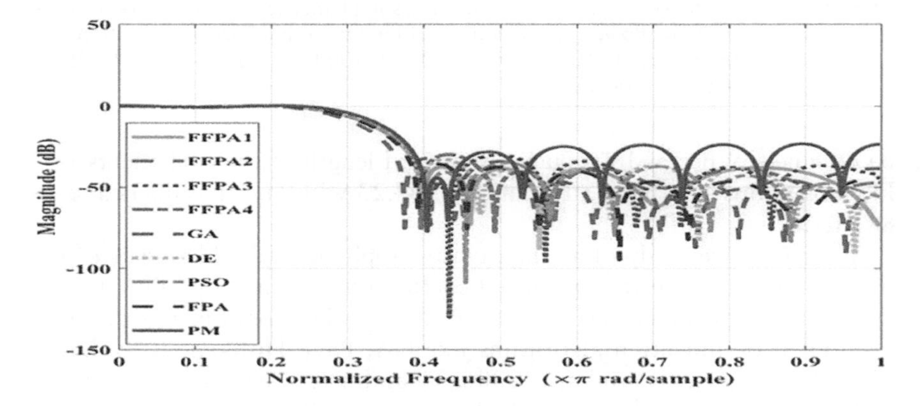

Figure 18.6 Comparison of magnitude response of filters designed by the optimized coefficients.

18.5 CASE STUDY

Present case study of our research work includes filtration of a noisy Phonocardiogram

(PCG) Signal. For proper diagnosis of diseases, any biomedical signal is needed to be noise free. In Phonocardiography high-fidelity recording of the sounds and murmurs made by the heart during a cardiac cycle caused by flow of blood through the heart are plotted using machine named phonocardiograph (Ganguly & Sharma, 2017). Usually heart produces the sound Lub & Dub, where Lub is the first sound S1 and S2 is the second sound Dub. The time between S1 and S2 is systole (Lub-------Dub), caused by the flow of blood from the heart to the lungs and body, flow of blood across the Pulmonic and Aortic valves (ausmed, 2018). For analysis a noise free PCG signal of a 40 years old human being negatively diagnosed with any heart disease is collected from Jeevan Rekha Diagnostic Pvt. Ltd., India is collected and then mixed with Gaussian white noise. Noisy PCG signal is then filtered using the implemented filters FFPA1, FFPA2, FFPA3, and FFPA4 (Table 18.2).

In Figure 18.7ag, frequency spectra of the original PCG signal, noisy PCG signal, and the filtered PCG signal are shown. Analysis of Figure 18.7a proves the presence of S1 and S2 in lower frequency range. Figure 18.7b shows the presence of noise in higher frequency range also. Figure 18.7c–g proves the efficiencies of different filters in filtering noise from higher frequency range. From Figure 18.8a–e, it can be observed that the filters implemented by PMA have reduced a lesser amount of noise in the frequency ranges of 7,000–7,500 Hz and 8,000–9,000 Hz compared to the filters designed using the coefficients optimized by FPA. In Figure 18.9a and b, column chart representations of SNR and correlation values of the filtered signals are shown, respectively. In Table 18.2, a comparative study of average error calculated by the following Equation (18.16) (Hamza Cherif et al., 2014) for different filters used to filter noisy PCG signal is shown.

Table 18.2 Comparative study of SNR, correlation value and average error (ε_{avg}) of filtered phonocardiogram signal filtered using different designed filters

Filters	*SNR*	*Correlation value*	ε_{avg}
PM	11.0577	0.9663	0.0324
FFPA1	11.3342	0.9763	0.0313
FFPA2	11.4646	0.9764	0.0308
FFPA3	11.6489	0.9767	0.0301
FFPA4	12.0718	0.9894	0.0297

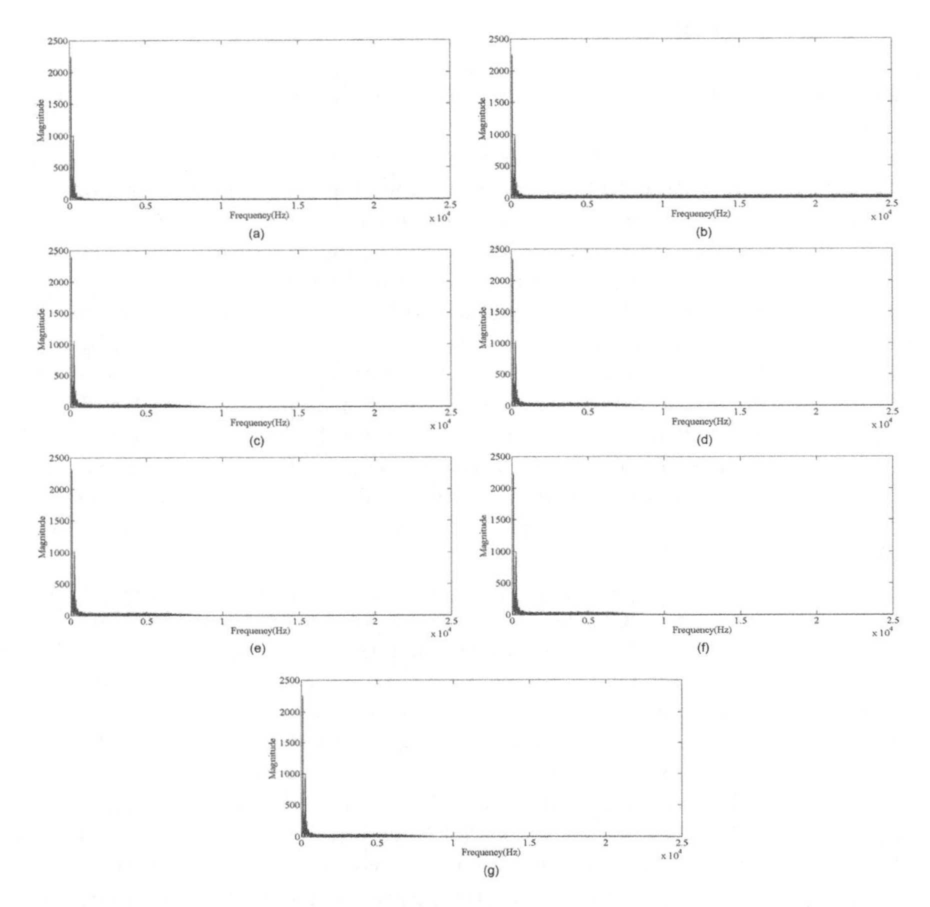

Figure 18.7 Frequency spectrum of (a) Original PCG signal. (b) Noisy PCG signal. (c) PCG signal filtered by PMA. (d) PCG signal filtered by FPA1. (e) PCG signal filtered by FPA2. (f) PCG signal filtered by FPA3. (g) PCG signal filtered by FPA4.

$$\varepsilon_{avg} = \frac{\sum_{i=1}^{N} \left| P_{oi} - P_{ri} \right|}{N} \tag{18.16}$$

p_o is the original PCG signal, p_{oi} is i^{th} sample of p_o. p_r is the synthesis PCG signal and p_{or} is i^{th} sample of p_r. It is proved that filters implemented using the coefficients optimized by FPA outpace the filter designed using PMA and also the IIR filters used in (Hamza Cherif et al., 2014) for filtering noisy PCG signal in terms of average error.

Figure 18.7a–g

Figure 18.8a–e

Figure 18.9a and b and Figure 18.10

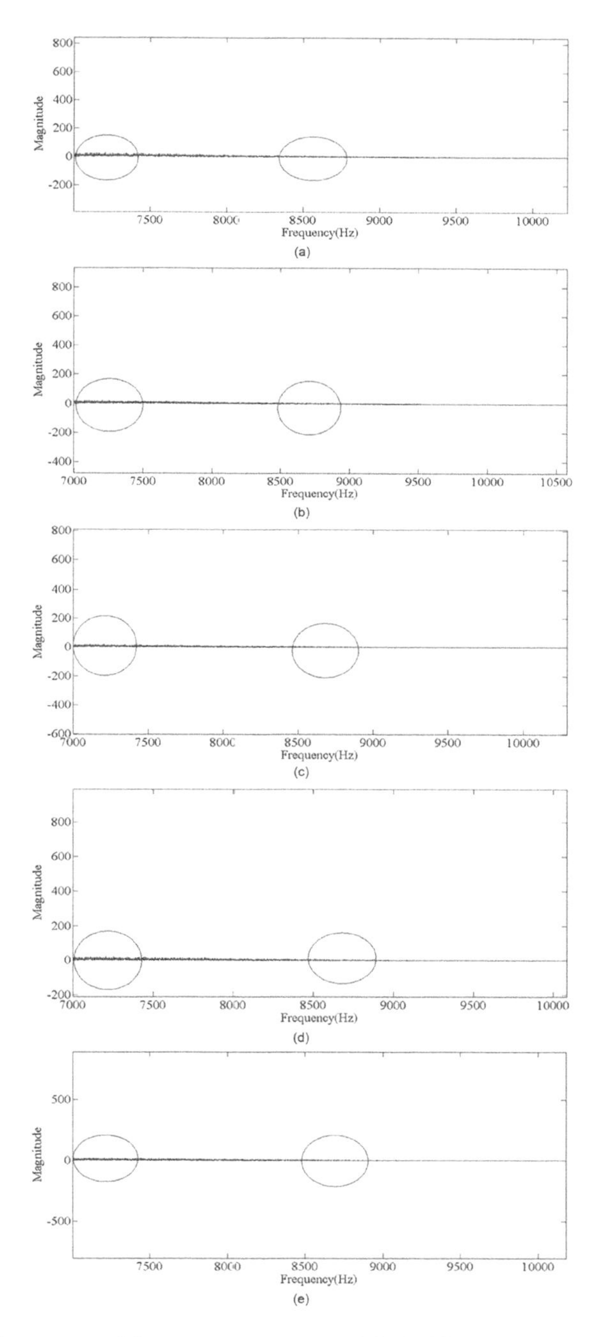

Figure 18.8 Near view of frequency spectrum of (a) PCG signal filtered by PMA. (b) PCG signal filtered by FPA1. (c) PCG signal filtered by FPA2. (d) PCG signal filtered by FPA3. (e) PCG signal filtered by FPA4.

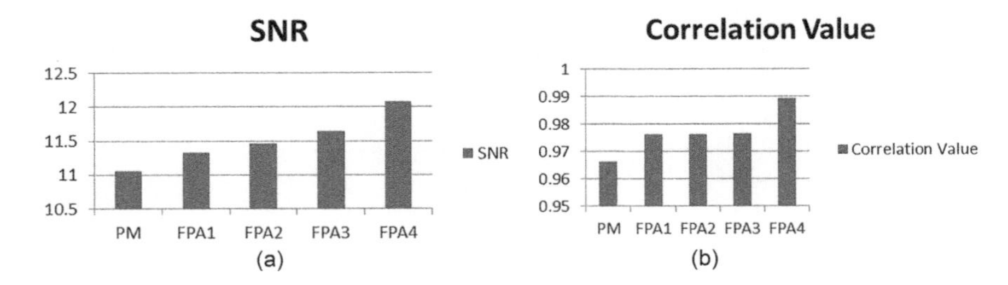

Figure 18.9 Column chart representation of (a) Comparison of SNR. (b) Correlation value of filtered PCG signals.

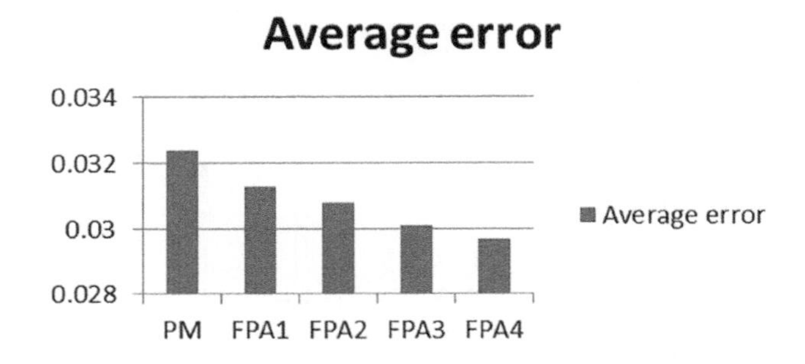

Figure 18.10 Column chart representation of average error of the filtered PCG signals.

18.6 CONCLUSIONS

This chapter proposes an approach to design FIR filter by optimizing filter coefficients by FFPA. Proper optimization of the filter coefficients leads to a reduction in number of adders in the circuit, thereby reducing the hardware cost. FFPAs also proved its efficacy in filter coefficient optimization in fewer generations compared to traditional flower pollination algorithms as well as other metaheuristic algorithms. Performance analysis of the designed hardware-efficient filters has been shown by filtering a noisy phonocardiogram signal.

BIBLIOGRAPHY

Aggarwal D., Reddy K. S. & Sahoo S. K. (2013). FIR filter design approach for reduced hardware with order optimization and coefficient quantization. *2013 International Conference on Intelligent Systems and Signal Processing (ISSP)*, Gujrat, India.

Ausmed. (2018). Retreived from https://www.ausmed.com/articles/heart-murmur-sounds/ (last accessed on 27th July 2022).

Chandra A. & Chattopadhyay S. (2012). Role of mutation strategies of differential evolution algorithm in designing hardware efficient multiplier-less low-pass FIR filter. *Journal of Multimedia*, 7: 353–363.

Das P., Naskar S. K. & Patra S. N. (2018). Hardware effiecient fir filter using global best steered quantum inspired cuckoo search algorithm. *Applied Soft Computing*, 71: 1–19.

Dhabal S. & Sengupta S. (2015). Efficient design of high pass FIR filter using quantum-behaved particle swarm optimization with weighted mean best position. *Third International Conference on Computer, Communication, Control and Information Technology*, Hooghly, IEEE, pp. 1–6.

Ganguly A. & Sharma M. (2017). Detection of pathological heart murmurs by feature extraction of phonocardiogram signals. *Journal of Applied and Advanced Research*, 2: 200–205.

Gold B. & Jordan K. (1969). A direct search procedure for designing finite duration impulse response filters. *Transaction on Audio and Electroacoustics*, 17: 33–36.

Hamza Cherif L., Mostafi M. & Debbal S. M. (2014). Digital filters in heart sound analysis. *International Journal of Clinical Medicine Research*, 1(3): 97–108.

Harris, F. J. (1978). On the use of windows for harmonic analysis with the discrete fourier transform. *Proceedings of the IEEE*, 66(1): 51–83.

Kaiser, J. F. (1966). Digital filters. In Kuo, F. F. & Kaiser, J. F. (eds.), *System Analysis by Digital Computer*, chap. 7. New York, Wiley.

Karaboga N. & Cetinkaya B. (2006). Design of digital FIR filters using differential evolution algorithm. *Circuits Systems Signal Processing*, 25: 649–660.

McClellan J. H. & Parks T. W. (1973). A unified approach to tf»e design of optimum FIR linear phase digital filters. *Transaction on Circuit Theory*, 20: 697–701.

Mitra S. K. (2013). *Digital Signal Processing - A Computer Based Approach* (3rd ed.). Mc Graw Hill, India.

Reddy K. S. & Sahoo S. K. (2015). An approach for FIR filter coefficient optimization using differential evolution algorithm. *International Journal of Electronics and Communications*, 69, 101–108.

Reddy S. K. (2015). High performance VLSI architecture for digital FIR filter design. Birla Institute of Technology & Science, Electrical & Electronics Engineering, Pilani.

Sasahara T. & Suyama K. (2015). An ACO approach for design of CSD coefficient FIR filters. *Asia-Pacific Signal and Information Processing Association Annual Summit and Conference (APSIPA)*, Hong Kong, IEEE, pp. 463–468.

Sharma S. (2009). *Theory and Design of Finite Impulse Response (FIR) Filters in Digital Signal Processing (with Matlab Programs)* (5th ed.). India: KATSON.

Singh T., Josan H. S. (2014). Design of low pass digital FIR filter using cuckoo search algorithm. *International Journal of Engineering Research and Applications*, 4: 72–77.

Solank S. (2012). Design FIR filter with signed power of two terms using MATLAB. *International Journal of Engineering & Technology* 1: 1–6.

Thapar S., Kaur P. & Aggarwal N. (2012). A low pass FIR filter design using genetic algorithm based artificial neural network. *International Journal of Computer Technology and Electronics Engineering*, 2: 99–103.

Tsutsumi S. & Suyama K. (2014). Design of FIR filters with discrete coefficients using ant colony optimization. *Electronics and Communications in Japan*, 132(7): 1066–1071.

Yang X-S. (2008). *Nature Inspired Meta-Heuristics Algorithm*. UK: Luniver Press.

Yang X-S. (2010). *A New Metaheuristic Bat-Inspired Algorithm*, https://arxiv.org/pdf/1004.4170.pdf (last accessed on 20 April 2018).

Yang X.-S. (2013). *Flower Pollination Algorithm for Global Optimization*, https://arxiv.org/pdf/1312.5673.pdf (last accessed on 22 April 2018).

Yang X-S. (2014). *Nature Inspired Optimization Algorithms* (1st ed.). London: Elsevier Insghits.

Yang X-S. (2011). Metaheuristic optimization: Algorithm analysis and open problems. In Pardalos, P. M. & Rebennack, S. (eds.). *Experimental Algorithms. SEA 2011, Lecture Notes in Computer Science*, Springer, Berlin, Heidelberg, vol. 6630, pp. 21–32.

Yang X-S. & Deb S. (2009). Cuckoo search via lévy flights. *World Congress on Nature & Biologically Inspired Computing*, Coimbatore, IEEE, pp. 210–214.

Yang X-S., & Deb S. (2010). Engineering optimization by cuckoo search. *International Journal of Mathematical Modeling and Numerical Optimization*, 1(4): 330–343.

Chapter 19

Voice recognition system using deep learning

B. K. Tripathy, Sudershan Sridhar, and Sharmila Banu K
VIT Vellore

CONTENTS

19.1 INTRODUCTION

Pattern recognition is the study of the manner in which machines use reasoning to learn and identify patterns of interest by observing the environmental background and categorizing them. The ability to understand speech depends on the quality of the pattern recognition algorithm used. A very attractive skill is the ability to understand speech using only visual information through a technique called lipreading. The applicability of this technique is in speech transcription where audio records are not available, like the archival silent films. It complements the audio understanding of speech. If there is no consistency between the audio and the lip movements, perception suffers greatly. This has necessitated rigorous research on lipreading over the past decade or so. Homophones put a constraint on the performance of lipreading. Homophones are words whose sounds are different, but the lip movements are similar. As a result, it is not possible to distinguish such words from their visual information, like lip movements. The other features that make lipreading a challenging problem are intraclass variations and adversarial imaging conditions. Intraclass variations such as accents, mumbling, speed of speaking, and conditions under adverse situations like motion, strong shadows, poor lighting, foreshortening, and resolution are difficult tasks [1].

DOI: 10.1201/9781003381167-19

A judicious increase in the number of hidden layers in an ANN [2] led to the development of deep neural networks (DNNs) [3]. Deep learning (DL) is a specific class of representational learning methods that use several levels of representations, like human beings [4]. DL has several applications [5]. Image processing and, in particular, their classification using DNN are presented in [6]. Following how artificial tongue visualized through computer vision and how artificial skin takes direction, it became necessary for computers to hear and speak as human beings do. To facilitate this audio signal processing, it is very much necessary [7]. DNNs have been extensively used for the development of healthcare systems [8]. Using the convolutional operator instead of the matrix operation used by standard feed-forward networks, convolutional neural networks (CNN) [9] were developed. There are several advantages to doing so. In the second step, bias is added to the convoluted image, followed by passing through so that the features of the image are detected. Next, the output is passed through a pooling function, which summarizes the information. Recurrent neural networks (RNN) are a specific type of DNN that saves the output of a specific layer and inputs it back to generate the output of that layer [10–13]. Long-short-term memory (LSTM) is a typical RNN that has feedback connections [14]. LSTM provides a convenient solution to the vanishing gradient and exploding gradient problems of RNN. The design of the architecture of the LSTM is such that it can model chronological sequences and their long-range dependencies more precisely than conventional RNNs.

In the context of word recognition, deep bottleneck features (DBF) are used to encode shallow input features such as linear discriminant analysis (LDA) and graphics interchange format (GIF). Similarly, DBF is used to encode the image for every frame, and the system is trained with an LSTM classifier to generate a word-level classification. Traditionally, Haar cascade is used to detect lips and then pass the model through CNN or LSTM-based model, which produces accuracy. In addition, accuracy can be improved by relatively reducing the mouth crop to the smallest possible size. Feature extraction plays a vital role in increasing accuracy [15–17]. In this chapter, we seek to identify the impact created by feature extraction in increasing final accuracy. In the next phase, most of the papers ensure that the entire clip is in sync with the audio channel; the results show the significance of fine-tuning the audio before training the model [18–20]. Not being careful about the selection of the optimum temporal window size harms the model. The analysis can be derived by plotting the word-vs-frame-count graph and choosing the right window size using the word boundary distribution [21]. After fine-tuning, pre-trained models like ImageNet and VGG16 models are used for better results. Removing the top layers from those huge models and combining them with layers of LSTM or Seq2seq encoder-decoder or Bidirectional LSTM (BLSTM) cells proves to be more efficient than training the model from scratch [15,18,19,22].

In this work, we seek to improve the accuracy by making tweaks at every phase. First, we employ Blazeface [23] and compare the results with MTCNN [22]. Second, we use silence detection techniques to split the coherent sentences and assign them to the corresponding word-clip pair. This technique makes the best use of the variable-length input for the models and gets the model more used to words of varied length. Finally, we use Bi-GRUs instead of the more complex Bi-LSTM to improve the generalization capability of the model.

19.2 RELATED WORKS

Lipreading has served as an alternative technique where the audio channel is unreliable [15]. However, recognizing the active speaker and speech labeling is also an alternative scope for using these techniques [20]. Our objective is to provide a scalable approach to comprehend words from continuous speeches, where the words are not separated but are impacted by what comes before and after them.

The prior works in lipreading are based on character-level recognition, where the model predicts characters based on frame-wise analysis called Connectionist Temporal Classification (CTC). The BLSTM-based models perform knowledge transfer and hence are preferred over traditional CTC methods. Based on the regions that activate the neurons of the NNs the most, the source is identified [24]. The important element to consider for architecture is the visual registration of lips. Alternatively, Haar cascades and HMM-based speech recognizers were very popular [25]. But recently, research has moved toward LSTM cells to establish more complex time-based relations and patterns. With high processing tensors and GPUs, more active object detection algorithms like MTCNN evolved [26]. As identifying words with longer pronunciation time require larger models with an enormous-sized dictionary, the research in [20] has been restricted to just 51 labels with variable lengths like "in" to "again". However, as with any word-level classification task, the setting is still distant from the real world, given that the word boundaries must be known beforehand [20].

19.3 DATASET

For this study, speech perception Lombard Grid, a bi-view audiovisual Lombard speech corpus, is used. The corpus includes 54 talkers, with 100 utterances per talker [27]. The statistics of this corpus include over 54 speakers, and each of them has a front recording of when they utter every word. They select their sentences from a pool of 51 words. The GRID corpus contains 51 different words in total. There is a uniform distribution

Table 19.1 Dataset label description

Words	*Category*
Bin, Lay, Place, Set	Command
Blue, Green, Red, White	Color
At, By, In, With	Preposition
A-Z (Excluding W)	Letter
0–9	Digit
Again, Now, Please, Soon	Adverb

among the six sentence parts. The transcript for the model is used as the file name for naming convention. The pattern is <code> <color> <preposition> <letter> <digit> <adverb>.

The speakers had the choice of using about four words for code: color, adverb, and preposition. But it is significantly less for the letters and digits. An unbalanced data set is used for training. For example, each color appears 240 times, whereas each letter from "a" to "z" appears 30 times [26]. By reading their respective sentence list in Lombard conditions and in plain together for 100 times, 100 utterances were produced by each talker. Five blocks of ten utterances were made from the collection of the utterances in each condition [26]. Each file consists of the speakers saying around six words on average about 2 seconds to run the clip. That gives us about 60 frames to extract this information. If we stick to splitting the clips using the conventional 15 frames for every word, the accuracy would be affected. In this paper, we use silence detection for identifying the split between these words. After we get segregated clips for all the respective words, we arrange them in a folder based on the word uttered and then shuffle them before we pass it to training. This mimics a real word scenario where the model should be trained to identify words with overlapping or in presence of other words. Dataset label description is provided in Table 19.1.

19.4 METHODS

We use a pipeline that essentially results in lipreading. In a single clip, the speaker says six words, and the clip names are named based on the transcript. In the GRID corpus, this process is omitted, because the sync is already observed. After this stage, we use a precise face detection algorithm that can actively track and understand how the movement occurs. We must reduce the unnecessary regions from being passed. Similarly, we need to reduce the bounding box for the lips so that the model is fed with more meaningful data to comprehend and make predictions. After the lip tracking is done, we align and mark the region required for tracking each word. This is different from other single clip learning because the crop will partially contain the state of the previous word so that the model differentiates between a no-change start versus a word uttered before it. The silence detection algorithm is used to align and crop videos. After we have managed to clip all the videos for the 51 labels, we then shuffle the data based on speakers. We preserve some of the clips for testing, both seen and unseen speakers. This filtering will help us make inferences on the model if the prediction is depending on whether the model has seen the speaker or not. We then shuffle the data and then feed it to a deep learning model. We perform transfer learning on VGG based on ImageNet weights and then pass the model to BGRU and BLSTM models to draw results. The pipeline of actions to train the model is provided in Figure 19.1.

19.4.1 Detection and fine-tuning

Multiple methods can be used for the detection of the face and cropping down to the mouth as each of the methods provides unique result. More accurate and stable cropping of the face leads to better accuracy. Haar cascades are generally less computationally expensive and faster but the crop is not accurate enough. As the Haar face cascade neither gave out an accurate boundary or coordinates of the mouth we had to find new methods to work on. The e MTCNN [26] queued up for too long. For 5,200 videos, the crop and saving of all the frames took about 52 hours. However, the use of Blazeface [24] reduced the computation time to 6 hours. Blazeface was selected instead of MTCNN (although it provided coordinates for both the left and right parts of the mouth), because of the computation power required. The accuracy was close to MTCNN and hence was used for results. In addition, Blazeface is applicable for real-time testing. Mostly, the mouth coordinates are fixed for the entire clip but here we run the model on every alternative frame and keep the alignment continued till the end. This ensures that there are no sudden movements and the lip is always in the center of the frame. A comparative analysis of these models is provided in Figure 19.2.

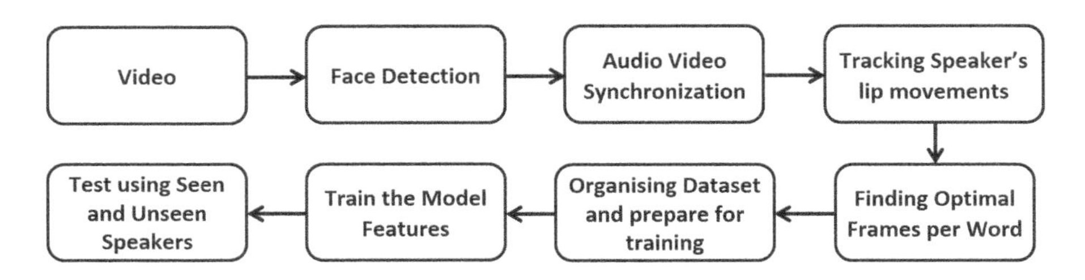

Figure 19.1 Pipeline to train the model.

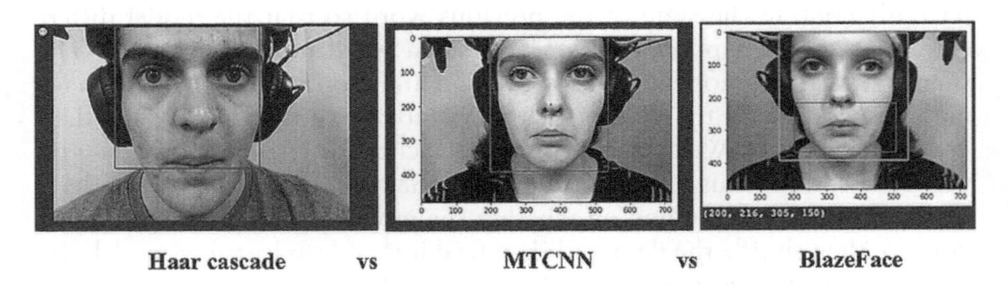

Figure 19.2 Comparison among Haar Cascade, MTCNN, and Blazeface.

19.4.2 Finding the optimal window size

Each file consists of the speakers saying around six words on average about 2 seconds to run the model on. That gives us about 60 frames to extract this information. This brings down the window size to about ten frames per word, but this is not accurate as some adverbs take longer to say than some digits. Hence, in order to find the optimal window time for all the clips, we used silence detection algorithms. The silence or pause quieter than the threshold meant there was a pause between the words. The minimum silence length was about 100 ms and the audio had to be quieter than -16 dBFS was considered as a pause. In Figure 19.3, the normalized spectral centroid for the waveform is presented.

This technique also allowed us to center the word and is preferred to prevent database bias. Altering the frames for even a period of time would have devastating effects on the model. This algorithm helps us by cropping the clip and giving a range of numbers for every word. The model could then be fed with varying length words as we are using Bi-GRU cells. We found this isolation technique very useful to improve accuracy.

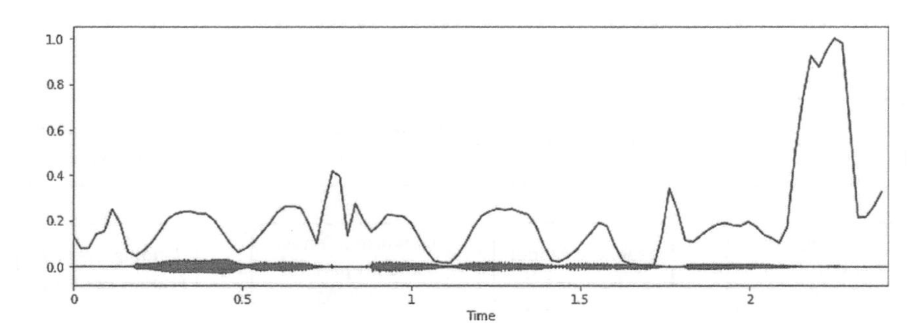

Figure 19.3 Normalized spectral centroids for the waveform.

19.4.3 Training the model

We are able to condense the clips into 4,800*12*128*128*3 for the training data and the labels were of the tensor, 4,800*51. We use one-hot representation for the training. A one-hot for four labels (A, B, C, D) of value A can be represented as [1, 0, 0, 0]. The activation is 1 for A and the rest are absent, this way we represent all the labels for the clips. We had pickled the data because the training time for building the model was very long. In addition, any range of about 5,000, resulted in memory issues for the RAM to store during training. Every frame is first converted into the corresponding tensor format that is supported by the transfer learning model. We use VGG16, ResNet, and MobileNet for testing out the accuracy and checking if they vary the accuracy. All of the models supported 128*128*3 for the size of the tensor and hence based on the varied length, the frames were condensed into corresponding tensors. We remove the last layers out of these models as it is used for classification and we replace them with our final layers. The kernel used is $5 \times 7 \times 7$. Figure 19.4 provides the framework of the proposed model.

At the end of the CNN model, the dense layer consists of two sets of learnable states. This can be used to pass to our second part of the model which could be GRUs, LSTMs, Bi-LSTMs, or Bi-GRUs. The dense layer has 1,024, 128 nodes each and includes a dropout layer in the middle. We use adam as the optimizer as it is generally known to do with these applications. Since training of the RNNs is more time complex, the training took over 8 hours to complete for the Bi-GRUs. The final layer consisted of a softmax layer with 51 units that gave o that gave out predictions based on

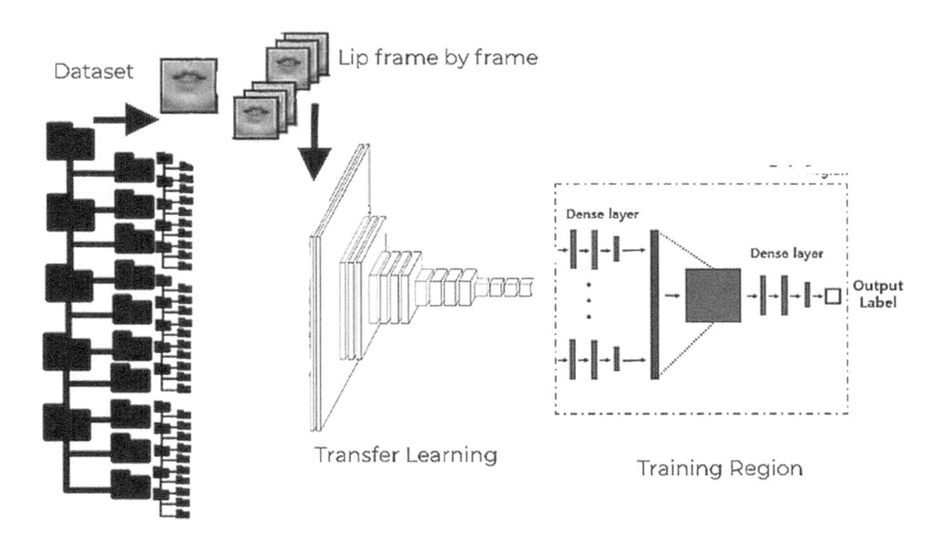

Figure 19.4 Framework of the model.

probability. The learning rate used was 0.001. The batch size was set to 16 as the RAM usage was a constraint. The validation radio was set to 0.2.

19.5 RESULTS AND ANALYSIS

Lipreading in isolation proves to be efficient as it enables comprehension over noisy channels. For the data preprocessing the base model was built with Haar cascade and then the crop was set for the entire clip. This resulted in not aligned lip clips and hence did affect the accuracy as seen in Table 19.2. The Haar cascade as observed does not perfectly align with the mouth. It does a good job at identifying the face but does not give lips bounding boxes with high accuracy. After which we move on to better object detection algorithms like the Blazeface and MTCNN paper. The data preprocessing would have taken about 50 hours if MTCNN were preferred. But since the model used was Blazeface, the GPU-based mobile prediction made the entire process quicker. This technique used for data preprocessing used up to 6 hours. After the entire clips were aligned and saved for training. The cropping and converting to black and white and creating a tensor took about another 8 hours. After which, the entire tensor, 3 GB in size, had to be saved dynamically in the RAM to be fed into the model for the number of epochs given.

From Table 19.2, it is evident that the improvisations incorporated like the optimum window selection and using Blazeface for object detection have created a difference and improved the accuracy by over 15%. But the datasets also had a few drawbacks. Traditionally, the frames per word are chosen to be 15 frames per word but using our optimum selection technique increases the accuracy by a lot. We also run this on top of a continuous speech and hence the state of the previous words contributes toward the next word. A language model could have also been fit to improve the accuracy even further. On evaluation with unseen speakers, the accuracy seems to be slightly lower than with the seen speakers. In this project, we were unable to better CNN models other than MobileNet due to the deficiency

Table 19.2 Comparison of models

Types of model	*Accuracy (%)*
VGG + Base model	58
ResNet + Base model	62
VGG + GRU	66
VGG + LSTM	69
VGG + Bi-LSTM	71
VGG + Bi-GRU	73

in time and memory. The number of labels was only 51 and the speakers were limited. But it was sufficient enough to prove that the varying in frame count per word caused a surge in the accuracy. However, the number of video clips for the alphabet and numerals was skewed, and hence the model had a very poor accuracy for most of the recognition of the alphabet. The accuracy was in the high 80s for the code, color, preposition, and adverbs types, as they were many examples and uniformly distributed. Speculations can also be made that because they consisted of more frames, the model was able to absorb the pattern. However, for the alphabet and numerals, the model had trouble fitting.

Additionally, the Bi-GRU also outperforms the Bi-LSTM model because of its simplicity at the core. Bi-LSTM is a little complex and makes it harder to establish relations for this application. For improving the finding for the optimal window, variations like 0 and –30 dB were tested but it was more adversaries for the GRID corpus. We also established that with the current data size, any more than 32 epochs also led to overfitting of the model. It is clearly visible how the algorithms when trained reach overfitting point where it clearly memorizes values instead of learning from the data. But when performing early stopping, we can figure out that the accuracy for the model for validation data maxes at around 75% which when compared to the testing set with new speakers and lips dropped down to around 73%. This could be solved by increasing the dataset size. More datasets like the LRW (Lipreading in the Wild) could be used to improve the model's potential. But the variation of the window for selection of the words combined with improvement in the lip object detection had an impact on the model's accuracy. Figure 19.5 provides trends observed during training.

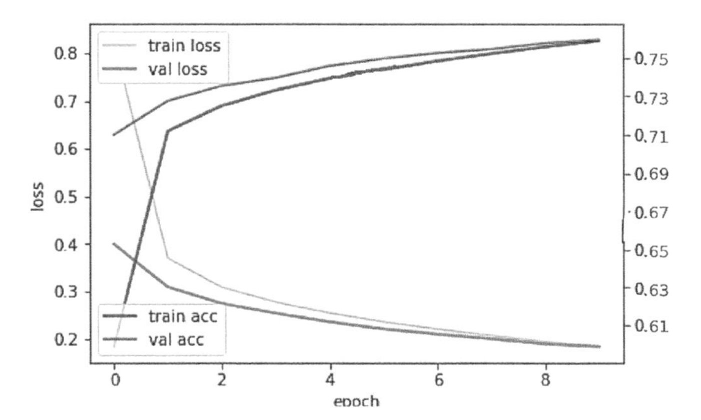

Figure 19.5 Trends observed during training.

19.6 CONCLUSIONS

In this work, we present some improvements to the preprocessing phase that greatly improves the model's accuracy. In the initial phase, Blazeface is used instead of Haar cascade and validated window selection techniques for better data segregation. Also, better lip tracking and choosing the frames based on word length established to have a significant impact on the accuracy. Execution time, complexity, and space requirement could be reduced as a result.

This modal can be extended to support multiple languages. Also, in the early stage audio can be fused with the automation of window selection using CNNs. Attachment of a language model before softmax activation can help to improve the accuracy. Similarly, CNNs can be replaced with the ResNet100 to enhance accuracy during the training phases.

REFERENCES

1. Chung, J. S., Zisserman, A. Lip reading in the wild. In: *Asian Conference on Computer Vision*, pp.87–103, Springer, Cham (2016).
2. Tripathy, B. K., Anuradha, J. *Soft Computing- Advances and Applications*. Cengage Learning publications, New Delhi (2015).
3. Bhattacharyya, S., Snasel, V., Hassanian, A. E., Saha, S., Tripathy, B. K. *Deep Learning Research with Engineering Applications*. De Gruyter Publications, Berlin (2020).
4. Adate, A., Tripathy, B. K., Arya, D., Shaha, A.: Impact of deep neural learning on artificial intelligence research. In: S. Bhattacharyya, A. E. Hassanian, S. Saha and B. K. Tripathy (Eds.), *Deep Learning Research and Applications*. De Gruyter Publications, Berlin, pp.69–84 (2020).
5. Adate, A., Tripathy, B. K. A survey on deep learning methodologies of recent applications. In: D.P. Acharjya, A. Mitra, and N. Zaman (Eds.), *Deep Learning in Data Analytics- Recent Techniques, Practices and Applications*, Springer Publications, Switzerland AG, pp.145–170 (2021). DOI: 10.1007/978-3-030-75855-4
6. Adate, A., Tripathy, B. K. Deep learning techniques for image processing. In S. Bhattacharyya, H. Bhaumik, A. Mukherjee and S. De (Eds.), *Machine Learning for Big Data Analysis*, De Gruyter, Berlin, Boston, pp.69–90 (2018).
7. Bose, A., Tripathy, B. K. Deep learning for audio signal classification. In: S. Bhattacharyya, A. E. Hassanian, S. Saha and B. K. Tripathy (Eds.), *Deep Learning Research and Applications*, De Gruyter Publications, Berlin, pp.105–136 (2020).
8. Kaul, D., Raju, H., Tripathy, B. K. Deep learning in healthcare. In: D.P. Acharjya, A. Mitra and N. Zaman (Eds.), *Deep Learning in Data Analytics- Recent Techniques, Practices and Applications*, Springer Publications, Switzerland, pp.97–115 (2021).

9. Maheswari, K., Shaha, A., Arya, D., Tripathy, B. K, Rajkumar, R. Convolutional neural networks: A bottom-up approach. In: S. Bhattacharyya, A. E. Hassanian, S. Saha and B. K. Tripathy (Eds.), *Deep Learning Research with Engineering Applications*, De Gruyter Publications, Berlin, pp.21–50 (2020).
10. Tripathy, B. K., Baktha, K. Investigation of recurrent neural networks in the field of sentiment analysis. In: *The Proceedings of IEEE International Conference on Communication and Signal Processing (ICCSP17)*, pp. 2047–2050, IEEE, Melmaruvathur (2017).
11. Adate, A., Tripathy, B. K. Understanding single image super-resolution techniques with generative adversarial networks. In: J. Bansal, K. Das, A. Nagar, K. Deep and A. Ojha (Eds.), *Soft Computing for Problem Solving, Advances in Intelligent Systems and Computing*, vol. 816, pp. 833–840. Springer, Singapore (2017). DOI: 10.1007/978-981-13-1592-3_66.
12. Debgupta, R., Chaudhuri, B.B., Tripathy, B.K. A wide resnet-based approach for age and gender estimation in face images. In: A. Khanna, D. Gupta, S. Bhattacharyya, V. Snasel, J. Platos and A. Hassanien (Eds.), *International Conference on Innovative Computing and Communications, Advances in Intelligent Systems and Computing*, vol. 1087, pp. 517–530, Springer, Singapore (2020). DOI: 10.1007/978-981-15-1286-5_44.
13. Tripathy, B. K., Parikh, S., Ajay, P., Magapu, C. Brain MRI segmentation techniques based on CNN and its variants, (Chapter-10). In: J. Chakki (Ed.), *Brain Tumor MRI Image Segmentation Using Deep Learning Techniques*, Elsevier publications (2022). DOI: 10.1016/B978-0-323-91171-9.00001-6.
14. Adate, A., Tripathy, B. K. S-LSTM-GAN: Shared recurrent neural networks with adversarial training. In: A. Kulkarni, S. Satapathy, T. Kang, and A. Kashan (Eds.), *Proceedings of the 2nd International Conference on Data Engineering and Communication Technology. Advances in Intelligent Systems and Computing*, Vol. 828, Springer, Singapore, pp.107–115 (2019).
15. Wand, M., Koutník, J., Schmidhuber, J. Lipreading with long short-term memory. In: *2016 IEEE International Conference on Acoustics, Speech and Signal Processing (ICASSP)*, pp.6115–6119. IEEE, Shanghai (2016).
16. Gupta, P., Bhachawat, S., Dhyani, K., Tripathy, B. K. A study of gene characteristics and their applications using deep learning, (Chapter 4). In: S.S. Roy and Y.-H. Taguchi (Eds.), H*andbook of Machine Learning Applications for Genomics, Studies in Big Data*, vol. 103, (2021). ISBN: 978-981-16-9157-7, 496166_1_En.
17. Bhardwaj, P., Guhan, T., Tripathy, B.K. Computational biology in the lens of CNN, (Chapter 5). In: S.S. Roy and Y.-H. Taguchi (Eds.), *Handbook of Machine Learning Applica-tions for Genomics, Studies in Big Data*, vol. 103 (2021). ISBN: 978-981-16-9157-7, 496166_1_En.
18. Afouras, T., Chung, J. S., Senior, A., Vinyals, O., Zisserman, A. Deep audio-visual speech recognition. *IEEE Transactions on Pattern Analysis and Machine Intelligence* 44(12), 8717–8727 (2018). DOI: 10.1109/TPAMI.2018.2889052

19. Afouras, T., Owens, A., Chung, J. S., Zisserman, A. Self-supervised learning of audio-visual objects from video. In: *Computer Vision–ECCV 2020: 16th European Conference, Glasgow, UK, August 23–28, 2020, Proceedings, Part XVIII 16*, pp.208–224. Springer International Publishing, Glasgow (2020).
20. Zhou, B., Khosla, A., Lapedriza, A., Oliva, A., Torralba, A. Learning deep features for discriminative localization. In: *Proceedings of The IEEE Conference on Computer Vision and Pattern Recognition*, pp.2921–2929. IEEE, Las Vegas, NV (2016).
21. Petridis, S., Stafylakis, T., Ma, P., Cai, F., Tzimiropoulos, G., Pantic, M. End-to-end audiovisual speech recognition. In: *2018 IEEE International Conference on Acoustics, Speech and Signal Processing (ICASSP)*, pp.6548–6552. IEEE, Calgary (2018).
22. Petridis, S., Li, Z., Pantic, M. End-to-end visual speech recognition with LSTMs. *In: 2017 IEEE International Conference on Acoustics, Speech and Signal Processing (ICASSP)*, pp.2592–2596. IEEE, New Orleans, LA (2017).
23. Bazarevsky, V., Kartynnik, Y., Vakunov, A., Raveendran, K., Grundmann, M. Blazeface: Sub-millisecond neural face detection on mobile gpus. *arXiv preprint arXiv*: 1907.05047 (2019).
24. Chiu, C. C., Sainath, T. N., Wu, Y., Prabhavalkar, R., Nguyen, P., Chen, Z., Kannan, A., Weiss, R.J., Rao, K., Gonina, E., Jaitly, N., Li, B., Chorowski, J., Bacchiani, M. State-of-the-art speech recognition with sequence-to-sequence models. In: *2018 IEEE International Conference on Acoustics, Speech and Signal Processing (ICASSP)*, pp. 4774–4778. IEEE, Calgary (2018).
25. Alghamdi, N., Maddock, S., Marxer, R., Barker, J., Brown, G. J. A corpus of audio-visual Lombard speech with frontal and profile views. *The Journal of the Acoustical Society of America*, 143(6), EL523–EL529 (2018).
26. Zhang, K., Zhang, Z., Li, Z., Qiao, Y. Joint face detection and alignment using multitask cascaded convolutional networks. *IEEE Signal Processing Letters* 23(10), 1499–1503 (2016).
27. Owens, A., Efros, A. A. Audio-visual scene analysis with self-supervised multisensory features. In: *Proceedings of the European Conference on Computer Vision (ECCV)*, Munich, pp.631–648 (2018).

Chapter 20

Modified Harris Hawk optimization algorithm for multilevel image thresholding

Soumyaratna Debnath, Abhirup Deb, and Sourav De
Cooch Behar Government Engineering College

Sandip Dey
Sukanta Mahavidyalaya

CONTENTS

20.1 INTRODUCTION

In the realm of nature and that of science, optimization [1] plays a pivotal role in the general course of events. Nature exemplifies its very own form of optimization through natural selection. Through genetic evolution [2], the features and the responses that are critical for the survival of a species, are filtered and optimized, through thousands of millennia, to culminate into the current plant and animal life that we see all around us today. These implicit constraints of selection form an integral part of a domain of computer science that deals with the solution and modeling of real-life problems. For the

DOI: 10.1201/9781003381167-20

past few decades, researchers have often faced the dilemma of creating an optimally designed solution for the problem. However, with the emergence of nature-oriented optimization methodologies, nowadays researchers are able to efficiently handle even multi-dimensional and multi-modal problems, provided in [3].

In the study of images, image segmentation and clustering is a critical domain that partitions an image into common areas of similarity. Segmenting and segregating an image with high accuracy is generally considered a task of extreme difficulty that poses a challenge of assembling the pixels, with similarity in their defined features [4], and separating those that are dissimilar accordingly. The principles of discontinuity and that odd Pixel-Similarity act as the driving proponents based on which, segmentation algorithms work. The segmentation algorithms, provided in [5,6], obeying the discontinuity principle tend to segment an image based on the abrupt changes in pixel intensity. However, those following the Similarity approach tend to segregate pixels based on their homogeneity. Within the existing realm of segmentation algorithms, thresholding [7] may be termed as the simplest and most renowned similarity-based approach, from an implementation point of view. It has been extensively used time and again for image processing applications including clustering [8], classification, and others [9].

Meta-heuristics are termed as approximate search stochastic algorithms that are highly useful in finding optimal and near-optimal solutions to optimization problems. Throughout computer literature, the aggregation behaviors of living organisms, especially in situations of high demand, have always been a source of high-interest and overwhelming inspiration for researchers. This in effect, has given birth to a vast array of nature-inspired meta-heuristic algorithms. The moderation to the Harris Hawks Optimization Algorithm proposed via this chapter is of a singular nature that aid in revolutionizing the approach to the problem. A variant of HHO, namely, Hybrid Parallel Harris Hawks Optimization (HPHHO) [10] deals with the re-entry of RLV aircraft into the defined flight trajectory and trajectory optimization problems pertinent to the domain of aerodynamics and aviation. Some other widely used variants include Quasi Reflected Harris Hawks Optimization (QRHHO) [11] and Hybrid Harris Hawks Nelder Mead Optimization (H-HHONM) [12].

20.2 HARRIS' HAWKS OPTIMIZATION

Harris' Hawk Optimization (HHO) [13], population-based, meta-heuristic method, was presented by Ali Asghar Heidari et al. The algorithm simulates the natural intellect of Harris' hawks to inspire and persuade the solutions of optimization problems, which are based on exploratory and

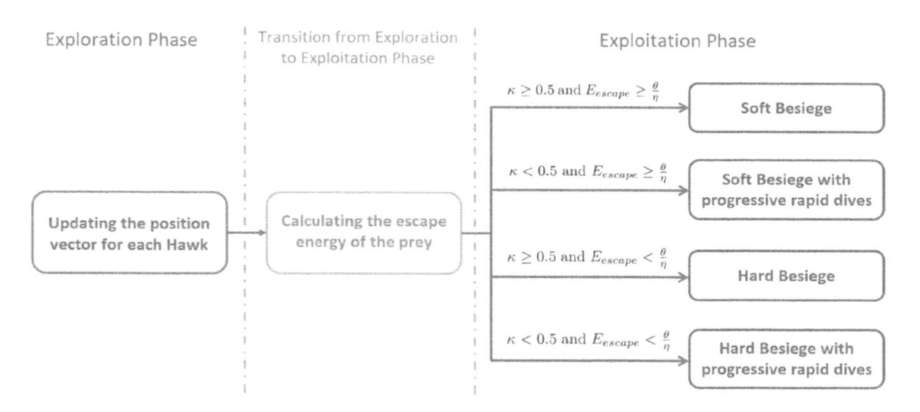

Figure 20.1 Different phases of HHO algorithm [13].

exploitative mechanisms, including mutation and crossover [13]. Different phases of the HHO algorithm are presented in Figure 20.1. The principal technique utilized by the Harris Hawks to capture their prey is the "Surprise Pounce" or "Seven Kills" strategy [13]. This strategy relies upon the following principle-several hawks try to attack the escaping rabbit, converging from different directions at different angles, simultaneously. The attack might be successful if the prey is caught by surprise in a very short duration of time, but occasionally the escaping capability of the prey is above the mark. In this scenario, the Seven-Kill strategy may include several successive miniature rapid dives in closed proximity to the prey, through a considerable duration of time. The Hawks are capable of exhibiting a plethora of persuasion techniques, relying on the circumstances. These tactics provide an advantage in that the Harris Hawks are capable of pursuing the rabbits until exhaustion [13].

20.3 QUANTITATIVE EVALUATION MEASURE

20.3.1 Modified Otsu

The quantitative evaluation measures have been employed using modified Otsu's method [14]. Considering monochrome image with pixel count as X where the pixel intensities lie within the range [0, Y] and consider g_x signifies the count of pixels with intensity x . Thus, the probability of occurrence of pixel intensity x in the image is given by Equation (20.1) [14].

$$p_x = \frac{g_x}{X} \tag{20.1}$$

With $Z-1$ number of thresholds i.e. $[d_1, d_2, ..., d_{Z-1}]$, an image can be spitted into Z classes, i.e. C_1 for $[1,..., d_1]$, C_2 for $[d_1+1,..., d_2]$,... C_Z for $[d_{Z-1}+1,..., Y]$. Thresholds $d_1^*, d_2^*,...,d_{Z-1}^*$ are classified selectively by optimizing the optimum thresholds [14].

$$d_1^*, d_2^*,...,d_{Z-1}^* = \text{ArgMax}\left\{(\sigma_D)^2 \{d_1, d_2,..., d_{Z-1}\}\right\} \tag{20.2}$$

where $(\sigma_D)^2 = \Sigma_{h=1}^{Z}\Omega_h\mu_h^2$, $\Omega_h = \Sigma_{x\in C_h} x \times \frac{p_x}{\Omega(h)}$, $\Omega(h)$ denotes the probability of pixel in h^{th} class, and μ_h denotes the mean value.

20.3.2 Entropy Yen

The entropic correlation proposed by Yen et al. is defined as Equation (20.3) [15].

$$W_{opt}(W) = \text{ArgMax}\left\{\Sigma_{l=1}^{x} C(W_l)\right\} \tag{20.3}$$

where x denotes the count of unique thresholds, i.e. $\{W_1, W_2,..., W_x\}$ and $C(W_m)$ is defined as in Equation (20.4.1–20.4.3) [15].

$$C(W_m) = -\log\left(\Sigma_{g=0}^{W_1}\left(\frac{p(g)}{p(W_1)}\right)^2\right), \text{ if } m = 1 \tag{20.4.1}$$

$$C(W_m) = -\log\left(\Sigma_{g=W_x}^{Y}\left(\frac{p(g)}{p(W_x)}\right)^2\right), \text{ if } m = x \tag{20.4.2}$$

$$C(W_m) = -\log\left(\Sigma_{g=W_m-1}^{W_m}\left(\frac{p(g)}{p(W_m)}\right)^2\right), \text{ otherwise} \tag{20.4.3}$$

where $p(g)$ signifies the odds of occurrence of the pixel with intensity g in a monochrome image, where the pixel intensities vary from 0 to Y. The odds of the area with the threshold value W_m is defined by Equation (20.5.1–20.5.3) [15].

$$p(W_m) = \Sigma_{g=0}^{W_1}(p(g)), \text{ if } m = 1 \tag{20.5.1}$$

$$p(W_m) = \Sigma_{g=W_x}^{Y}(p(g)), \text{ if } m = x \tag{20.5.2}$$

$$p(W_m) = \Sigma_{g=W_m-1}^{W_m}(p(g)), \text{ otherwise} \tag{20.5.3}$$

20.4 PROPOSED METHODOLOGY

20.4.1 Exploration phase

In this phase of the modified HHO, the hawks wait, persevere, and track the sites to detect the presence of prey (rabbits in this case). The Harris Hawks are observed as the possible solutions from which the optimal candidate solutions are encouraged in each step. In the modified HHO, this is modeled by Equation (20.6) which has been mathematically optimized with reduced randomization and systematic engineering of observed parameters, to obtain a more accurate representation based on a single random variable α.

$$M(\theta+1)=\left\{ \begin{array}{l} M_{rabbit}(\theta)-M_{\mu}(\theta)+\alpha\left(LB+\frac{\eta-\frac{\theta}{\eta}}{\theta^{2}\times\eta}(UB-LB)\right), \text{if } \alpha\geq 0.5 \\ M_{rabbit}(\theta)-M_{\mu}(\theta)-\alpha\left(LB+\frac{\eta-\frac{\theta}{\eta}}{\theta^{2}\times\eta}(UB-LB)\right), \text{if } \alpha<0.5 \end{array} \right\} \quad (20.6)$$

where $M(\theta+1)$ is the position vector of hawks in the next iteration, $M_{rabbit}(\theta)$ is the position of rabbit, $M(\theta)$ is the current position vector of hawks, η is the maximum number of generations, and α is selected randomly in the range [0, 1]. $M_{\mu}(\theta)$ is the average position of the hawks in the range [LB, UB], given by Equation (20.7).

$$M_{\mu}(\theta)=\frac{1}{N}\Sigma_{i=1}^{N}M_{i}(\theta) \quad (20.7)$$

where $M_i(\theta)$ signifies the position of individual hawks, where θ denotes the current iteration and N signifies the total number of hawks.

20.4.2 Transition from exploration to exploitation phase

The modified HHO algorithm follows the original HHO [13] algorithm in this scenario and transitions from the Exploration to the Exploitation Phase, formulated upon the prey's entropy. The escape energy is modeled as in Equation (20.8).

$$E_{escape}=2E_{0}\left(1-\frac{\theta}{N}\right),\ E_{0}=2\times \text{rand}-\frac{\theta}{N} \quad (20.8)$$

Here, E_{escape} indicates the Entropy of the prey, E_0 is initial energy, and *rand* returns a random number in the range (0,1).

20.4.3 Exploitation phase

The Hawks' attack their intended prey, in this case, the rabbits, by performing and enacting their Seven-Kill strategy. The prey displays some curious escaping behavior based on which the chasing strategy changes. The four possible strategies are employed. Supposing that, λ denotes the escaping opportunity for a prey and $\lambda = \frac{\eta \times \theta}{4 \times (\theta^2 + \eta)}$, δ is a randomly selected in [0, 1], and $\kappa = \frac{\lambda + \delta}{1 + \delta}$, then, $\kappa < 0.5$ denotes that the prey will successfully escape, and $\kappa \geq 0.5$ denotes that the prey will be caught. The Hawks either follow the hard besiege or soft besiege strategy to catch the prey by encircling the prey from different directions. This entire scenario depends on the energy retained by the prey. As the rabbits lose their escaping energy, the hawks intensify their besiege procedure. In this respect, in the modified strategy, soft besiege occurs when $E_{escape} \geq \frac{\theta}{\eta}$ and hard besiege occurs when $E_{escape} < \frac{\theta}{\eta}$.

Soft besiege: In case of soft besiege, it is considered that the rabbits have enough energy for trying random jumps in order to escape, but eventually, they cannot. In the case of modified HHO, soft besiege is governed $\kappa \geq 0.5$ and $E_{escape} \geq \frac{\theta}{\eta}$. Soft besiege is modeled by Equation (20.9).

$$M(\theta + 1) = \begin{Bmatrix} 2\lambda + E_{escape}\left(T \times M_{rabbit} - 2\delta \times M(\theta)\right), \text{ if } \lambda \geq \delta \\ \cdot 2\lambda - E_{escape}\left(T \times M_{rabbit} + 2\delta \times M(\theta)\right), \text{ if } \lambda < \delta \end{Bmatrix} \tag{20.9}$$

$$\Delta M(\theta) = M_{rabbit} - M(\theta) \tag{20.10}$$

$M_{rabbit}(\theta)$ signifies position vector of the rabbit and $M(\theta)$ denotes the current location in θ^{th} location, and $T = \frac{(\lambda + \delta) \times \theta}{\eta - \theta + 1}$ represents the random jump strength of the rabbit throughout the escaping procedure, where η is the maximum number of generations of the simulation. The equation is systematically modeled with the introduction of the parameters λ and δ in order to decrease the randomness and promote systematic upgradation of the factors governing the rabbits' escape chances.

Hard besiege: In the case of hard besiege, it is considered that the rabbits are exhausted, and thus have a low escaping energy. In the case of modified HHO, it is dictated by $\kappa \geq 0.5$ and $E_{escape} < \frac{\theta}{\eta}$. In addition, it is considered that the hawk encircles the prey and performs surprise pounce. The situation is modeled by Equation (20.11).

$$M(\theta+1)=\left\{\begin{matrix} 2\lambda\times M_{rabbit}(\theta)+2\delta\times E_{escape}(\delta M(\theta), \text{ if } \lambda\geq\delta) \\ \cdot 2\lambda\times M_{rabbit}(\theta)-2\delta\times E_{escape}(\delta M(\theta)), \text{ if } \lambda<\delta \end{matrix}\right\} \tag{20.11}$$

The equations governing the hard besiege is engineered with the introduction of the parameters λ and δ in order to promote the accuracy of the random pounce, thus improving the quality of the hawks in terms of fitness, and hunting ability.

Soft besiege with progressive rapid dives: If $\kappa < 0.5$ and $E_{escape} \geq \frac{\theta}{n}$, the prey still has sufficient energy and capabilities, to escape. In order to devise a mathematical model for the escaping patterns of the prey, the Levy Flight Function has been utilized both in the modified and the original HHO [13] algorithm. This imitates the irregularities of the motion observed during the erratic movements of the prey while being attacked. The Hawks possess the natural ability to select the best possible diving strategy in situations of intensity. The decisive action for the next move is evaluated based on the following modified equation:

$$P=\left\{\begin{matrix} 2\lambda\times M_{rabbit}(\theta)-E_{escape}(T\times M_{rabbit}-2\delta\times M(\theta)), \text{ if } \lambda+\delta\geq 1 \\ \cdot 2\lambda\times M_{rabbit}(\theta)+E_{escape}(T\times M_{rabbit}+2\delta\times M(\theta)), \text{ if } \lambda+\delta<1 \end{matrix}\right\} \tag{20.12}$$

This equation has been carefully modeled to replicate the behavior of the rabbits more accurately by the introduction of parameters λ and δ, in order to better quantify the random behavioral movements of the Rabbits. If the possible results of the dive are not as expected, the Hawks switch to irregular, abrupt, and rapid dives using Equation (20.13).

$$Q = P + S \times F_{levi}(D) \tag{20.13}$$

where D=Solution dimension, S is a $1\times D$ vector selected randomly and Levy Flight Function, F_{levi}, evaluated using the following formula:

$$F_{levi}(x) = 0.01\times\frac{u\times\sigma}{|v|^{\frac{1}{\beta}}}, \sigma = \left(\frac{\Gamma(1+\beta)\times\sin\left(\frac{\pi\beta}{2}\right)}{\Gamma\left(\frac{1+\beta}{2}\right)\times\beta\times 2^{\left(\frac{\beta-1}{2}\right)}}\right)^{\frac{1}{\beta}} \tag{20.14}$$

where u and v are random variables in range (0,1) and β=0.5. The ultimate procedure for modifying the position of the Harris Hawks is dictated through the following equation:

$$M(\theta+1)=\left\{\begin{matrix} P, \text{if } F(P)<F(M(\theta)) \\ \cdot Q, \text{if } F(Q)<F(M(\theta)) \end{matrix}\right\} \tag{20.15}$$

where P and Q are obtained using the rules in Equations (20.12 and 20.13).

Hard besiege with progressive rapid dives: Considering that the rabbits do not have enough energy, the Hawks orchestrate the hard besiege stage by performing a surprise pounce upon the prey. In the case of modified HHO, the situation was driven by $\kappa < 0.5$ and $E_{escape} < \frac{\theta}{\eta}$. The hard besiege with progressive rapid dives is modeled by Equation (20.16).

$$M(\theta+1)=\left\{\begin{matrix} P, \text{if } F(P)<F(M(\theta)) \\ \cdot Q, \text{if } F(Q)<F(M(\theta)) \end{matrix}\right\} \tag{20.16}$$

where P and Q are obtained using the rules in Equations (20.17 and 20.18). These equations have been carefully modeled to replicate the behavior of the rabbits more accurately by the introduction of parameters λ and δ.

$$P=\left\{\begin{matrix} 2\lambda\times M_{rabbit}(\theta)+2\delta\times E_{escape}(T\times M_{rabbit}(\theta)+M_{\mu}(\theta)), \text{if } \delta\geq\lambda \\ \cdot 2\lambda\times M_{rabbit}(\theta)-2\delta\times E_{escape}(T\times M_{rabbit}(\theta)-M_{\mu}(\theta)), \text{if } \delta<\lambda \end{matrix}\right\} \tag{20.17}$$

$$Q=P+S\times F_{levi}(D) \tag{20.18}$$

where $M_{\mu}(\theta)$ is obtained using Equation (20.4).

20.5 EXPERIMENTAL RESULTS

20.5.1 Details of the experimental data

For experimental purposes, two natural test images namely, Barbara and Couple [16], having pixel dimensions 512×512, have been considered. Multilevel thresholding has been performed upon test images for labels 6, 8, and 10, using modified Otsu's [14] and Yen's [17] as fitness functions. The original test image has been referred to in Figures 20.1 and 20.2.

20.5.2 Performance measures

Effectiveness of every method is visually and quantitatively examined condition to the methods mentioned below.

- The optimum fitness values (B_r) among different runs.
- Convergence time (B_{tm}) at best case.

Figure 20.2 Original test images. (a) Barbara and (b) Couple [16].

- The mean (μ_a) and standard deviation (σ) of all runs.
- The mean time for convergence (T_{at}) and mean time of execution (T_{rt}).
- Kruskal-Wallis test values [18] between MHHO & HHO [13], MHHO & PSO [19], and MHHO & DE [15].
- PSNR values at best case.
- Curves for convergence pertaining to every participating method.

20.5.3 Results

Modified Otsu's [14] method is applied as an optimization function. The specific parameters of the PSO [19] and DE [15] methods are furnished in Table 20.1. The size of the swarm and the maximal number of iterations for every method including MHHO and HHO is fixed at 50 and 100, over 30 independent runs.

The proposed MHHO is introduced to unravel the maximal threshold values for the test image. For the contingent study, the classical PSO [19], classical DE [15], and original HHO [13] are run upon the aforementioned test image. Table 20.1 presents the maximum fitness, optimum thresholds (Tr), and the time of convergence (at the best case) as obtained by the MHHO,

Table 20.1 Parameter setting for PSO [19] and DE [15] methods

Algorithm	*Parameter*	*Value*
DE	Scaling factor	0.5
	Crossover probability	0.5
PSO	Inertia factor	0.3
	Acceleration coefficient (c1)	1
	Acceleration coefficient (c2)	1

Table 20.2 For D=6, 8 and 10, optimum thresholds (T_r), optimum fitness (B_r) and time of convergence at the optimal case (B_{tm}) of MHHO, HHOs [13], PSO [19] and DE [15] algorithms for Barbara [16] using modified Otsu's method

Methods	*D*	T_r	B_r	B_{tm}
MHHO	6	49, 86, 126, 170, 207, 218	5,118.411	0.735
MHHO	8	43, 66, 93, 121, 151, 179, 212, 239	5,162.929	1.252
	10	35, 59, 80, 99, 125, 149, 172, 194, 217, 230	5,183.949	1.105
HHO	6	43, 80, 121, 167, 206, 244	5,116.701	0.822
HHO	8	39, 76, 93, 111, 137, 176, 206, 214	5,153.168	1.883
	10	40, 57, 73, 93, 116, 137, 167, 195, 212, 229	5,178.989	1.695
PSO	6	54, 95, 110, 127, 165, 207	5,084.920	1.730
PSO	8	42, 80, 100, 119, 150, 180, 210, 217	5,129.360	1.330
	10	39, 70, 86, 112, 122, 133, 154, 173, 191, 209	5,168.580	1.750
DE	6	45, 88, 118, 131, 175, 209	5,108.810	1.002
DE	8	39, 79, 96, 124, 157, 172, 217, 219	5,149.210	1.301
	10	38, 73, 82, 114, 122, 145, 190, 203, 211, 239	5,166.520	1.550

HHO [13], PSO [19], and DE [15] for the eighth and tenth levels of image thresholding. The average fitness (μ_{av}), mean time of convergence (T_{av}), mean execution time (T_{rt}), and standard deviation (σ) of each test case are presented in Table 20.2. The P SNR values are mentioned in each case. The P SNR values reported here are in favor of the modified HHO algorithm. The efficacy of the MHHO algorithm is confirmed with a subject to a popular statistical test, namely, the Kruskal-Wallis test [18] (95% significance). The test results confirm the superiority of the modified method. Figure 20.3 presents the threshold images for each test case at D=8 and 10. Finally, the convergence curves (at the best cases) visually establish the quality of the proposed MHHO are presented in Figure 20.4 for D=8 and 10. In light of the above facts and figures, the following conclusions can be obtained.

- Tables 20.2–20.7 present the test results that adhere to the supremacy of the MHHO algorithm subject to optimum fitness, the optimum time of convergence, mean time of convergence, and mean execution time.
- With reference to Tables 20.6 and 20.7, smaller PSNR values of MHHO are testimony to the fact that it is encouraging better thresholding capabilities compared to the other two algorithms.
- The Kruskal-Wallis test [4] results, computed between MHHO & HHO [1], MHHO & PSO [3], and MHHO & DE [20] are mentioned in Tables 20.8 and 20.9. This indicates the supremacy of the proposed MHHO algorithm.

Table 20.3 For $D = 6$, 8 and 10, optimum thresholds (T_r), optimum fitness (B_r) and time of convergence at the optimal case (B_{tm}) of MHHO, HHOs [13], PSO [19] and DE [15] algorithms for Barbara [16] using Yen's method

Methods	*D*	T_r	B_r	B_{tm}
MHHO	6	39, 81, 121, 168, 209, 222	19.660	2.250
	8	38, 66, 96, 132, 159, 191, 220, 222	24.001	0.674
	10	29, 52, 79, 102, 121, 146, 174, 199, 220, 234	27.878	1.245
HHO	6	36, 78, 129, 161, 208, 228	19.600	3.339
	8	33, 65, 105, 142, 179, 194, 219, 235	23.789	1.077
	10	32, 59, 77, 106, 124, 149, 176, 197, 223, 235	27.809	3.678
PSO	6	41, 84, 97, 123, 165, 200	19.640	2.620
	8	37, 68, 90, 121, 149, 188, 197, 227	23.870	1.440
	10	44, 65, 87, 110, 132, 144, 156, 176, 204, 223	27.580	2.150
DE	6	48, 78, 95, 134, 156, 220	19.643	2.310
	8	27, 74, 86, 117, 139, 197, 206, 239	23.906	1.044
	10	34, 52, 89, 121, 133, 154, 167, 173, 214, 239	27.620	1.510

Table 20.4 For $D=6$, 8 and 10, optimum thresholds (T_r), optimum fitness (B_r) and time of convergence at the optimal case (B_{tm}) of MHHO, HHOs [13], PSO [19] and DE [15] algorithms for Couple [16] using modified Otsu's method

Methods	*D*	T_r	B_r	B_{tm}
MHHO	6	48, 87, 120, 146, 178, 200	1,870.379	1.510
	8	42, 70, 98, 120, 141, 161, 188, 237	1,912.580	1.525
	10	38, 64, 89, 106, 125, 136, 151, 167, 191, 198	1,931.129	0.870
HHO	6	45, 86, 120, 144, 181, 219	1,868.838	2.719
	8	42, 75, 106, 127, 150, 168, 184, 213	1,908.367	1.688
	10	38, 66, 80, 98, 117, 133, 149, 174, 194, 204	1,929.679	1.071
PSO	6	51, 89, 120, 145, 178, 202	1,870.110	1.690
	8	39, 70, 98, 120, 140, 162, 182, 217	1,910.510	1.950
	10	36, 52, 72, 99, 112, 131, 148, 169, 202, 221	1,927.250	1.740
DE	6	45, 78, 116, 138, 188, 205	1,855.610	1.663
	8	27, 77, 89, 135, 158, 182, 192, 219	1,911.021	1.680
	10	46, 59, 83, 110, 124, 147, 178, 188, 198, 235	1,930.210	1.021

- Figures 20.3 and 20.4 depict thresholded images using the modified HHO algorithm.
- Figures 20.5 and 20.6 depict the curves for convergence for each test case.

Table 20.5 For $D=6$, 8 and 10, optimum thresholds (T_r), optimum fitness (B_r) and time of convergence at the optimal case (B_{tm}) of MHHO, HHOs [13], PSO [19] and DE [15] algorithms for Couple [16] using Yen's method

Methods	*D*	T_r	B_r	B_{tm}
MHHO	6	43, 88, 131, 169, 199, 202	19.875	1.906
	8	41, 69, 107, 136, 166, 196, 235, 249	24.472	0.886
	10	20, 40, 66, 97, 125, 143, 176, 202, 235, 252	28.572	2.287
HHO	6	42, 79, 120, 157, 202, 227	19.776	2.520
	8	47, 92, 117, 148, 169, 199, 235, 251	24.322	4.184
	10	23, 55, 85, 108, 125, 154, 191, 205, 235, 248	28.245	5.771
PSO	6	43, 83, 124, 167, 198, 212	19.850	2.410
	8	45, 83, 108, 141, 175, 188, 203, 235	24.440	1.740
	10	22, 67, 93, 121, 141, 160, 187, 191, 211, 237	28.300	2.780
DE	6	39, 85, 134, 187, 198, 226	19.760	2.010
	8	38, 87, 109, 131, 178, 189, 213, 239	24.437	1.843
	10	28, 56, 91, 131, 142, 161, 182, 202, 237, 245	28.451	2.378

20.6 CONCLUSION AND FUTURE DIRECTION

This chapter has aimed to present an improved version of the recent Harris Hawks Optimization Algorithm. In the proposed Modification of HHO [13], an overall improvement to the approach has been proposed, that eliminates the degree of uncertainty in the Mathematical Equations, through a reduction in the number of parameters. These subtleties have resulted in a remarkable improvement in the Exploitation Phase and its sub-phases of hard besiege and soft besiege. The Performance of modified HHO has been critically and extensively evaluated upon a natural grayscale test image, available in the public domain. Rigorous statistical analysis has been carried out to support the efficiency of the algorithm. Through such analysis, it can be inferred that modified HHO stands out to be the best-performing algorithm in its class of optimization algorithms. An emphasis is also being laid upon exploring the proposed variants of the Harris Hawks Optimization Algorithm namely, Hybrid Parallel Harris Hawks Optimization (HPHHO) [10], Quasi-Reflected Harris Hawks Optimization (QRHHO) [11] and Hybrid Harris Hawks Nelder Mead Optimization (H-HHONM) [12]. The researchers involved are striving toward pushing this particular avenue of interest toward more real-life use cases, through future works. The immediate objective is to extend and adapt the architecture of the proposed modified HHO algorithm to maneuver complex real-life situations.

Table 20.6 For D=6, 8 and 10, average fitness (μ_a), standard deviation (σ), mean times of convergence (T_{at}), mean execution time (T_{rt}) and *PSNR* of MHHO, HHO [1], PSO [3] and DE [20] for Barbara and Couple [16] using modified Otsu's [8] method

Images	*Methods*	*D*	μ_a	σ	T_{at}	T_{rt}	*PSNR*
Barbara	MHHO	6	5,110.11	2.93	1.56	3.16	36.120
		8	5,153.29	2.24	1.23	3.55	36.605
		10	5,175.46	3.30	1.69	3.46	36.461
	HHO	6	5,089.91	8.57	1.86	3.67	36.877
		8	5,138.37	11.83	2.26	3.80	37.035
		10	5,166.10	8.78	1.76	4.09	37.015
	PSO	6	5,078.52	4.01	2.45	3.68	39.135
		8	5,118.49	5.60	2.39	3.63	38.558
		10	5,138.84	4.45	1.91	3.69	38.342
	DE	6	5,098.12	5.04	2.35	3.81	39.351
		8	5,131.44	6.67	1.79	3.53	37.248
		10	5,155.83	7.57	1.95	4.01	37.152
Couple	MHHO	6	1,864.87	5.11	1.46	3.36	38.229
		8	1,905.5	5.96	1.48	3.39	38.100
		10	1,924.98	4.35	1.77	3.63	38.102
	HHO	6	1,853.34	11.05	1.73	3.72	38.705
		8	1,892.59	9.58	1.84	3.93	38.906
		10	1,916.54	6.71	2.04	2.97	39.048
	PSO	6	1,860.65	8.05	1.67	3.69	40.666
		8	1,898.93	6.81	1.71	3.66	40.285
		10	1,915.59	7.60	2.29	3.69	40.455
	DE	6	1,862.51	8.35	1.69	3.24	39.456
		8	1,896.23	6.61	1.62	3.57	39.184
		10	1,918.89	6.60	1.89	2.79	39.455

Table 20.7 For D=6, 8 and 10, average fitness (μ_a), standard deviation (σ), mean times of convergence (T_{at}), mean execution time (T_{rt}) and $PSNR$ of MHHO, HHO [1], PSO [3] and DE [20] for Barbara and Couple [16] using Yen's [8] method

Images	*Methods*	*D*	μ_a	σ	T_{at}	T_{rt}	$PSNR$
Barbara	MHHO	6	19.62	0.04	1.70	3.40	37.046
		8	23.84	0.11	1.86	3.59	36.605
		10	27.45	0.29	2.45	3.19	36.606
	HHO	6	19.38	0.13	1.98	3.88	38.836
		8	23.36	0.21	1.48	3.97	38.270
		10	27.05	0.37	2.61	5.04	38.682
	PSO	6	19.58	0.07	2.55	3.72	39.748
		8	23.68	0.11	2.45	3.63	39.693
		10	27.24	0.57	2.87	4.15	38.703

(*Continued*)

Table 20.7 (Continued) For D=6, 8 and 10, average fitness (μ_a), standard deviation (σ), mean times of convergence (T_{at}), mean execution time (T_{rt}) and $PSNR$ of MHHO, HHO [1], PSO [3] and DE [20] for Barbara and Couple [16] using Yen's [8] method

Images	*Methods*	*D*	μ_a	σ	T_{at}	T_{rt}	$PSNR$
	DE	6	19.47	0.65	2.15	3.56	37.748
		8	23.48	0.21	2.85	3.73	38.193
		10	27.39	0.63	2.71	4.52	37.513
Couple	MHHO	6	19.8	0.04	2.19	3.15	37.893
		8	24.2	0.09	1.87	3.69	37.793
		10	28.02	0.12	2.62	3.08	37.724
	HHO	6	19.58	0.14	1.88	3.11	38.741
		8	23.92	0.24	2.01	4.27	38.214
		10	27.78	0.31	3.79	4.07	38.632
	PSO	6	18.71	0.06	2.62	4.58	39.892
		8	24.01	0.15	2.77	3.66	39.823
		10	27.2	0.19	3.35	3.63	39.880
	DE	6	18.89	0.08	2.59	3.98	38.861
		8	23.90	0.14	2.81	3.54	38.113
		10	26.92	0.17	2.95	3.51	38.880

Table 20.8 For $D=6$, 8 and 10, results of Kruskal-Wallis test [18] between MHHO & HHO [13], MHHO & PSO [19] and MHHO & DE [15] for Barbara and Couple [16] using modified Otsu's [14] method

		MHHO & HHO	*MHHO & PSO*	*MHHO & DE*
Images	*D*	*p-value*	*p-value*	*p-value*
Barbara	6	< 0.001	< 0.001	< 0.001
	8	< 0.001	< 0.001	< 0.001
	10	< 0.001	< 0.001	< 0.001
	6	< 0.001	< 0.001	< 0.001
Couple	8	< 0.001	< 0.001	< 0.001
	10	< 0.001	< 0.001	< 0.001

Table 20.9 For $D=6$, 8 and 10, results of Kruskal-Wallis test [18] between MHHO & HHO [13], MHHO & PSO [19] and MHHO & DE [15] for Barbara and Couple [16] using Yen's [14] method

		MHHO & HHO	*MHHO & PSO*	*MHHO & DE*
Images	*D*	*p-value*	*p-value*	*p-value*
Barbara	6	< 0.001	< 0.001	< 0.001
	8	< 0.001	< 0.001	< 0.001
	10	< 0.001	< 0.001	< 0.001
	6	< 0.001	< 0.001	< 0.001
Couple	8	< 0.001	< 0.001	< 0.001
	10	< 0.001	< 0.001	< 0.001

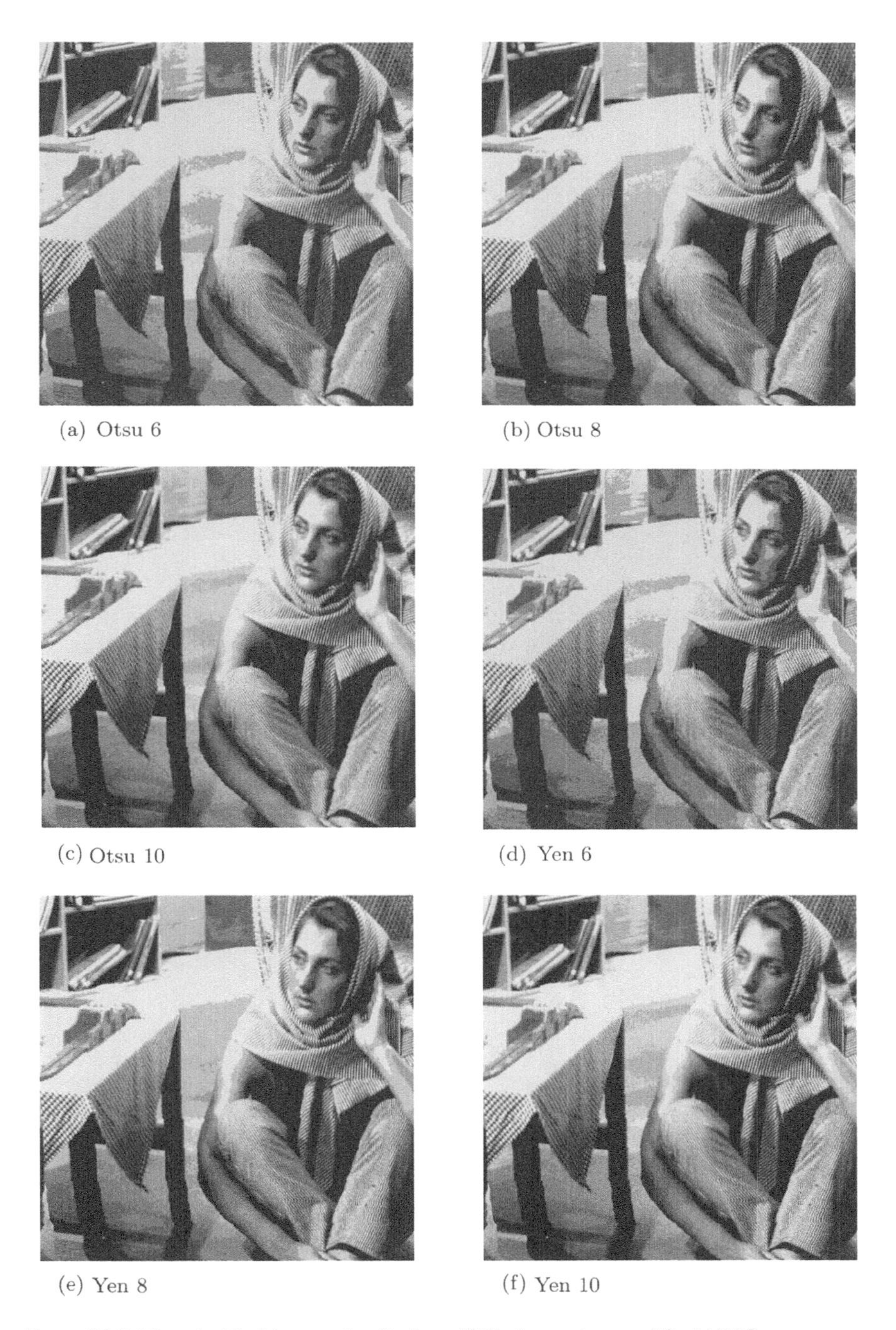

Figure 20.3 Thresholded images for Barbara [16] after using modified HHO.

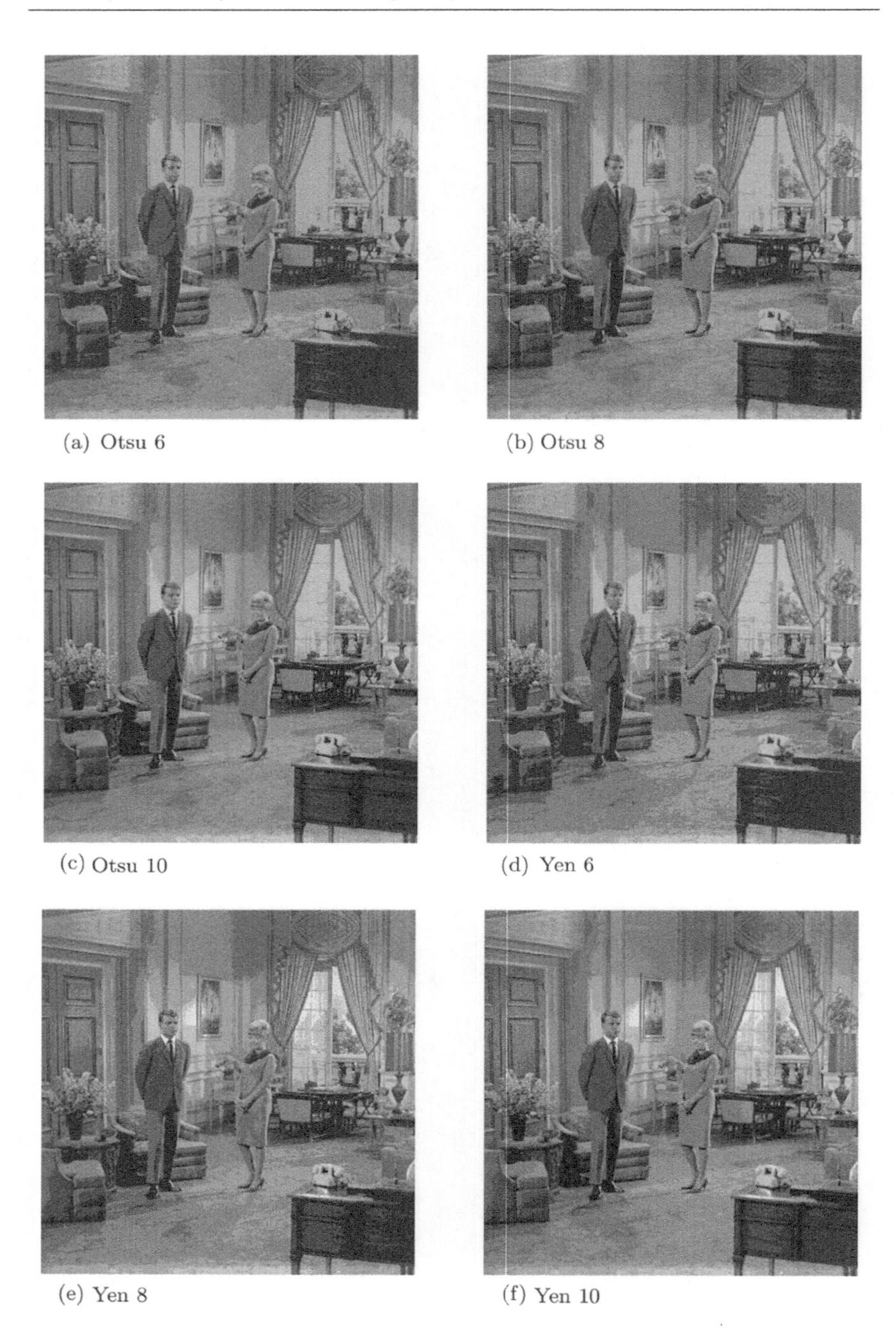

(a) Otsu 6

(b) Otsu 8

(c) Otsu 10

(d) Yen 6

(e) Yen 8

(f) Yen 10

Figure 20.4 Thresholded images for couple [16] after using modified HHO.

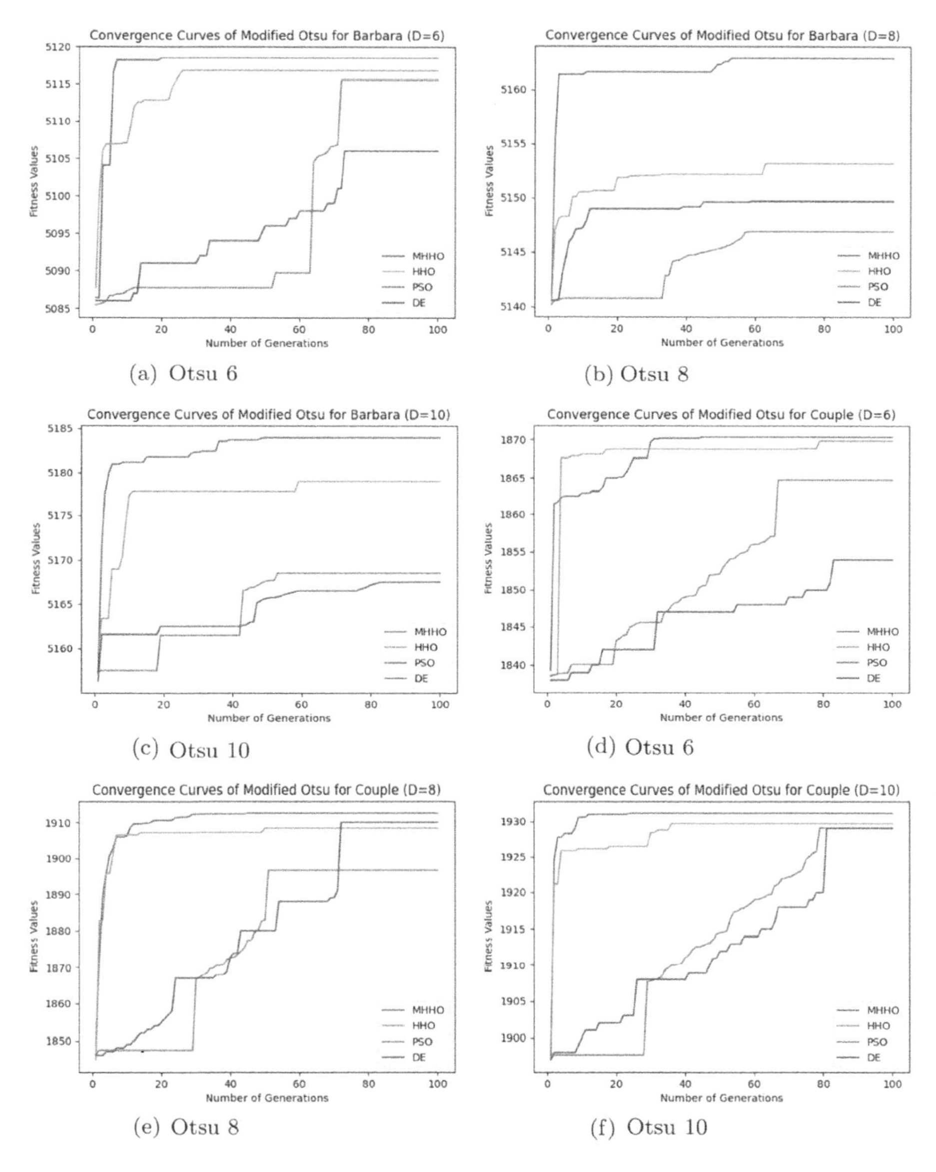

Figure 20.5 Convergence curves. (a)–(c) Barbara [16], (d)–(f) Couple [16] using modified Otsu's method [8].

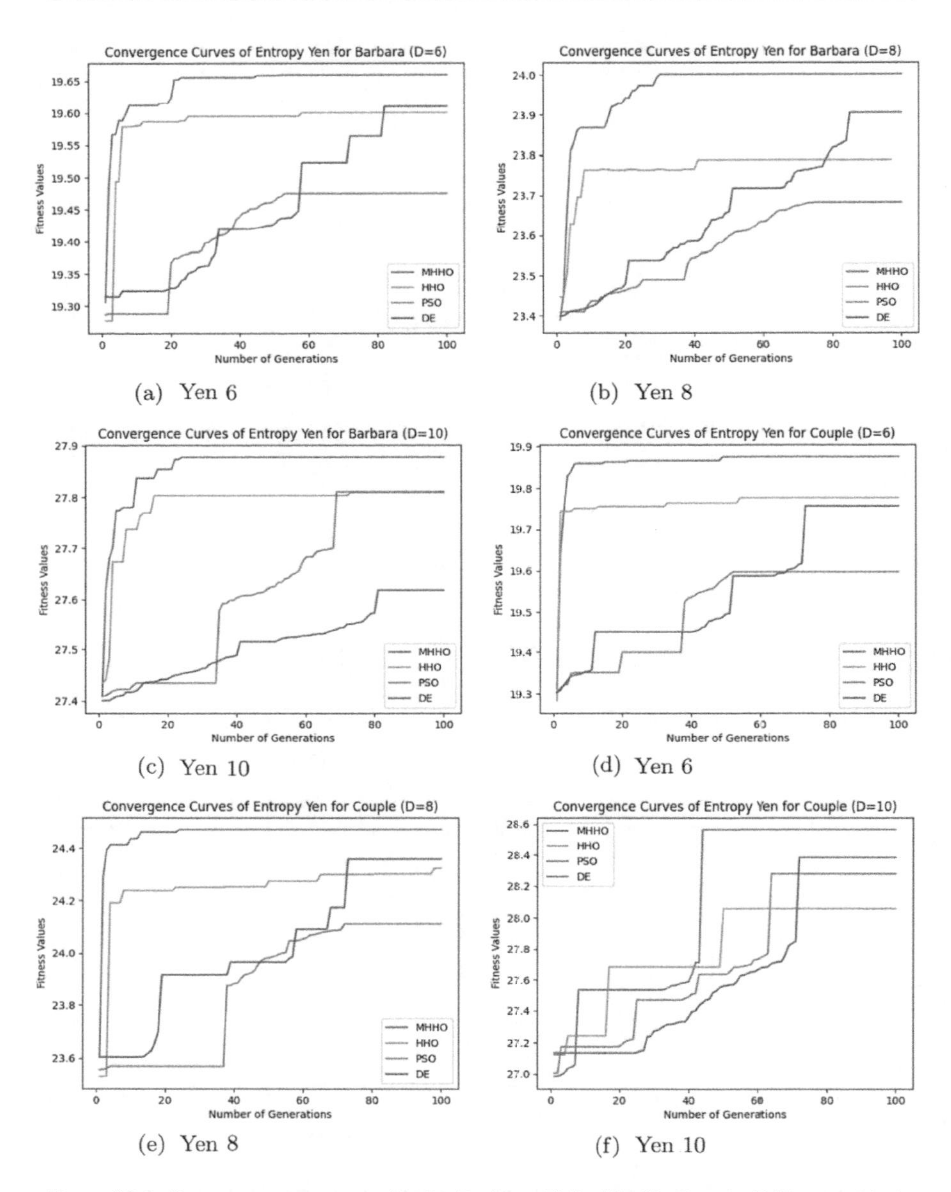

(a) Yen 6 (b) Yen 8 (c) Yen 10 (d) Yen 6 (e) Yen 8 (f) Yen 10

Figure 20.6 Convergence curves. (a)–(c) Barbara [16], (d)–(f) Couple [16] using Yen's method [8].

REFERENCES

1. De, S., Bhattacharyya, S., Chakraborty, S. Color image segmentation using parallel OptiMUSIG activation function. *Applied Soft Computing* 12(10), 3228–3236 (2012).
2. Holland, J. Scientific American. *Genetic Algorithms* 267, 66–72 (1992).
3. Bansal, J.C., Sharma, H., Jadon, S.S., Clerc, M. Spider monkey optimization algorithm for numerical optimization. *Memetic Computing* 6(1), 31–47 (2014).
4. De, S., Bhattacharyya, S., Dutta, P. Efficient grey-level image segmentation using an optimised musig (OptiMUSIG) activation function. *International Journal of Parallel, Emergent and Distributed Systems* 26(1), 1–39 (2011).
5. Dey, S., Bhattacharyya, S., Maulik, U. New quantum inspired meta-heuristic techniques for multi-level colour image thresholding. *Applied Soft Computing* 46, 677–702 (2016).
6. Dey, S., Bhattacharyya, S., Maulik, U. Efficient quantum inspired meta-heuristics for multi-level true colour image thresholding. *Applied Soft Computing* 56, 472–513 (2017).
7. Dey, S., De, S., Deb, A., Debnath, S. Multilevel image segmentation using modified red deer algorithm pp. 362–368 (2021). https://doi.org/10.1109/Confluence51648.2021.9377112.
8. De, S., Dey, S., Debnath, S., Deb, A. A new modified red deer algorithm for multi-level image thresholding. *2020 Fifth International Conference on Research in Computational Intelligence and Communication Networks (ICRCICN)*, India, pp. 105–111 (2020).
9. Ng, H. Pattern recognition letters. *Automatic Thresholding for Defect Detection* 27(14), 1644–1649 (2006).
10. Su, Y., Dai, Y., Liu, Y. *A Hybrid Parallel Harris Hawks Optimization Algorithm for Reusable Launch Vehicle Reentry Trajectory Optimization with No-Fly Zones* (2021). DOI:10.21203/rs.3.rs-554106/v1
11. Fan, Q., Chen, Z., Xia, Z. A novel quasi-reflected harris hawks optimization algorithm for global optimization problems. *Soft Computing* 24(19), 14825–14843 (2020).
12. Yıldız, A.R., Yıldız, B.S., Sait, S.M., Bureerat, S., Pholdee, N. A new hybrid harris hawks-nelder-mead optimization algorithm for solving design and manufacturing problems. *Materials Testing* 61(8), 735–743 (2019).
13. Heidari, A.A., Mirjalili, S., Faris, H., Aljarah, I., Mafarja, M., Chen, H. Harris hawks optimization: Algorithm and applications. *Future Generation Computer Systems* 97, 849–872 (2019).
14. Liao, P.S., Chen, T.S., Chung, P.C. A fast algorithm for multilevel thresholding. *Journal of Information Science and Engineering* 17, 713–727 (2001).
15. De, S., Bhattacharyya, S., Dutta, P. A differential evolution algorithm based automatic determination of optimal number of clusters validated by fuzzy intercluster hostility index. In: *2009 First International Conference on Advanced Computing*, Chennai, India, pp. 105–111 (2009).
16. Public-Domain Test Images for Homeworks and Projects. (2020) https://homepages.cae.wisc.edu/ ece533/images/ (accessed September 3, 2020).

17. Yen, J.-C., Chang, F.-J., Chang, S. A new criterion for automatic multilevel thresholding. *IEEE Transactions on Image Processing* 4(3), 370–378 (1995).
18. Kruskal, W.H., Wallis, W.A. Use of ranks in one-criterion variance analysis. *Journal of the American Statistical Association* 47(260), 583–621 (1952).
19. De, S., Haque, F. Multilevel image segmentation using modified particle swarm optimization. In: Bhattacharyya, S., Bhaumik, H., De, S., Klepac, G. (eds.) *Intelligent Analysis of Multimedia Information*, pp. 106–142. IGI Global (2016). DOI: 10.4018/978-1-5225-0498-6
20. Alabool, H. M., et al. Harris hawks optimization: A comprehensive review of recent variants and applications. *Neural Computing and Applications* 33, 8939–8980 (2021).

Chapter 21

An automatic probabilistic framework for detection and segmentation of tumor in brain MRI images

K Bhima, A. Jagan, and K. Dasaradh Ramaiah
B V Raju Institute of Technology

CONTENTS

21.1 INTRODUCTION

Future treatment planning based on brain imaging are planned using MRI images [1–3] to diagnose the patient. Conventional method for tumor detection in MRI images is a monotonous and overwhelming task due to invariant characteristics in scanner images. Due to the rise of generated input, using MRIs to investigate and estimate the accurate tumor area is a challenging task, and doing so takes time. Moreover, MRI images are produced in diverse sequences [6,7]. Therefore, an automatic and unified framework for segmentation methodology is desperately required for tumor diagnosis and future treatment planning.

The BraTS challenge denotes the dataset with multiple MRI sequences available. Clinically, typically presented MRI sequences include T1, T1C, T2, and FLAIR.

MRI images [1,10,11], which are produced by scanner, are shown in Figure 21.1 with varied characteristics. Figure 21.2 show the process of segmentation of brain tumor [26] and all acquired MRI images are preprocessed with bilateral filter [14,15] in order to produce quality results and to create further process quite easy for accurate detection [4,13]. To assess and analyze brain tumor, an automatic probabilistic model based on EM-GM is offered.

DOI: 10.1201/9781003381167-21

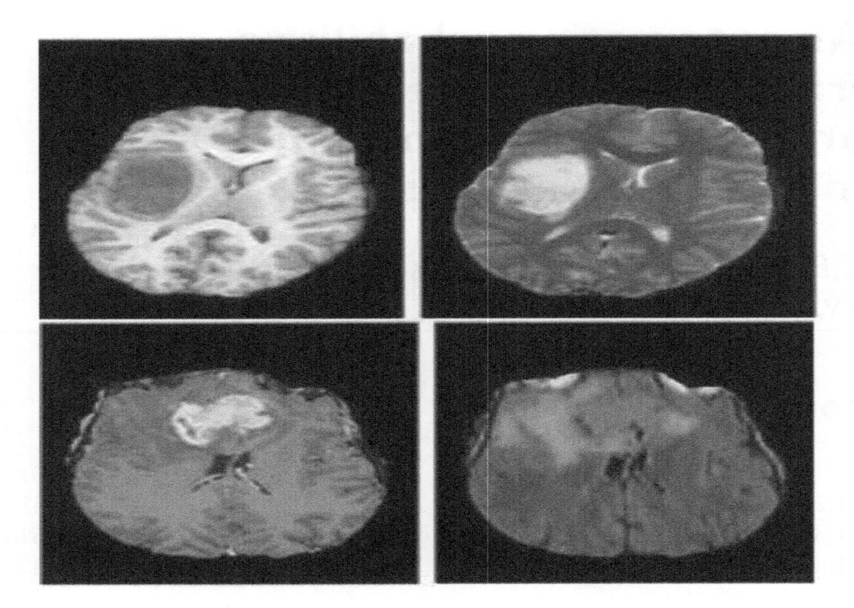

Figure 21.1 Brain MRI images from BRATS.

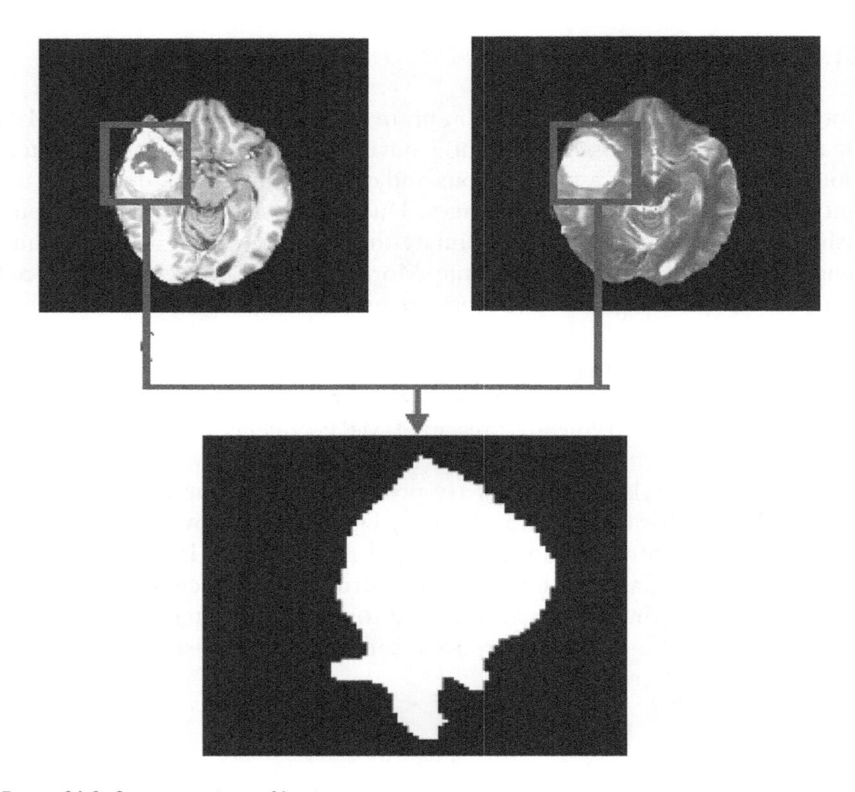

Figure 21.2 Segmentation of brain tumor.

21.2 RELATED WORK

The probability of error in MRI with conventional segmentation of tumor in MRI is unusually high due to varied characteristics of MRI [9–12]. Due to the increase in number of generated MRI images, the process for investigation of tumor is delayed. Henceforth, there is a requirement of automatic segmentation methodology to process and generate accurate segmentation results. Numerous techniques [21–23] were proposed to validate MRI brain disorders and to discover brain anatomy in order to analyze a large amount of data, but competent results were not produced in the survey [18–20].

The presence of the tumor in MRI can be detected using the fully automatic segmentation method called "watershed segmentation," which also shows the average segmentation accuracy for MRI imaging.

K – Mean technique is a widely used method for clustering of brain MRI images. It presents better segmentation results for datasets that are well separated and K – Mean technique is best suitable for large datasets and complex images.

FCMC - Fuzzy clustering method is established on fuzzy knowledge with region growing method. Fuzzy clustering is an unsupervised technique, and this algorithm precisely extracts tumor from MRI image modalities but has low accuracy for tumor segmentation. The shortcoming of this algorithm is difficulty of handling outlier points in MRI images segmentation.

Gaussian mixture technique demonstrates worthy segmentation results but yields medium accuracy for segmentation of MRI image modalities. Conditional Random Field with Global Classification method presents average segmentation accuracy for few datasets.

Figure 21.3 shows various Classification and Clustering techniques[24,25], which are used for MRI images, such as Neural networks (NNs), k-nearest neighbors (k-NN), Clustering methods, Expectation maximization (EM) methods; Bayesian approach, Random forest (RF) methods, Support Vector Machine (SVM), and Random fields. Expectation Maximization Gaussian

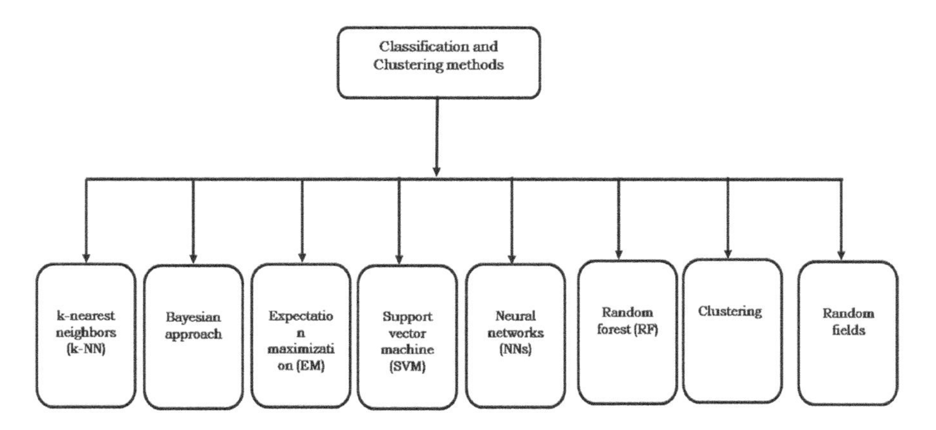

Figure 21.3 Classification and clustering methods to MRI images.

Mixture Technique (GMM) is used for noisy and low-contrast MRI images but less segmentation accuracy.

The fuzzy c-means technique was developed to overcome the issues in the MRI segmentation process, but segmentation presents inaccurate results for noisy MRI images, difficult to select initial condition, and undesirable segmentation results due to intensity in-homogeneity prevalent in MRI images. Although intensity in-homogeneity in MRI and the presence of noise are key challenges to gain enriched segmentation results, numerous segmentation techniques have been developed despite these challenges with inadequate accuracy. Probabilistic EMGMM [5–8] was presented for the classification and exploration of MRI into tumors and non-tumors. Nevertheless, existing techniques of finite mixtures for MRI detection retain a key challenge that is sensible to noise. The presented automatic probabilistic framework is a standard approach to extract precise tumor regions in MRI.

21.3 PROPOSED FRAMEWORK AND METHODOLOGY

An automatic probabilistic framework is recommended to develop a novel approach to accurate tumor extraction in MRI. The significant method constituents of the presented probabilistic framework are MRI images classification and tumor detection. A novel framework is developed for MRI images for automatic classification and segmentation in large datasets. The automatic probabilistic EMGMM remains most appropriate for a large set of input MRI. Novel method is established to compute Gaussian mixtures factors and finite mixture that is most flexible probabilistic model used for MRI segmentation. The key objective of an automatic probabilistic framework is amalgamation of Gaussian mixtures in EM to challenge noisy MRI images.

Figure 21.4 represents the workflow of automatic segmentation of proposed work with automatic probabilistic EMGMM.

The proposed method is most reliable and accurate for brain tumor detection based on a probabilistic approach to improve segmentation results. An automatic probabilistic framework is used to determine the maximization likelihood function along with missing data. Still, the presence of intensity in-homogeneity and noise in MRI are foremost limitations to find superior segmentation accuracy. The probabilistic EMGMM is cyclic process to determine maximum-likelihood, estimation of data points, and unsupervised setting. An automatic probabilistic framework has two steps: computing expectation and maximization. The repetition of steps is continuous till convergences. An automatic probabilistic framework is presented for recursively complete optimization parameter on unobserved label data.

The quality of MRI images is enhanced with pre-processing technique using bilateral filter for smoothing edge and image features for accurate extraction and analysis of tumor region. Maximizing the likelihood function is determined with base parameters, i.e weights in every component,

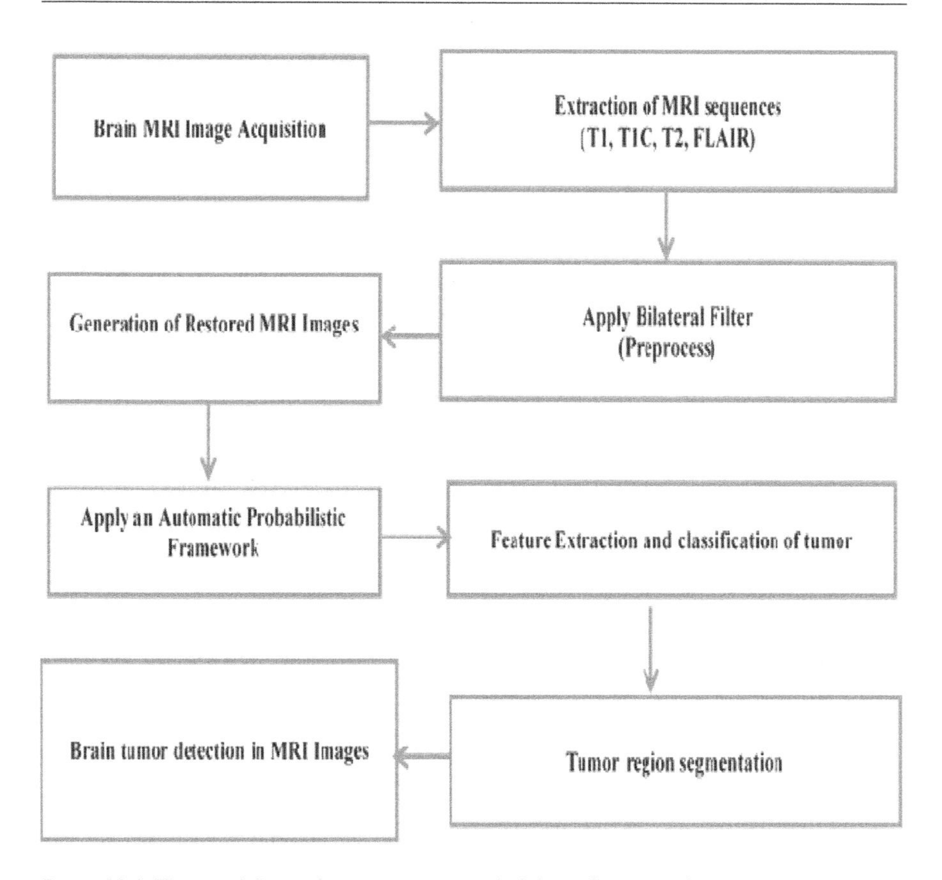

Figure 21.4 The workflow of an automatic probabilistic framework.

covariances, and means. EMGMM framework is used to extract and analyze tumor features in MRI images.

The precise and automatic segmentation of MRI images is realistic when a framework is evaluated on huge and assorted, and open challenging datasets. The proposed method may enhance the performance on varied data.

Figure 21.5 presents the flowchart of proposed an automatic probabilistic framework, which shows the entire process followed for precise tumor detection.

The following are main procedures in an automatic probabilistic framework:

- The Framework Initialize mixing coefficients, means, and covariance's parameters to discover initial parameters of likelihood.
- Expectation mechanism is used to determine posterior probability by current values.

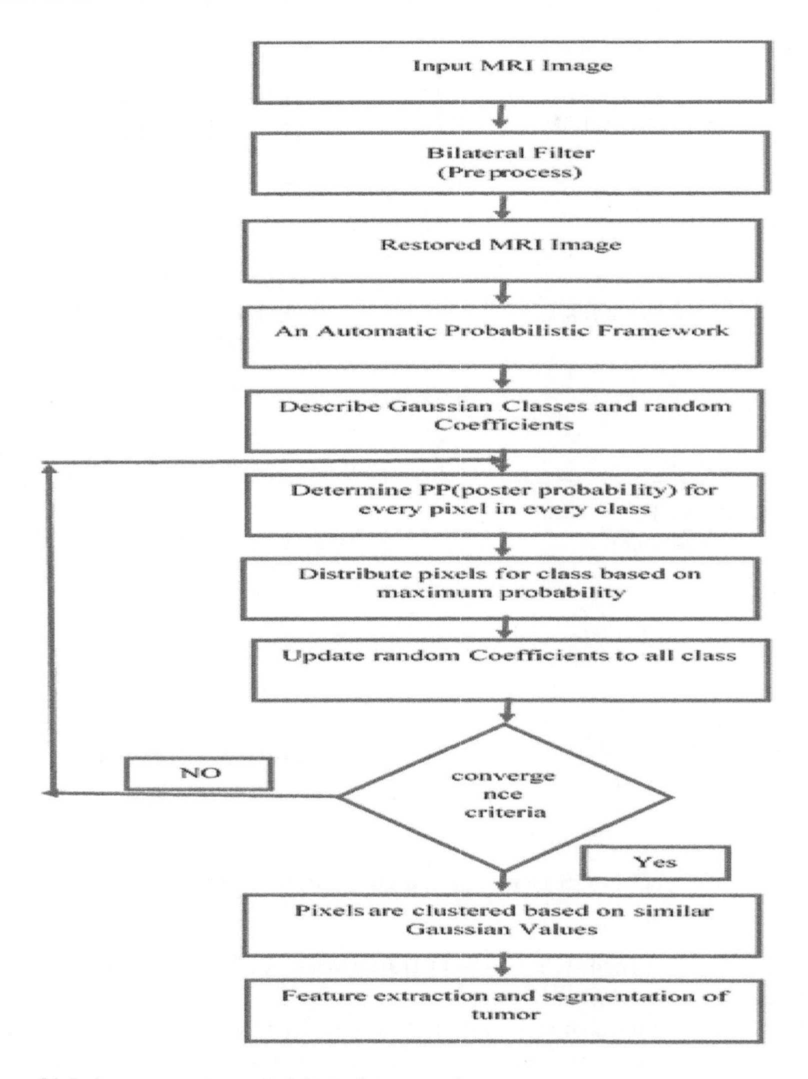

Figure 21.5 An automatic probabilistic framework.

- The maximization mechanism is used to recalculate values by current posterior and revised values.
- Finally, compute likelihood value and process is repeated till convergence standard is achieved.
- An automatic probabilistic framework is presented for recursively complete optimization parameter on unobserved label data.

21.4 RESULTS AND DISCUSSION

An automatic probabilistic framework is developed and examined to evaluate the performance [8–11,16,17] of work with standard datasets, which has ground truth and original MRI images. An automatic probabilistic framework in this chapter is demonstrated on challenge MRI datasets, i.e., BraTS15, BraTS16, BraTS17, and BraTS18. Clinically, typically presented MRI sequences include T1, T1C, T2, and FLAIR. The inconsistency in input MRI images dataset may influence predictable results of segmentation algorithms. In addition, the proposed work tested on large assorted datasets with varied sequences.

- Segmentation Accuracy:

$$\frac{\text{TruePositive} + \text{TrueNegative}}{\text{TruePositive} + \text{TrueNegative} + \text{FalsePositive} + \text{FalseNegative}} * 100$$

- An automatic probabilistic framework is quantitatively evaluated on challenge datasets and the superior segmentation accuracy signifies precise segmentation approach.

  Figure 21.6 presents illustrations of 40 assorted input MRI images from **BraTS15, BraTS16, BraTS17 and BraTS18,** which are used for assessment of the proposed probabilistic framework.

Figure 21.7 shows an improvement to MRI image by bilateral filter. In turn to assessment of the proposed method, two different approaches are used: classification accuracy and segmentation accuracy of an automatic probabilistic framework on multimodal image dataset. The open BraTS (Multimodal Brain Tumor Image Segmentation) MRI images dataset and clinical MRI images dataset through comparable imaging modalities of BraTS datasets used for assessment. The proposed probabilistic framework was evaluated on tumor and non-tumor data.

Figure 21.8 shows the examples of brain images from the BRATS database for T1 image. Table 21.1 and Figure 21.9 signify the comparative results analysis on an automatic probabilistic framework of ten patients' T1 MRI images chosen from BraTS15, BraTS16, BraTS17, and BraTS18 for tumor segmentation.

Figure 21.10 shows the examples of brain images from the BRATS database for T1C image. Table 21.2 and Figure 21.11 signify the comparative results analysis of an automatic probabilistic framework on ten patients' T1C brain MRI images chosen from BraTS15, BraTS16, BraTS17, and BraTS18 for tumor segmentation.

Figure 21.12 shows the examples of brain images from the BRATS database for T2 images. Table 21.3 and Figure 21.13 signify the comparative

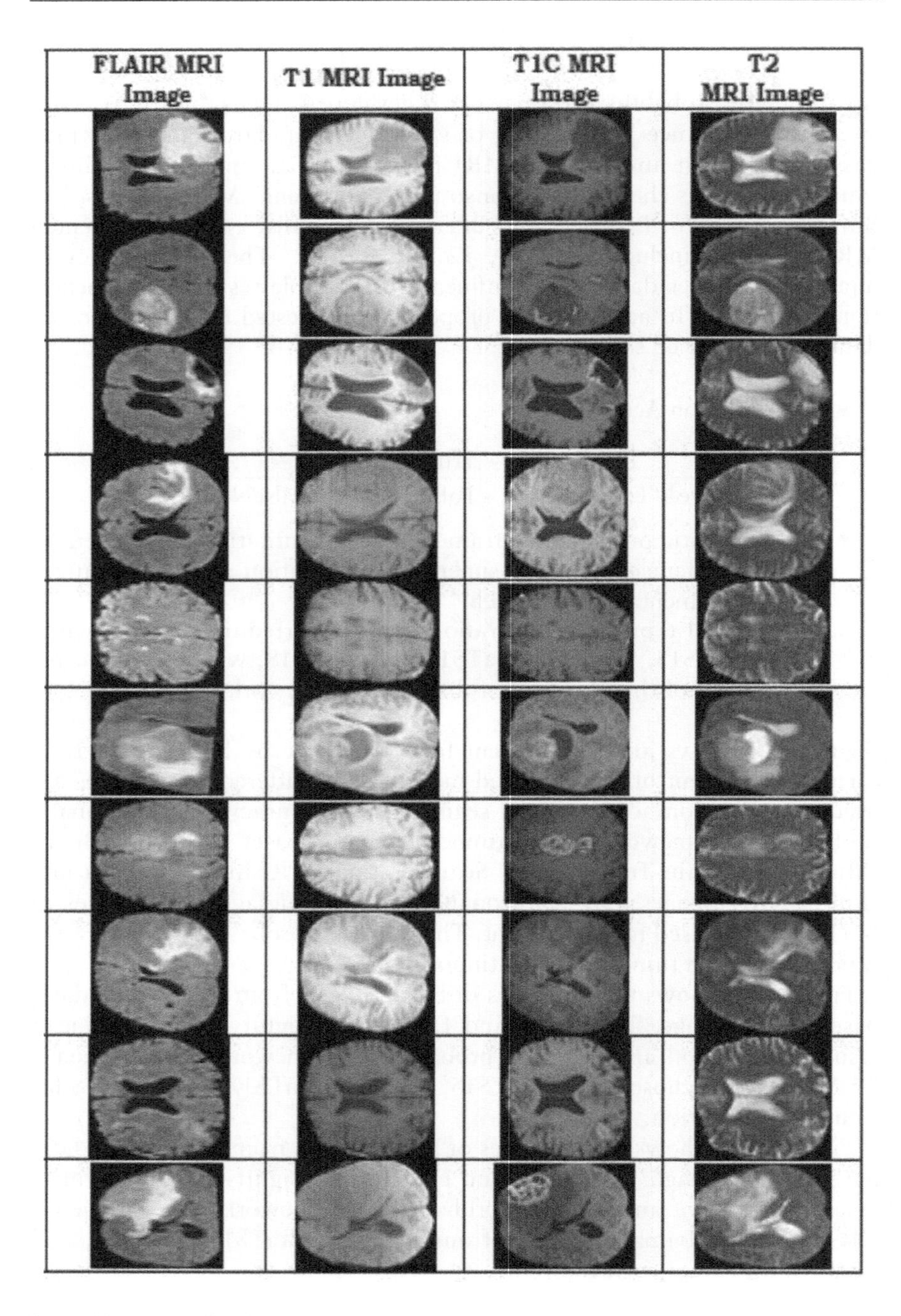

Figure 21.6 T1, T1C, T2, and FLAIR images.

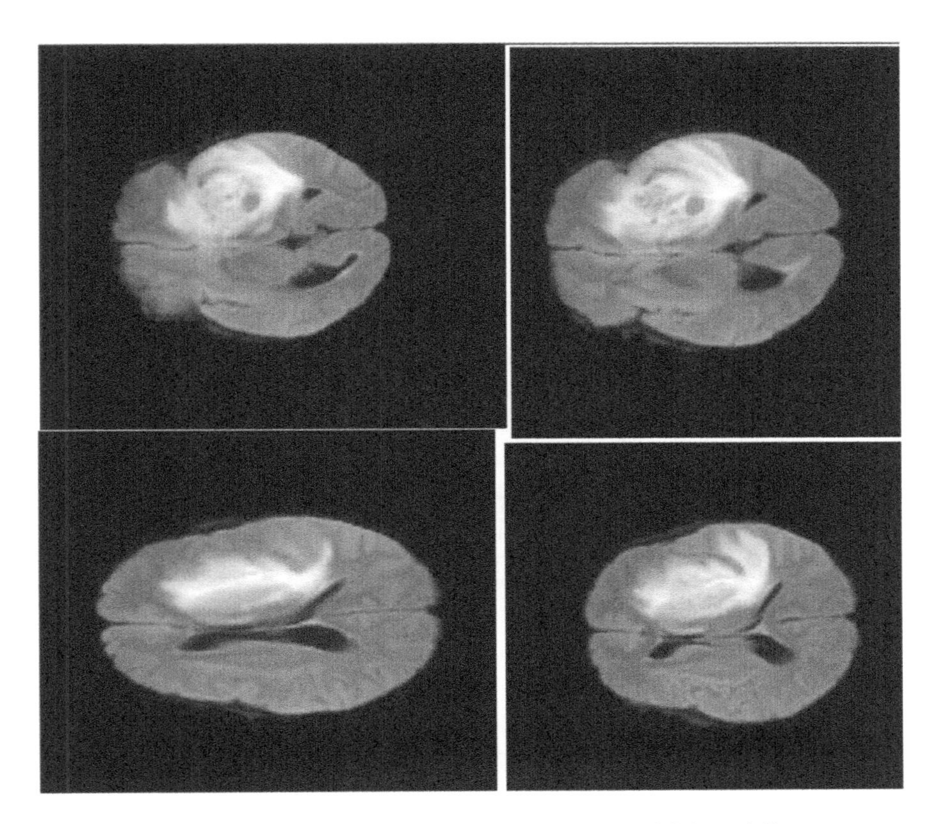

Figure 21.7 Example of the input data after pre-processing with bilateral filter.

results analysis of an automatic probabilistic framework on ten patients' T2 brain MRI images chosen from BraTS15, BraTS16, BraTS17, and BraTS18 for tumor segmentation.

Figure 21.14 shows the examples of brain images from the BRATS database for FLAIR images. Table 21.4 and Figure 21.15 signify the comparative results analysis of an automatic probabilistic framework on ten patients FLAIR brain MRI images chosen from BraTS15, BraTS16, BraTS17, and BraTS18 for tumor segmentation.

The proposed automatic probabilistic framework is evaluated, and segmentation results were obtained against the ground truth images with testing the segmentation accuracy. Figure 21.15 details the MRI images used for evaluation. Segmentation accuracy determines the overlap among mentioned MRI plus segmented image. As shown in Figures 21.6–21.9,

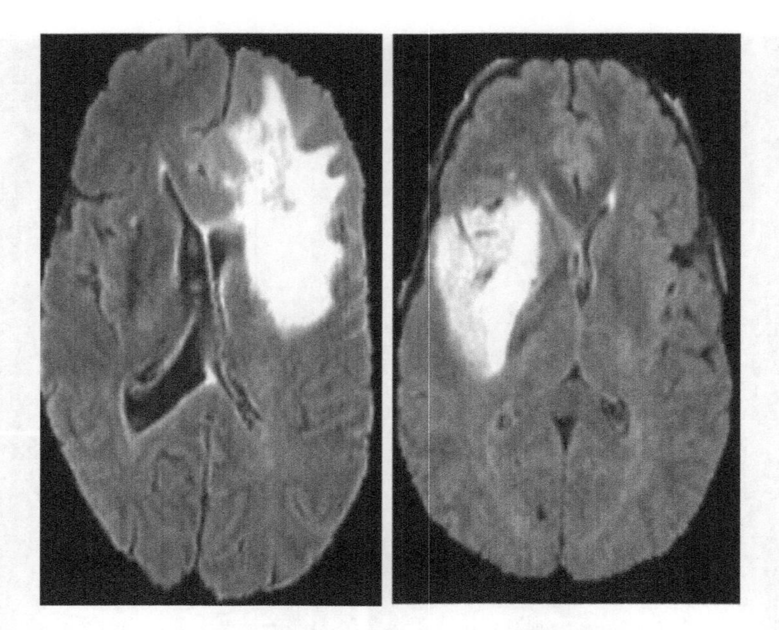

Figure 21.8 Examples of brain images from the BRATS database: T1 image.

Table 12.1 Evaluation results for automatic probabilistic framework for T1 MRI

Input MRI dataset	*Classification accuracy (%)*	
	Existing method	*Proposed method*
T1_001	96.99	97.13
T1_002	97.29	98.29
T1_003	97.64	98.64
T1_004	97.81	98.85
T1_005	98.30	98.96
T1_006	98.24	99.07
T1_007	97.38	98.13
T1_008	98.52	99.17
T1_009	95.15	96.54
T1_010	98.46	99.03

the final segmentation results are summarized and present promising segmentation results as compared to existing method. An automatic probabilistic framework was evaluated on BraTS15, BraTS16, BraTS17 and

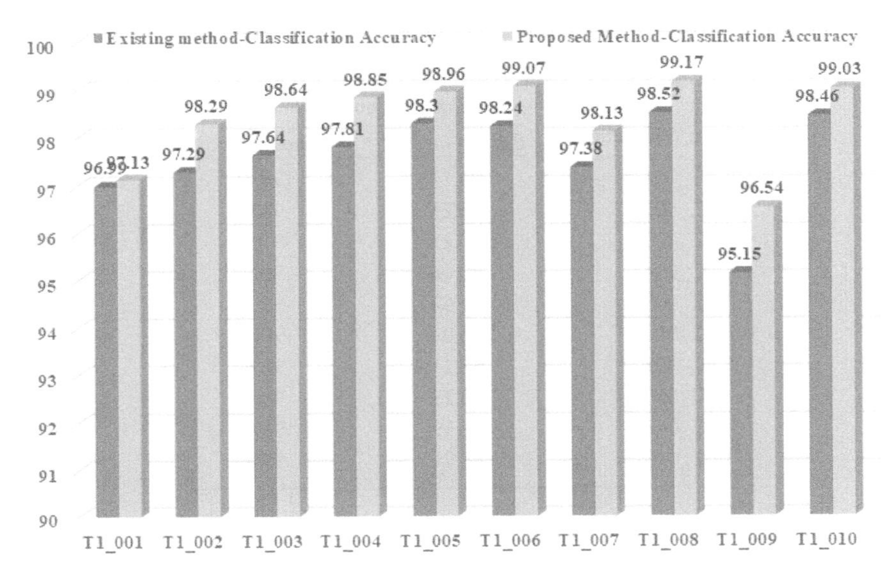

Figure 21.9 Analysis of automatic probabilistic framework on T1.

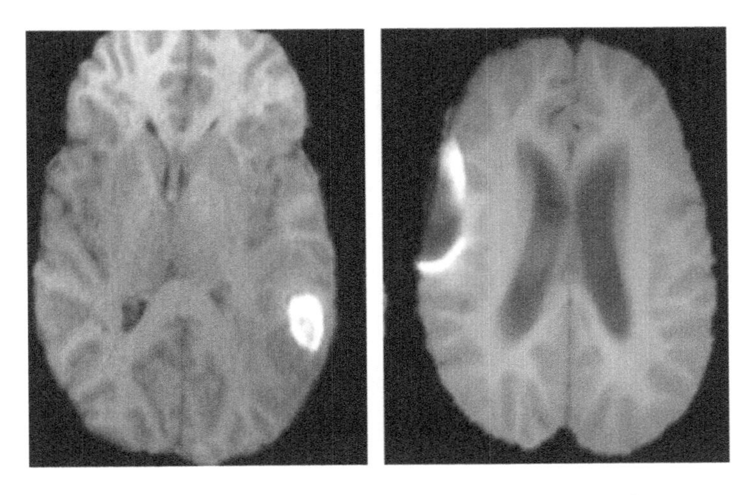

Figure 21.10 Examples of brain images from the BRATS database: T1C image.

BraTS18 MRI images, acquired results are further compared by obtainable methods and the superior results for segmentation of brain tumor are s.

Table 21.2 Evaluation results for automatic probabilistic framework for T1C MRI

Input MRI dataset	*Classification accuracy (%)*	
	Existing method	*Proposed method*
T1C_001	98.17	99.15
T1C_002	98.21	99.03
T1C_003	97.23	97.92
T1C_004	98.06	98.97
T1C_005	97.62	98.14
T1C_006	98.04	98.87
T1C_007	97.21	98.62
T1C_008	98.23	98.41
T1C_009	95.39	96.72
T1C_010	98.51	99.27

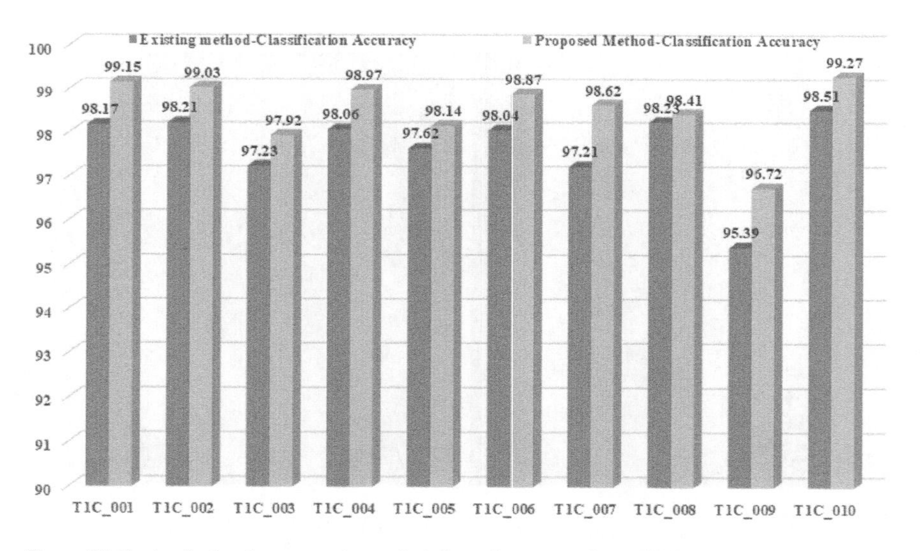

Figure 21.11 Analysis of automatic probabilistic framework on T1C.a

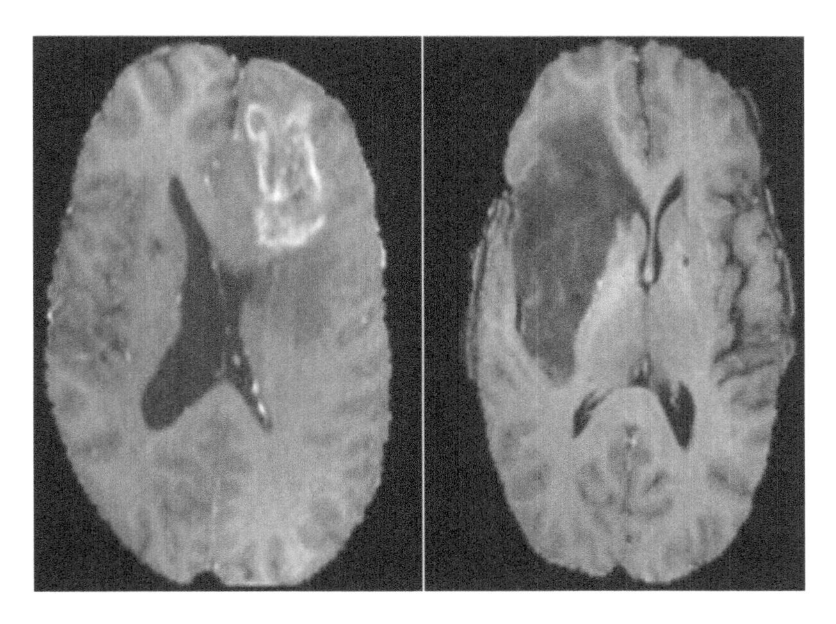

Figure 21.12 Examples of brain images from the BRATS database: T2 image.

Table 21.3 Evaluation results for automatic probabilistic framework for T2 MRI

Input MRI dataset	*Classification accuracy (%)*	
	Existing method	*Proposed method*
T2_001	98.51	99.16
T2_002	97.32	98.14
T2_003	98.27	99.01
T2_004	97.51	98.43
T2_005	98.17	99.11
T2_006	96.13	97.04
T2_007	97.44	98.51
T2_008	97.41	98.13
T2_009	98.21	99.08
T2_010	97.35	98.23

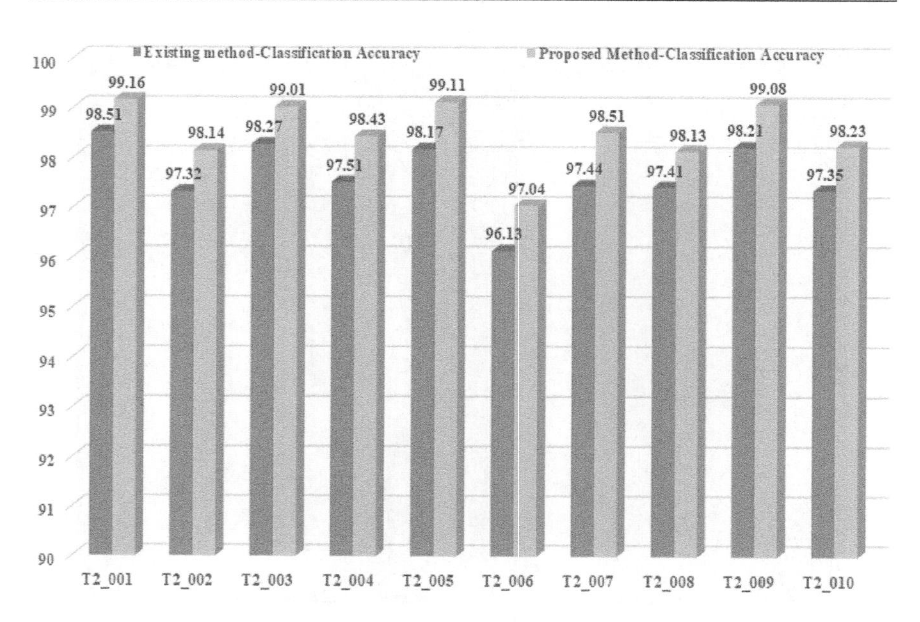

Figure 21.13 Analysis of automatic probabilistic framework on T2.

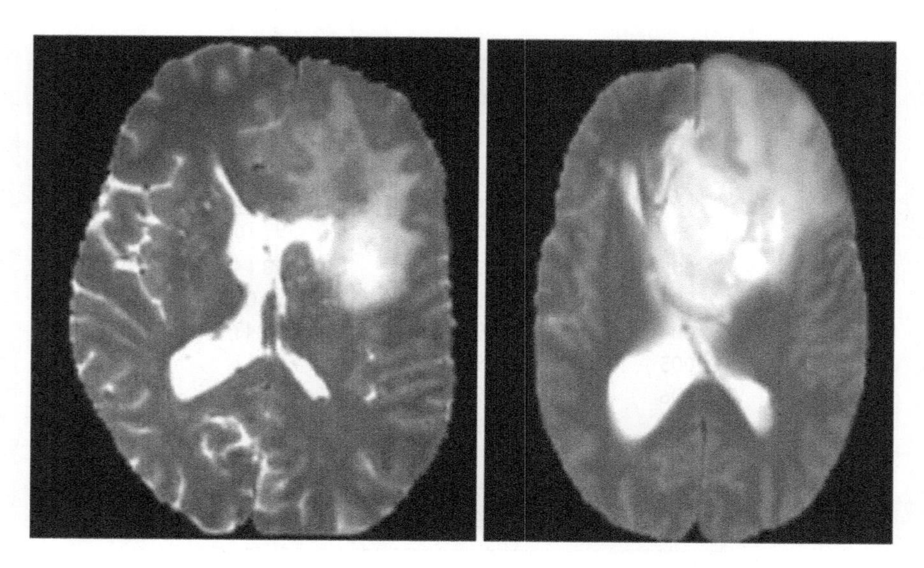

Figure 21.14 Examples of brain images from the BRATS database: FLAIR image.

Table 21.4 Evaluation results for automatic probabilistic framework for FLAIR MRI

Input MRI dataset	*Classification accuracy (%)*	
	Existing method	*Proposed method*
FLAIR_001	97.83	98.39
FLAIR_002	97.24	98.21
FLAIR_003	98.25	98.79
FLAIR_004	98.03	98.91
FLAIR_005	97.25	98.11
FLAIR_006	96.25	97.87
FLAIR_007	97.13	98.17
FLAIR_008	97.63	98.42
FLAIR_009	98.25	99.05
FLAIR_010	97.17	98.36

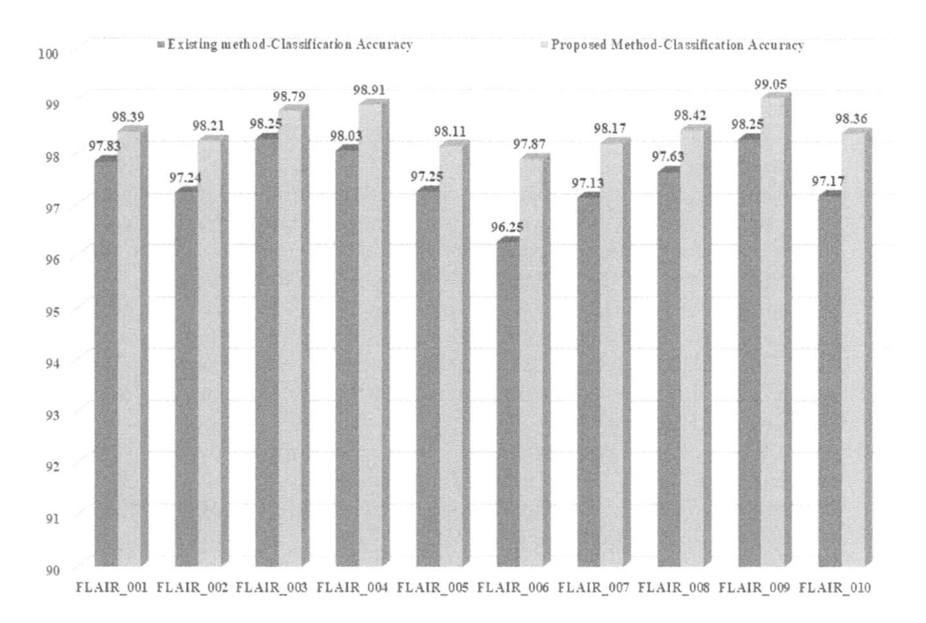

Figure 21.15 Analysis of automatic probabilistic framework on FLAIR.

21.5 CONCLUSION

This chapter submits an automatic probabilistic framework for tumors extraction in MRI images. The proposed technique presents the improved form of EMGM method. The EMGMM was developed to classify and assess a large set of MRI dataset images. An automatic probabilistic framework in this chapter has been evaluated on BraTS15, BraTS16, BraTS17, and BraTS18 challenge MRI datasets. In this work, an automatic probabilistic framework is quantitatively compared with existing segmentation techniques on synthetic MRI images. The proposed automatic segmentation of MRI images is realistic when a framework is evaluated on huge and assorted and open challenging datasets. The proposed method may enhance the performance on varied data.

REFERENCES

1. Menze, B. H., et al. The multimodal brain tumor image segmentation benchmark (BRATS). *IEEE Transactions on Medical Imaging* 34(10), 1993–2024, 2015.
2. Kalaiselvi, T, Sriramakrishnan, P. Rapid brain tissue segmentation process by modified FCM algorithm with CUDA enabled GPU machine. *International Journal of Imaging Systems and Technology* 28(3), 163–174, 2018.
3. Sharma, A., Hamarneh, G. Missing MRI pulse sequence synthesis using multi-modal generative adversarial network. *IEEE Transactions on Medical Imaging* 39(4), 1170–1183, 2020.
4. Kalaiselvi, T., Kumarashankar, P., Sriramakrishnan, P. Three-phase automatic brain tumor diagnosis system using patches based updated run length region growing technique. *Journal of Digital Imaging, Society for Imaging Informatics in Medicine*, 2019. doi: 10.1007/s10278-019-00276-2.
5. Pravitasari, A. A., Qanita, S. F., Iriawan, N., Fithriasari, K., Irhamah, Purnami, S. W., Ferriastuti, W. MRI-based brain tumor segmentation using Gaussian and hybrid Gaussian mixture model-spatially variant finite mixture model with expectation-maximization algorithm. *Malaysian Journal of Mathematical Sciences* 14(1), 77–93, 2020.
6. Song, Y, Ji, Z, Sun, Q. An extension Gaussian mixture model for brain MRI segmentation. *2014 36th Annual International Conference of the IEEE Engineering in Medicine and Biology Society*, pp. 4711–4714, 2014. doi: 10.1109/EMBC.2014.6944676. PMID: 25571044.
7. Nguyen, T. Gaussian mixture model based spatial information concept for image segmentation. Electronic Theses and Dissertation 438, 2011. https://scholar.uwindsor.ca/etd/438.
8. Kumar, N. S., Satoor, S., Buck, I. Fast parallel expectation maximization for Gaussian mixture model on GPUs using CUDA. *11th IEEE International Conference on High Performance Computing and Communications*, 103–105, 2009. doi: 10.1109/HPCC.2009.45.
9. Khalil, M, Ayad, H, Adib, A. Performance evaluation of feature extraction techniques in MR-brain image classification system. *Procedia Computer Science* 127, 218–225, 2018.

10. Wang, G., Li, W., Ourselin, S., Vercauteren, T. Automatic brain tumor segmentation using cascaded anisotropic convolutional neural networks. In *International MICCAI Brain-lesion Workshop*, pp. 178–190. Springer, Cham, 2017.
11. Sriramakrishnan, P, Kalaiselvi, T, Rajeswaran, R. Modified local ternary patterns technique for brain tumour segmentation and volume estimation from MRI multi-sequence scans with GPU CUDA machine. *Biocybernetics and Biomedical Engineering* 39(2), 470–487, 2019.
12. Bhima, K., Jagan, A. Analysis of MRI based brain tumor identification using segmentation technique. *2016 International Conference on Communication and Signal Processing (ICCSP)*, pp. 2109–2113, Melmaruvathur, 2016. doi: 10.1109/ICCSP.2016.7754551.
13. Bhima, K., Jagan, A. An improved method for automatic segmentation and accurate detection of brain tumor in multimodal MRI. *International Journal of Image, Graphics and Signal Processing (IJIGSP)* 9(5), 1–8, 2017. doi: 10.5815/ijigsp.2017.05.01.
14. Ryan, P. C., David, H. *Laidlaw: Bilateral Filtering of Multiple Fiber Orientations in Diffusion MRI.* Springer-Verlag, 2015. http://link.springer.com/book/10.1007/978-3-319-11182-7.
15. Bhonsle, D., Chandra, V., Sinha, G.R. Medical image denoising using bilateral filter. *I.J. Image, Graphics and Signal Processing* 6, 36–43, 2012. doi: 10.5815/ijigsp.2012.06.06.
16. Hassen, O. A., et al. Nature-inspired level set segmentation model for 3D-MRI brain tumor detection. *Computers, Materials & Continua* 68(1), 961–981, 2020. doi: 10.32604/cmc.2021.014404.
17. Menze, B. H., Jakab, A., Bauer, S., et al. The multimodal brain tumor image segmentation benchmark (BRATS). *IEEE Transactions on Medical Imaging* 34(10), 1993–2024, 2015.
18. Muir, C. S., Storm, H. H., Polednak, A. Brain and other nervous system tumours. *Cancer Surveys* 19(20), 369–392, 1994.
19. Bauer, S., Wiest, R., Nolte, L. P., Reyes, M. A survey of MRI-based medical image analysis for brain tumor studies. *Physics in Medicine and Biology* 58(13), R97–R129, 2013.
20. Jiang, Y., Zhao, K., Xia, K., et al. A novel distributed multitask fuzzy clustering algorithm for automatic MR brain image segmentation. *Journal of Medical Systems* 43(5), 118:1–118:9, 2019.
21. Ranjbarzadeh, R., Bagherian Kasgari, A., Jafarzadeh Ghoushchi, S., et al. Brain tumor segmentation based on deep learning and an attention mechanism using MRI multi-modalities brain images. *Scientific Reports* 11, 10930, 2021. doi: 10.1038/s41598-021-90428-8.
22. Tjahyaningtijas, H. P. A. Brain tumor image segmentation in MRI image. *IOP Conference Series: Materials Science and Engineering* 336, 012012, 2018.
23. Bakas, S., Zeng, K., Sotiras, A., Rathore, S., Akbari, H., Gaonkar, B., Rozycki, M., Pati, S., Davatzikos, C. GLISTRboost: Combining multimodal MRI segmentation, registration, and biophysical tumor growth modeling with gradient boosting machines for glioma segmentation. *Lecture Notes in Computer Science* 9556, 144–155, 2016. doi: 10.1007/978-3-319-30858-6_13.
24. Erasa, M., Meena, K. A phenomenological survey on various types of brain diseases using soft computing techniques. *International Journal of Civil Engineering and Technology* 8, 1209–1220, 2017.

25. Nyo, M., Mebarek-Oudina, F., Hlaing, S., Khan, N. Otsu's thresholding technique for MRI image brain tumor segmentation. *Multimedia Tools and Applications*, 2022. doi: 10.1007/s11042-022-13215-1.
26. Gordillo, N., Montseny, E., Sobrevilla, P. State of the art survey on MRI brain tumor segmentation. *Magnetic Resonance Imaging* 31(8), 1426–1438, 2013.

Chapter 22

Comparative study of generative adversarial networks for sensor data generation-based remaining useful life classification

Anindya Chatterjee, Indrajit Kar, Rik Das, Sudipta Mukhopadhyay, Jaiyesh Chahar, and Pravar Kulbhushan

Siemens Technology Services Pvt. Ltd.

CONTENTS

22.1 INTRODUCTION AND MOTIVATION

For estimating the breakdown of modern time equipment which are Multiphysics in nature, Remaining Useful Life (RUL) is significantly applicable [1]. A multiphysics system represents devices working on various concepts of physics including electrical and mechanical components, chemical reactions, electromagnetic waves together, etc. Most of the complex systems in industry are depreciating assets of excessive cost, which makes it necessary to design and develop data-driven artificial intelligence (AI) techniques for Prognostics and Health Management (PHM) of the machinery. Prognostics-based tools can be used by any organization to find out RUL of any depreciable system. The probable length of time before any machine that is repaired or replaced is termed as RUL. This time estimation is useful for engineers to schedule necessary maintenance activities by understanding the remaining time to failure and thus mitigating the challenge of sudden downtime. Hence, the estimation model created using RUL has significant contribution in providing a confidence bound for prediction of failure [2].

DOI: 10.1201/9781003381167-22

RUL prognostics-based techniques [3] are divided into data-driven analytics-based method and mechanism-driven analytics-based method. Effectiveness of any PHM [4] system is governed by RUL. RUL must be estimated very accurately as the results generated play a critical role in operation.

Motivations behind this work are:

- Addressing the dearth of training data is instrumental in enhancing the classification accuracy for RUL estimation.
- Automating RUL prediction by implementing machine learning algorithms to address lack of domain expertise [5] in designing a complex system-based solution.

The authors have addressed these major concerns by proposing the data-driven approach that is facilitated by using time series-based ACGAN and image-based ACGAN. The propositions have proven to be useful in quantifying the uncertainties associated with RUL computation and prognostics in general.

Therefore, the objective of this work is

- To generate data from different ACGANs similar to the original data.
- To improvise the RUL estimation and classification using generated data.

In this research work, time series data is transformed into spectrograms and recurrence plot images. The images are used to train convolution neural network (CNN)-based image ACGAN. After the data generation from ACGANs, bidirectional LSTM neural network and CNN are used for sequential classification of RUL on actual data and generated data. The classification accuracies on original data, image ACGAN data, and time ACGAN data are evaluated. The C-MAPSS dataset is provided by NASA [6] and is used as a test bed in this work. The improvement in accuracies of RUL estimation and classification using datasets generated with proposed ACGANs is observed.

The structure of this chapter initiates with Introduction followed by related works and proposed techniques. The research outcomes are discussed in "Results and discussion" section. The future scope of work is stated in "Conclusion" section.

22.2 RELATED WORK

Implementing RUL-governed Prognosis Health Management (PHM) can prevent radical failure [7]. Model-based, sensor-based, and hybrid approaches can predict a machine's RUL. Data-driven methods are a good way to extract information from historical and online statistics [8].

Descriptor definition is used to extract features from high-dimensional PHM data. Deep convolution neural network (DCNN) model extracts high-level features [9,10]. Reference [11] proposes a time window approach for RUL prediction using an attention-based DCNN. RUL estimation typically uses sequential data. Sequential data used in RUL estimation can be processed by RNN and LSTM [12–14]. Reference [15] introduces a semi-supervised RUL estimation architecture using Restricted Boltzman Machine (RBM). A semi-supervised learning approach using VAE-based nonlinear embedding is introduced [16]. Reference [17] introduces generative adversarial networks (GANs) for image generation [18]. Deep Convolution GAN DCGAN [19,20] generates data well. RUL must be estimated to develop new measures for evaluating engine performance degradation.

Auxiliary classifier developments in deep learning have facilitated GAN implementation [21]. ACGAN was used to extract features and simplify RUL class distributions. Input images are reshaped using a native variant of ACGAN. Every GAN-generated image has a class label. GANs generate images using noise and class labels.

In this work, the authors propose generating synthetic data using time series ACGAN and image-based ACGAN for the first time to improve existing RUL prediction, with promising results.

22.3 PROPOSED TECHNIQUE

This research uses GANs to generate RUL data. Time series and image data are generated. Early experiments showed classification loss, but model output quality improves with more classes. To make RUL calculations more accurate, they've been divided into three classes, so alarms can be raised when remaining useful life crosses a threshold. Natural images have multidimensional, hard-to-quantify features. Improved deep learning models like GANs have helped image compression [22], denoising [23], and super-resolution [24]. ImageNet model synthesis [21] is difficult because of more classes, but GANs perform well by memorizing many examples. A GAN is a combination of two deep learning models trained in opposition to one another. Generator GEN takes a random noise vector z as an input and outputs an image (I) as introduced in reference [17]. The equation for the output image (I) is

$$I_{fake} = \text{GEN}\left(Z\right) \tag{22.1}$$

Synthetic image generated from GEN along with a real image is fed as an input to discriminator DIS which gives a probability distribution (P) as the output in reference [17].

$$P\left(S|I\right) = \text{DIS}\left(X\right) \tag{22.2}$$

Error is backpropagated. For maximizing the log-likelihood, which is assigned to the correct source, DIS is trained. As shown in reference [22], the second term of the equation is minimized when the GAN is trained. The expression for the proposition is given in Equation (22.3)

$$LD = E[\log P(S = \text{real} \mid I_{fake})] + E[\log P(S = \text{fake} \mid I_{fake})] \tag{22.3}$$

Conditioning of GANs can be achieved by adding auxiliary information [17]. Smaller dataset classes can be compensated using ACGANs-based approach and that helped them significantly to improve their results. An image ACGAN is used to reshape the input images to make them suitable for the training. Every generated image will have a class label associated with it. GANs use both noise and class labels to generate synthetic images (Figure 22.1).

ACGAN uses NASA Turboprop data. This dataset includes Turboprop engine sensor data for anomalies. ACGAN is trained using data from all

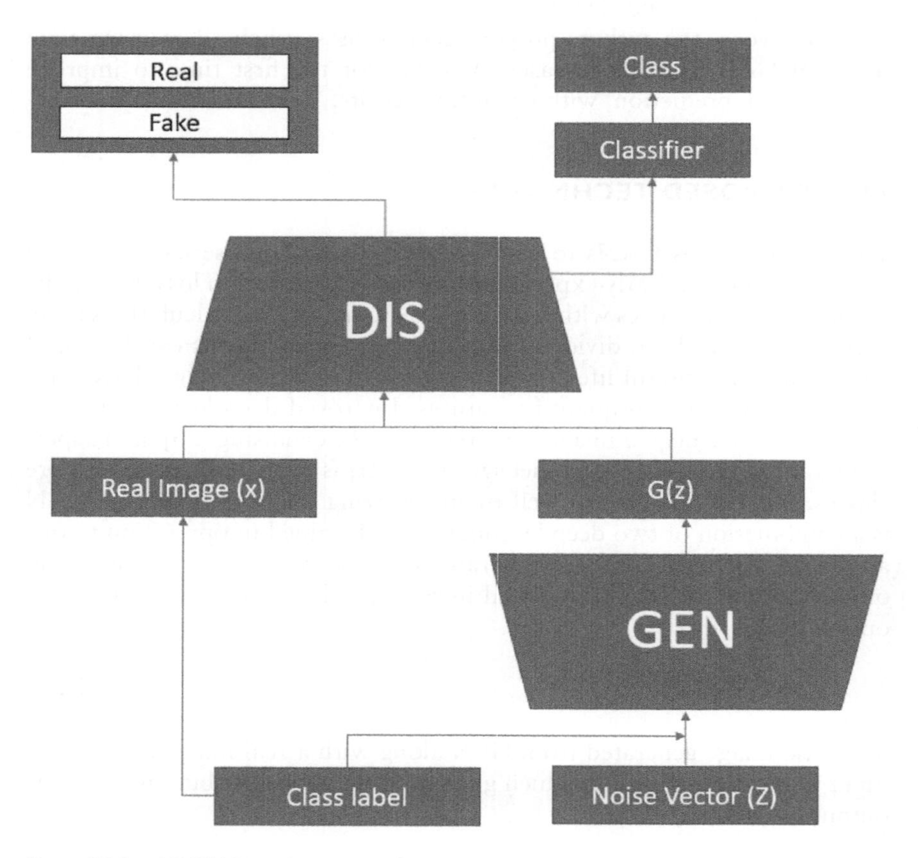

Figure 22.1 ACGAN working procedure.

three engine sets. State space trajectory describes a dynamical system. Recurrence Quantification Analysis (RQA) measures a dynamical system's recurrence duration and number. The graph can be passed to any CNN [25] for feature extraction and processing [26]. This chapter trains auxiliary classifier GANs using recurrence plots. For time series RUL classification, a trained discriminator will classify the generated images. Time GAN extends generative adversarial networks [27]. An efficient generative model for time series data should maintain temporal dynamics and consider original variable relationships with time. GAN's time framework. The method uses a time series ACGAN. Instead of images, ACGAN passes time series data. Non-deterministic output improves generalization and training data quality and quantity. ACGAN uses NASA Turboprop data to train. The time ACGAN model's output data matched the original data. Neural network training should avoid overfitting.

22.4 RESEARCH METHODOLOGY

Temporal data has more information than non-temporal data sampled at a single time step. Single window sliding is used in the proposed method to efficiently extract multivariate temporal information. Different window sizes generate a 2D matrix for the deep learning model's input. The proposed deep learning pipeline receives this segmented signal matrix. Unsupervised learning is used to train ACGAN, which will help with RUL image classification. ACGAN's generator will use a Decoder to hide input data. In the generator, convolution network parameters are carefully chosen to match the training set's output tensor. The time complexity for training neural network with n layers with i, j, k, ... nodes with t training samples and m epochs would be O(mt∗(ij+jk+kl+...)). This is necessary because both go to the discriminator to be evaluated (Figure 22.2).

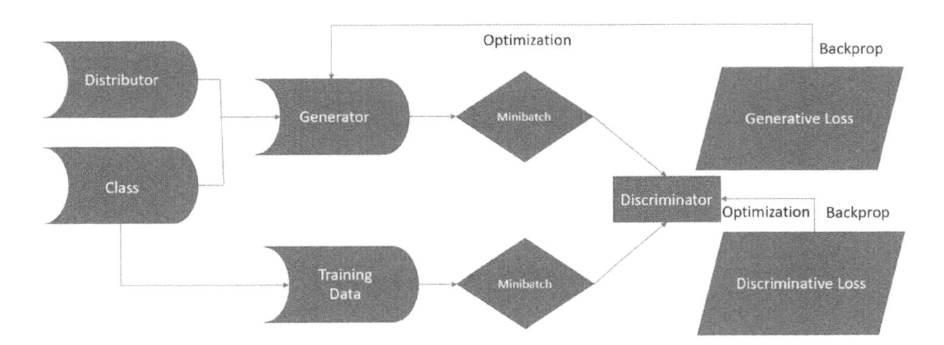

Figure 22.2 Architecture of ACGAN.

- To handle multiphysics data, three convolutional layers are stacked with 2-dimensional convolution filters and single padding to extract features from generator sensor observations.
- The first convolutional layer has 128 3×3 filters. These stride-1 convolution filters extract pixel relationships.
- Leaky ReLU is used as an activation function to generate more dense matrices so gradients don't saturate and errors can be better backpropagated.
- Discriminator is coded like generator. Last convolution layer has tanh activation. Discriminator sorts of input and generated data.
- When generator can capture input data distribution, discriminator can't tell fake from real images. Discriminator network uses 25% dropout to prevent overfitting network.

22.5 EXPERIMENTAL SETUP

CMAPPS dataset is used for building ACGAN. Dataset contains the simulation of faulty sensors and their deterioration information when they are exposed to different operating conditions. Schematic representation of sensors in CMAPSS is shown in Figure 22.3.

Altitude, Mach-Number, and Throttle revolver angle determine engine operating modes. Subsets may have different numbers of engines with varying operational cycles. All engines vary in manufacturing and initial wear and tear. Over time, the engine degrades. From subsets FD001 and FD003, sensors 2,3,4,7,8,9,11,12,13,14,15,17,19,20,21 are chosen for feature selection. Min-max data normalization follows. Piecewise model with linear degradation generates labels. The proposed architecture is shown in Figure 22.4 using a flowchart:

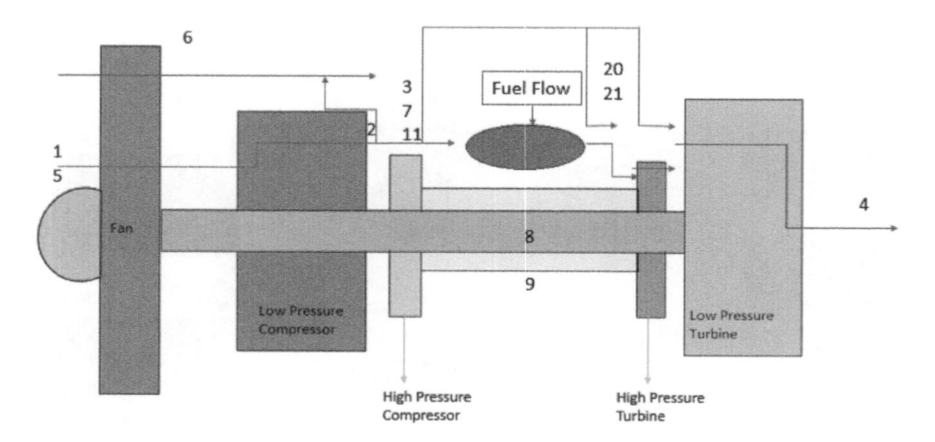

Figure 22.3 Schematic representation of sensors in CMAPSS.

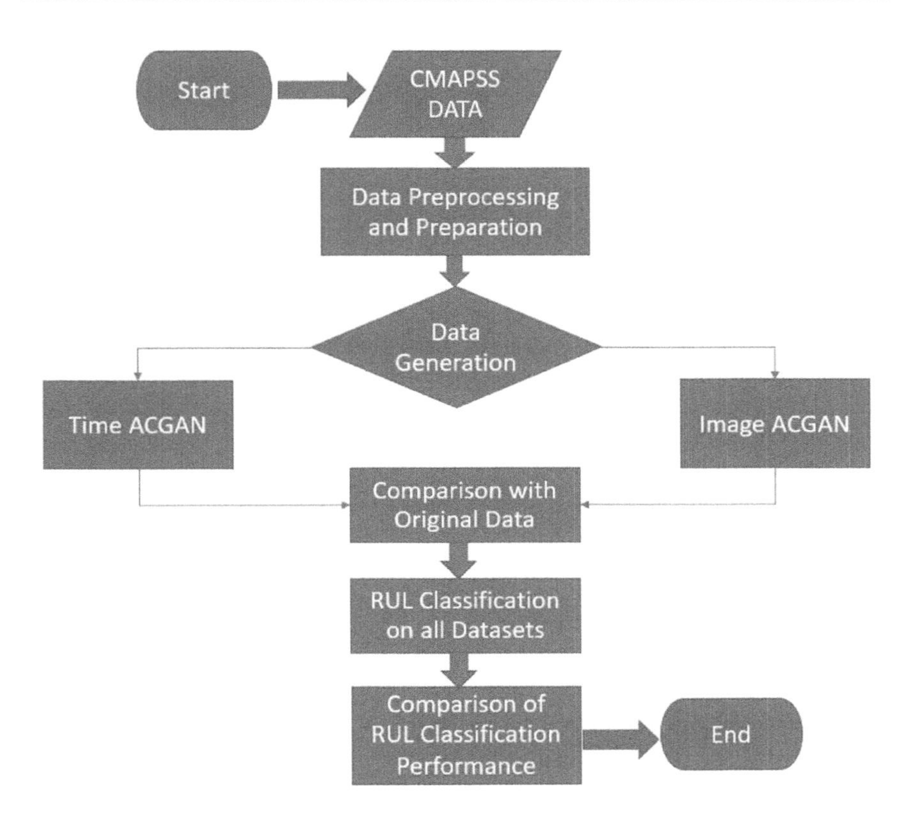

Figure 22.4 Proposed architecture.

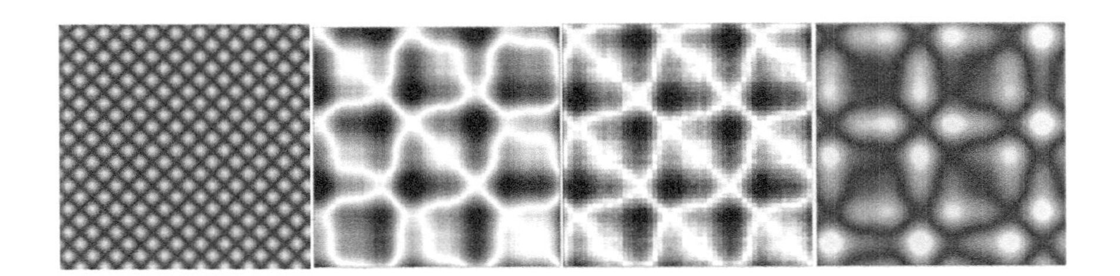

Figure 22.5 Recurrence plots generated from ACGAN network.

Some of the generated recurrence plot samples from the ACGANS are shown in Figure 22.5.

22.6 RESULT AND DISCUSSION

Experiments run on a 9th-generation Intel® Core i7, 32 GB RAM, NVIDIA Quadro T1000 GPU, and Windows 10 system. TensorFlow [28] uses anaconda's Python 3.7. Proposed ACGANs are evaluated by comparing generated data with the original dataset and RUL classification using bidirectional LSTM and CNN. Two methods are compared to see which is more accurate (Figures 22.6–22.8).

To compare ACGAN-generated datasets to original data. Higher similarity means the model can understand the data and generate similar data for RUL classification. Since the series lengths are the same, Pearson correlation coefficient and Mahalanobis distance are used to compare them. *Dp* and *DM* denote Pearson correlation [29] and Mahalanobis distance [30], respectively in Equations (22.3) and (22.4).

$$D_p(x_i, x_j) = \frac{\mathring{a}\left(x_i - \overline{x_i}\right)\left(x_j - \overline{x_j}\right)}{\sqrt{\mathring{a}\left(x_i - \overline{x_i}\right)^2 \mathring{a}\left(x_j - \overline{x_j}\right)^2}} \tag{22.4}$$

$$D_M\left(x_i, x_j\right) = (x_i - x_j)W^{-1}(x_i - x_j)^T \tag{22.5}$$

The third metric is a comparison between subsets of time epochs between the generated series and actual series as in Equation (22.5). A dynamic time warping (DTW) technique can be used to compare time series of unequal

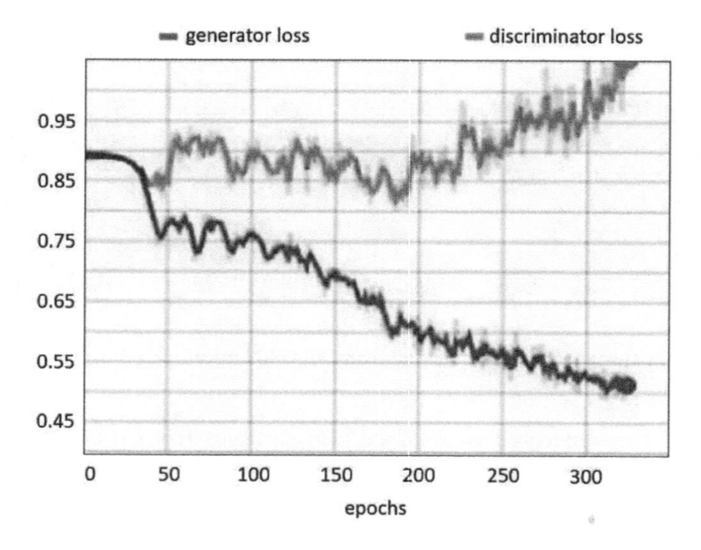

Figure 22.6 ACGAN loss – Generator (Blue) and discriminator (Red).

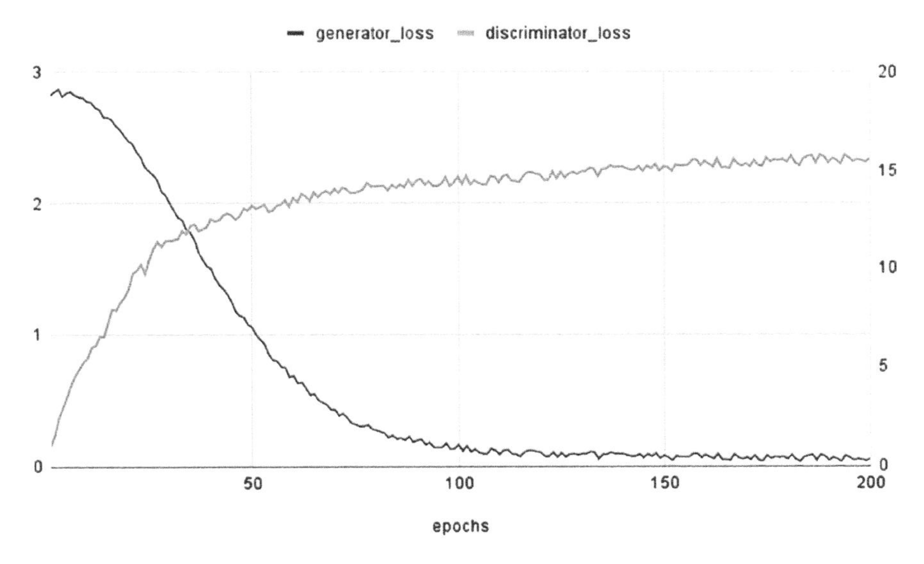

Figure 22.7 Time ACGAN loss – Generator (Blue) and discriminator (Red).

lengths. For DTW, a distance matrix [31] is formed between two time series n and m as below:

$$M(d)\left(\left(x_i, x_j\right)\right) = \left(d\left(x_k^i, x_l^j\right)\right)_{ij} \tag{22.6}$$

for $1 \le k \le n$ and $1 \le l \le m$, where $d\left(x_k^i, x_l^j\right)$ is the distance of the kth point to the series x_i from the lth point of the series x_j. and the aim is to minimize the DTW distance.

Table 22.1 shows the derived value of comparative metrics of the generated series using image ACGANs and time ACGANs with respect to actual data.

The data computed in Table 22.1 has shown the Pearson correlation value for image ACGAN is more compared to time ACGAN. This denotes that similarity of generated data using image ACGAN is higher to the original sensor dataset when compared to that of time series ACGAN.

Image ACGAN has lower Mahalnobis and DTW distances. Timeseries ACGAN data is less accurate than image ACGAN data.

Classification compares original data to image and time ACGAN data. After 15 cycles, a device is in the first class; after 45, the second. If the prediction doesn't fall in either class, a third class indicates a high RUL. Classifying actual, image, and time ACGAN data. All data are converted to spectrograms for evaluation. Evaluating bidirectional LSTM neural network accuracy and epochs. The image ACGAN dataset had better accuracy after 17.5 k epochs, whereas Time ACGAN overfits after 14.5 k epochs.

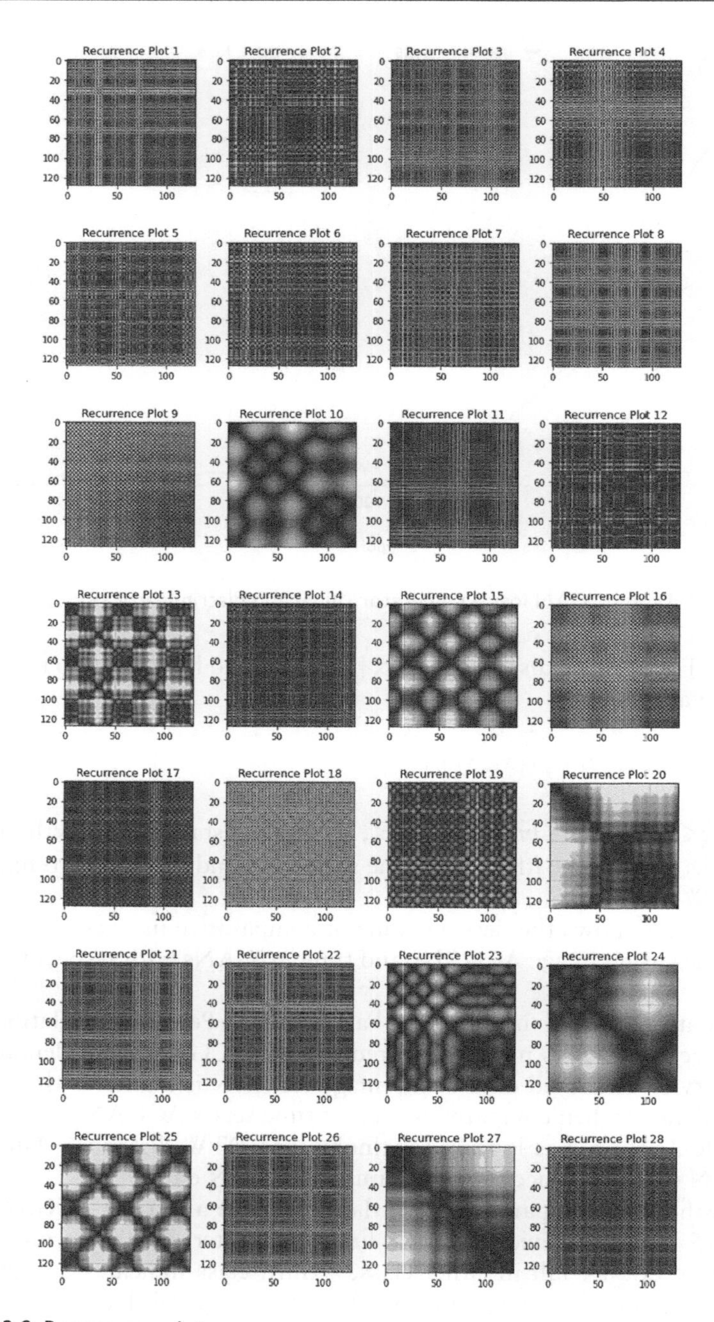

Figure 22.8 Recurrence plots.

Table 22.1 Comparative analysis of generated series with respect to original series

Metric	*Image ACGAN*	*Time ACGAN*
Pearson correlation	0.82	0.69
Mahalanobis distance (mean of 1st 5,000 points)	0.70	1.08
Mean DTW distance (5 observations)	67.3	109.21

The original data trained on the same model performed poorly as shown in Figures 22.9–22.11.

Figures 22.9–22.11 shows superior classification accuracy using image ACGAN data generation for RUL prediction. The classification accuracy for data generated using image ACGAN and time series ACGAN is even better than the original sensor data used for RUL prediction (Figures 22.9–22.11). Figure 22.10 shows Time Series ACGAN overfitting during training. The goals of this work were achieved by successfully generating data using ACGANs that are useful for RUL estimation and classification and outperform original sensor data.

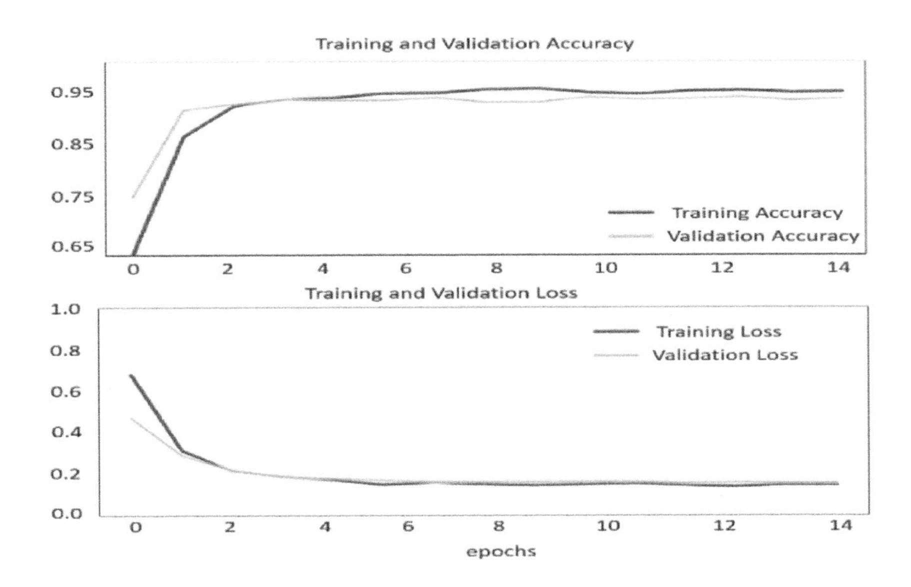

Figure 22.9 Accuracy and loss vs Epoch (K) – Actual data.

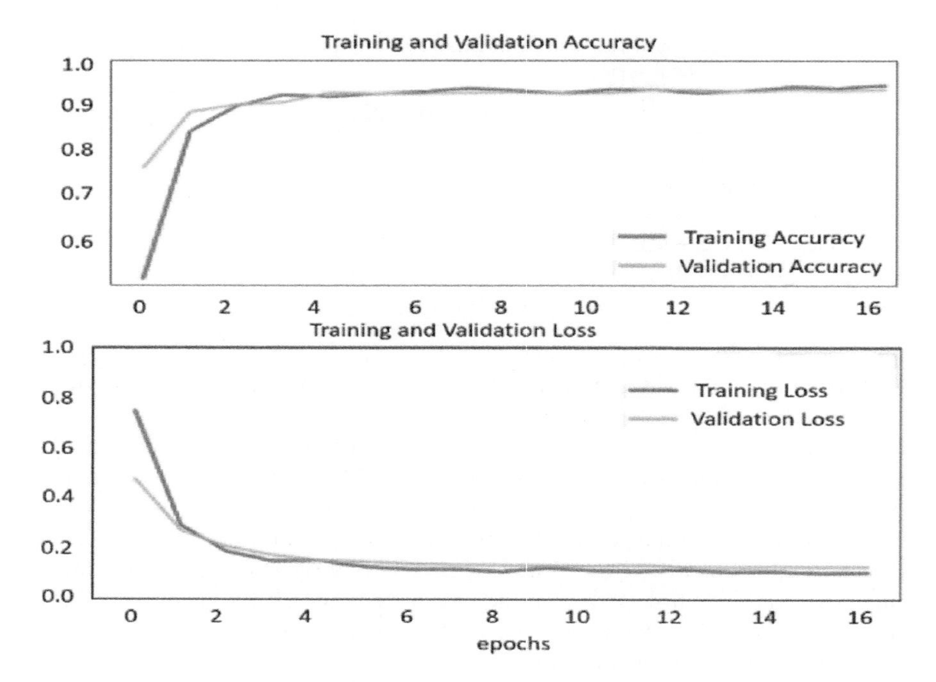

Figure 22.10 Accuracy and loss vs Epoch (K) – Time ACGANs.

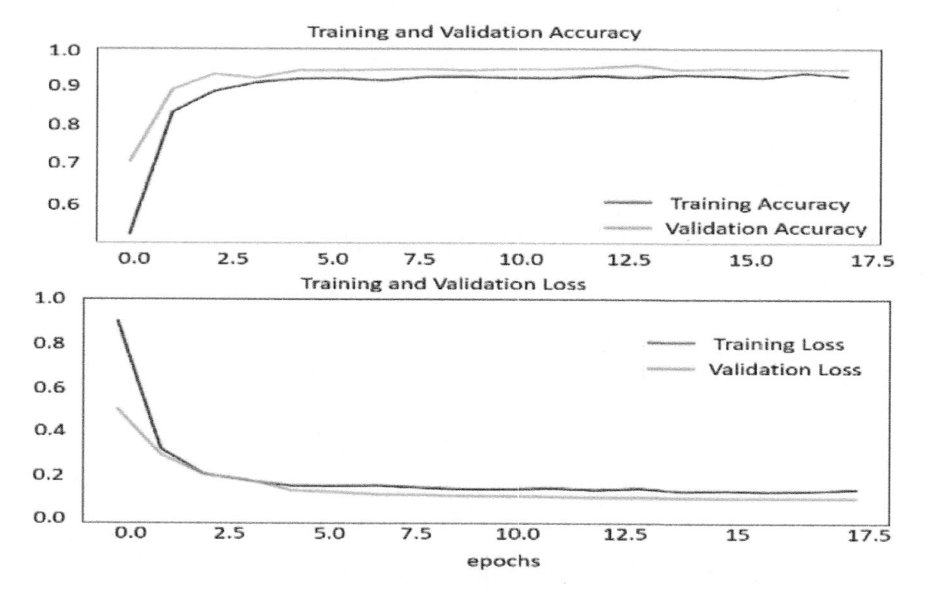

Figure 22.11 Accuracy and loss vs Epoch (K) – Image ACGANs.

22.7 CONCLUSION

In this chapter, two novel ACGAN-based approaches are demonstrated for RUL classification. The research results have established the superiority of image data-based ACGAN over time series data-based ACGAN in generating data resembling the original data used for RUL prediction. Experimental results have shown improvement in RUL prediction for both estimation and classification accuracies using the proposed techniques of data generation. It is also observed that the generated dataset with image ACGAN has higher similarity to actual sensor data in comparison to data generated using Time ACGAN. The superior performance of generated data over original sensor data has facilitated the possibilities to predict RUL in absence of original data.

REFERENCES

[1] J. Thilmany. "Multiphysics: All at once." *Mechanical Engineering* 132(2), 39–41, 2010. https://doi.org/10.1115/1.2010-Feb-5.
[2] S. Hong, Z. Zhou, E. Zio, and B. Wang. "An adaptive method for health trend prediction of rotating bearings." *Digital Signal Processing* 35, 117–123, 2014.
[3] Zeqi Zhao, Bin Liang, Xueqian Wang, and Weining Lu. "Remaining useful life prediction of aircraft engine based on degradation pattern learning. *Reliability Engineering System Safety* 164, 74–83, 2017.
[4] Francesca Calabrese, et al. "Predictive maintenance: A novel framework for a data-driven, semi-supervised, and partially online prognostic health management application in industries." *Applied Sciences* 11(8), 3380, 2021.
[5] Qi Li, Zhanbao Gao, Diyin Tang, and Baoan Li. "Remaining useful life estimation for deteriorating systems with timevarying operational conditions and condition-specific failure zones." *Chinese Journal of Aeronautics* 29(3), 662–674, 2016.
[6] A. Saxena and K. Goebel. "Turbofan engine degradation simulation data set." *NASA Ames Prognostics Data Repository*. NASA Ames Research Center, Moffett Field, CA, 2008. http://ti.arc.nasa.gov/project/prognostic-data-repository.
[7] Biggio, Luca, and Iason Kastanis. "Prognostics and health management of industrial assets: Current progress and road ahead." *Frontiers in Artificial Intelligence* 3 (2020): 578613.
[8] Jay Lee, Hung-An Kao and Shanhu Yang. "Service innovation and smart analytics for industry 4.0 and big data environment." *Procedia Cirp* 16, 3–8, 2014.
[9] LeCun, Yann, Bernhard Boser, John Denker, Donnie Henderson, Richard Howard, Wayne Hubbard, and Lawrence Jackel. "Handwritten digit recognition with a back-propagation network." *Advances in Neural Information Processing Systems* 2 (1989).
[10] Yanming Guo, et al. "Deep learning for visual understanding: A review." *Neurocomputing* 187, 27–48, 2016.

[11] Amgad Muneer, et al. "Deep-learning based prognosis approach for remaining useful life prediction of turbofan engine." *Symmetry* 13(10), 1861, 2021.
[12] Sepp Hochreiter and Jürgen Schmidhuber. "Long short-term memory." *Neural Computation* 9(8), 1735–1780, 1997.
[13] Zheng, Shuai, Kosta Ristovski, Ahmed Farahat, and Chetan Gupta. "Long short-term memory network for remaining useful life estimation." In *2017 IEEE International Conference on Prognostics and Health Management (ICPHM)*, pp. 88–95. IEEE, 2017.
[14] Yuting Wu, et al. "Remaining useful life estimation of engineered systems using vanilla LSTM neural networks." *Neurocomputing* 275, 167–179, 2018.
[15] André Listou Ellefsen, et al. "Remaining useful life predictions for turbofan engine degradation using semi-supervised deep architecture." *Reliability Engineering & System Safety* 183, 240–251, 2019.
[16] Andre S. Yoon, et al. "Semi-supervised learning with deep generative models for asset failure prediction." *arXiv preprint arXiv:1709.00845*, 2017.
[17] Goodfellow, Ian, Jean Pouget-Abadie, Mehdi Mirza, Bing Xu, David Warde-Farley, Sherjil Ozair, Aaron Courville, and Yoshua Bengio. "Generative adversarial networks." *Communications of the ACM* 63, no. 11 (2020): 139–144.
[18] Emily Denton, et al. "Deep generative image models using a laplacian pyramid of adversarial networks." *arXiv preprint arXiv:1506.05751*, 2015.
[19] Alec Radford, Luke Metz, and Soumith Chintala. "Unsupervised representation learning with deep convolutional generative adversarial networks." *arXiv preprint arXiv:1511.06434*, 2015.
[20] Guisheng Hou, et al. "Remaining useful life estimation using deep convolutional generative adversarial networks based on an autoencoder scheme." *Computational Intelligence and Neuroscience 2020*, 2020. https://www.hindawi.com/journals/cin/2020/9601389/.
[21] Odena, Augustus, Christopher Olah, and Jonathon Shlens. "Conditional image synthesis with auxiliary classifier gans." In *International conference on machine learning*, pp. 2642–2651. PMLR, 2017.
[22] Wu, Lirong, Kejie Huang, and Haibin Shen. "A gan-based tunable image compression system." In *Proceedings of the IEEE/CVF Winter Conference on Applications of Computer Vision*, pp. 2334–2342, 2020.
[23] Hyoung Suk Park, Jineon Baek, Sun Kyoung You, Jae Kyu Choi, and Jin Keun Seo. "Unpaired image denoising using a generative adversarial network in X-ray CT." *IEEE Access* 7, 110414–110425, 2019.
[24] Ledig, Christian, Lucas Theis, Ferenc Huszár, Jose Caballero, Andrew Cunningham, Alejandro Acosta, Andrew Aitken et al. "Photo-realistic single image super-resolution using a generative adversarial network." In *Proceedings of the IEEE Conference on Computer Vision and Pattern Recognition*, pp. 4681–4690, 2017.
[25] Y. Le Cun, B. Boser, J. S. Denker, R. E. Howard, W. Habbard, L. D. Jackel, and D. Henderson. *Handwritten Digit Recognition with a Back- Propagation Network*, pp. 396–404. Morgan Kaufmann Publishers Inc., San Francisco, CA, USA, 1990.

[26] Kumar, Gaurav, and Pradeep Kumar Bhatia. "A detailed review of feature extraction in image processing systems." In *2014 Fourth International Conference on Advanced Computing & Communication Technologies*, pp. 5–12. IEEE, 2014.
[27] Jinsung Yoon, Daniel Jarrett, and Mihaela van der Schaar. "Time- series generative adversarial networks." In H. Wallach, H. Larochelle, A. Beygelzimer, F. d'Alch´ e-Buc, E. Fox, and R. Garnett, editors. *Advances in Neural Information Processing Systems*, vol. 32. Curran Associates, Inc., 2019.
[28] TensorFlow. Large-scale machine learning on heterogeneous systems, 2015. Software available from tensorflow.org.
[29] E. I. Obilor and E. C. Amadi. "Test for significance of Pearson's correlation coefficient." *International Journal of Innovative Mathematics, Statistics & Energy Policies* 6(1), 11–23, 2018.
[30] K. Song, N. Wang, and H. Wang. "A metric learning-based Univariate time series classification method. *Information* 11(6), 288, 2020.
[31] Y. Li, H. Chen, and Z. Wu. "Dynamic time warping distance method for similarity test of multipoint ground motion field." *Mathematical Problems in Engineering*, 2010.

Chapter 23

Toward a framework for implementation of quantum-inspired evolutionary algorithm on noisy intermediate scale quantum devices (IBMQ) for solving knapsack problems

Ravi Saini, Ashish Mani, and M.S. Prasad
Amity University Noida

Siddhartha Bhattacharyya
Rajnagar Mahavidyalaya

CONTENTS

DOI: 10.1201/9781003381167-23

23.1 INTRODUCTION

Quantum computing brings together two most important technologies of the 20th century viz., Quantum Mechanics and Computing. Quantum mechanics was introduced as a fundamental theory for physical properties at subatomic levels through the work carried out by leading physicists like Max Planck, Erwin Schrödinger, Werner Heisenberg, Paul Dirac (Zubairy, 2020) and many others, while the revolutionary paper on universal turing machine is commonly considered as the birth of computing (Bowen, 2016). The idea of using quantum effects for computing was introduced by Richard Fyedman (1986) and David Deutsch (1985).

Pioneering works in the development of theoretical quantum algorithms for prime factorization (Shor, 1994) and exhaustive search (Grover, 1998) established the benefits derived from these conceptual quantum machines over the classical computers to solve real world problems. However, practical utilization of quantum algorithms is severely restricted by the number of qubits available on present machines and their high susceptibility to noise, interference, and decoherence. The present-era quantum machines are designated as Noisy Intermediate Scale Quantum (NISQ) (Preskill, 2018). Here the term "intermediate scale" refers to machines with qubits restricted to a few hundreds and "Noisy" indicates that we can only have imperfect control on these qubits.

Parallelly, attempts were made to integrate some of the principles of quantum mechanics into the framework of evolutionary algorithms on classical computers. These quantum-inspired evolutionary algorithms (QEA) (Han & Kim, 2002) are a class of population-based multi-model estimation of distribution algorithms (Platel et al., 2009), which have been successfully applied to solve a wide variety of search and optimization problems. QEA uses Q-bit representation, unitary rotation gates, and measurement operators, which are simulated on classical digital computers (Mani et al., 2014). The NISQ device has been used to showcase selective quantum supremacy in sampling probability distributions, which are hard to simulate in classical digital computers (Frank Arute, 2019). These devices can be used to implement genotype qubit representation, rotation gates and for performing measurement operations in QEA more efficiently as compared to classical digital computers. It is conjectured that NISQ devices may help in inherently better estimation of the multi-model probability distribution in QEA, which may lead to faster search and optimization as compared to implementation on digital computers.

In the recent literature, there has been a persistent effort by academia and industry to develop algorithms for NISQ machines. An approach to use Estimation of Distribution Algorithm (EDA) on quantum machine to solve Traveling Salesman Problem was proposed in (P. Soloviev Vicente, 2021). Further, in (Vitiello, 2021) a new Hybrid Quantum Genetic Algorithm (HQGA) has been proposed for using GA on NISQ machines.

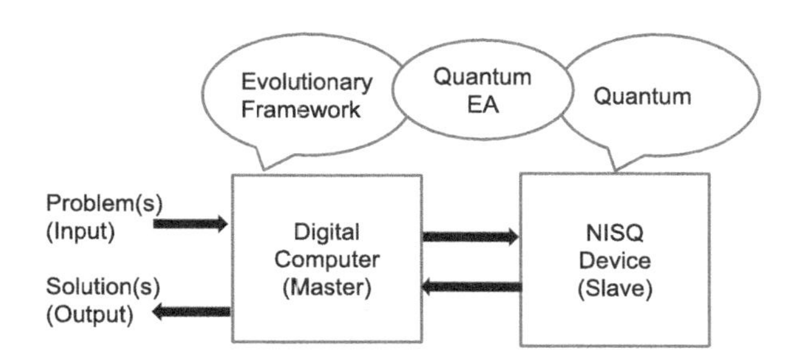

Figure 23.1 Block diagram - Proposed hybrid classical-NISQ computing system for quantum evolutionary algorithms.

This chapter proposes a hybrid classical-quantum implementation of QEA on digital computer and NISQ devices shown in Figure 23.1, with the evolutionary framework running on a digital computer and the quantum-inspired representation and operators running on low-depth quantum circuits of NISQ devices.

Application of evolutionary algorithms (EAs) with such an approach was proposed as quantum-inspired evolutionary algorithms (QIEAs) (Narayanan & Moore, 1996) to solve combinatorial problems. QIEAs were further expanded to show their superior performance to solve class of NP-Hard problems on classical computers in comparison to other EAs (Han & Kim, 2002; Mani et al., 2016).

In this chapter, a novel approach is presented to adapt EA for solving medium size knapsack problem on presently available IBM quantum machines using a circuit model. Section 23.2 of this chapter provides an overview of QIEA and the knapsack problem. Section 23.3 explains the proposed adaptation to run QIEA on an IBM Q machine. Section 23.4 elucidates the experimentation framework and parameters used to a solve set of knapsack taken on real quantum machines. Section 23.5 discusses the results of computation and simulation. Section 23.6 gives conclusion and future scope of research.

23.2 OVERVIEW

23.2.1 Motivation

In a recent review of quantum-inspired metaheuristics (Ross, 2020), it has been proposed that some QIEA may be adapted to be used on real quantum machines. The availability of NISQ machines through the internet cloud provides an opportunity for such an adaptation and test the efficacy of quantum-inspired EAs on quantum hardware. The primary motivation for this research was to formulate a practical implementation of QEA on quantum machines and evaluate results obtained from this adaptation vis-à-vis implementation on classical machines.

23.2.2 Knapsack problem

Knapsack problem derives its name and motivation from a situation faced by mountaineers, which is to select a set of items to maximize the utility while maintaining the weight of knapsack below a specified threshold (Assi & Haraty, 2018). Mathematically, it is an optimization problem where, from a set of n items each with positive profit p_j and a positive weight w_j, a subset of items is to be selected such that the profit is maximized and the weight is maintained below the capacity C of knapsack.

$$\max \Sigma^{n}{}_{j=1} p_j x_j \qquad : \forall(j) \qquad (23.1)$$

subject to following constraints:

$$\sum_{j=1}^{n} wj\, xj \leq C \qquad : \forall(\mathrm{j}) \qquad (23.2)$$

$$x_j = 0 \; or \; 1 \qquad : \forall(j) \qquad (23.3)$$

where:
n = Number of items
p_j = Profit of item j
w_j = Weight of item j
C = Maximum capacity of knapsack

$$x = \begin{cases} 0 \text{ if item } j \text{ is chosen to be in knapsack} \\ 1 \text{ if item } j \text{ is not chosen to be in knapsack} \end{cases}$$

As defined above, the problem is simple to understand, but a brute force solution for the problem is of the order of (2^n), which means that to find an optimal solution by exhaustive search would take about 30 years on a classical computer for 60 items problem, assuming that the machine can run 1 billion vectors in 1 second (Assi & Haraty, 2018). The proof of knapsack problem being p-complete and having reducibility among combinatorial problems (Karp, 1974) means that if a polynomial time algorithm is found for this problem, the same can be extended to other combinatorial problems. Based on the classification of knapsack problems (Kellerer et al., 2004), the following instances of problems were selected from the github repository (Onoue, n.d.).

- Uncorrelated instances (UC): There is no correlation between the weight and profit of an item, and p_j and w_j for each item is chosen randomly.
- Weakly correlated instances (WC): Here, after selecting the weight for each item randomly, the profit is selected so as to maintain high correlation between the weight and profit for each item.
- Strongly correlated instances (SC): In this class of problems, the profit is proportional to the weight with some additional fixed value. These problems are hard to solve and are comparable to real-life situations.
- Uncorrelated with similar weights (UCSW): Here, the profit for each item is selected over a larger range, while the weights are selected from a smaller range, with no correlation between the profit and weight.

23.2.3 QIEA

All EAs are probabilistic and maintain a population of solutions as individuals. This population is operated upon using EA operators to search for an optimal solution. In QIEA, the population is maintained as quantum individuals in form of Q-bits, which are probabilistic like qubits. Akin to a bit in a classical computer, quantum information is stored and processed in a qubit. A classical bit can store and process information only in two states viz., 0 or 1. While a qubit can have information as a superposition of 0 and 1, with a general representation as follows:

$$|\psi> \; = \alpha|0> \; + \beta|1> \tag{23.4}$$

$$\left(|\alpha|^2 + |\beta|^2\right)^{1/2} = 0 \tag{23.5}$$

where:
$|\alpha|^2$: Probability of $|\psi>=0$
$|\beta|^2$: Probability of $|\psi>=1$
$\alpha, \beta \in C$

The intrinsic probabilistic representation in QIEA provides a higher representation power ensuring greater diversity and lower chance of premature convergence.

23.2.4 QIEA for knapsack

For the purpose of adaptation to an actual quantum machine, we selected the QIEA for knapsack problem proposed by Han and Kim (2002). In this algorithm, a quantum population Q(t) is maintained with each quantum individual having n Q-bits representing a linear superposition of solutions. The quantum individuals are observed to generate classical population $P(t)$

with each individuals having n bits. The classical individuals are evaluated based on fitness function and the quantum individuals are updated based on the performance of respective classical individuals. A set of best solutions for each individual is stored in $B(t)$ and the best solution is stored as b. The proposed algorithm in Han and Kim (2002) with *MAX_GEN* generations is reproduced as Algorithm 23.2.3.1, with minor modifications.

Algorithm 23.2.3.1: QIEA for Knapsack

```
1:  procedure QIEA knapsack ()
2:          t ← 0
3:          Q(t) ← INITIALIZE ()
4:          P(t) ← OBSERVE (Q(t))
5:          P(t) ← REP AIR (P(t))
6:          B(t) ← Best solution among P(t)
7:          while t<MAX_GEN do
8:                  t ← t+1
9:                  P(t) ← OBSERVE (Q(t -1))
10:                 P(t) ← REPAIR (P(t))
11:                 EVALUATE (P(t))
12:                 UPDATE (Q(t))
13:                 B(t) ← Best solution among P(t) and B(t -1)
14:                 b ← Best solution among B(t)
15:                 if (migration−period) then
16:                         migrate b or b t j to B(t) globally or locally respectively
17:                 end if
18:                 end while
19: end procedure
```

For a detailed discussion on Algorithm 23.2.3.1, readers are requested to refer to Han and Kim (2002), however a brief introduction of routines used in the algorithm are elucidated below.

- Initialize(): The quantum population is initialized with each Q-bit of every quantum individual having 0.5 probability of collapsing to state 0 and 1 on being observed. This is done by the setting values in Equation (23.6) for each Q-bit as:

$$\alpha = \beta = (1/2)^{1/2} \tag{23.6}$$

- Observe (Q): In this routine, each Q-bit of every individual q in quantum population (Q) is observed to generate a classical bit (C-bit) for each individual p in population (P). The operation of the routine is presented in Algorithm 23.2.3.2.

Algorithm 23.2.3.2: Observe (Q)

```
1:  procedure Observe (Q)
2:          for each individual q in Q do
3:                  for each Q-bit (=α|0i+β|1i)in q do
4:                          if (random[0, 1)<|β|²) then
5:                                  C-bit of ind. p in P ← 1
6:                          else
7:                                  C-bit of ind. p in P ← 0
8:                          end if
9:                  end for
10:         end for
11: end procedure
```

- Repair (P): The repair routine is applied to every solution p in the classical population (P) to ensure validity with respect to capacity (C) constraint of knapsack problem (refer Equation 23.2). We use only random repair procedure, which is presented in Algorithm 23.2.3.3.

Algorithm 23.2.3.3: Repair (P)

```
1:  procedure Repair (P)
2:          for each individual p in P do
3:                  while Σ^n_{j=1} w_j x_j>C do
4:                          i ← random Integer (1, n)
5:                          x_i ← 0
6:                  end while
7:          end for
8:  end procedure
```

- Evaluate (P): Fitness function of net profit as per Equation (23.1) is calculated for every chromosome p in the classical population P. Algorithm 23.2.3.4 shows the procedure for this routine.

Algorithm 23.2.3.4: Evaluate (Q)

```
1:  Procedure Evaluate (P)
2:          for each individual p in P do
3:                  fitness[p] ← Σ^n_{j=1} p_j x_j
4:          end for
5:  vend procedure
```

- Update (Q): This routine updates Q-bits of every individual q in quantum population (Q) based on the fitness of respective classical individual p and the fitness of the best solution b. The update routines in Han and Kim (2002) are explained using a table of angles. However, we present a simpler representation of the same procedure applicable to the knapsack problem in Algorithm 23.2.3.5. The angle of rotation of Q-bit is selected as 0.01π and -0.01π as per the proposal in Han and Kim (2002). Rotation is applied using procedure *ROTATE* based on the quadrant in which the Q-bit is presently located when plotted on a polar plot.

Algorithm 23.2.3.5: Update (Q)

1: Procedure Update(Q, P, b)
2: for each (q and p) in (Q and P) do
3: if fitness(p) < fitness(b) then
4: for each (Q-bit, C-bit, and b_i) in (q, p, and b) do
5: if (C-bit== 0 and b_i== 1) then
6: Q-bit ← ROTATE (Q-bit, 0.01π)
7: else if (C-bit== 1 and b_i== 0) then
8: Q-bit ← ROTATE (Q-bit, -0.01π)
9: end if
10: end for
11: end if
12: end for
13: end procedure
14: procedure Rotate(Q-bit, Δθ)
15: if (Q-bit (= α|0>+β|1) in first/third quadrant) then
16: $\text{Q}-\text{bit} \leftarrow \begin{bmatrix} \cos(\Delta\theta) - \sin(\Delta\theta) \\ \sin(\Delta\theta) + \cos(\Delta\theta) \end{bmatrix} \begin{bmatrix} \alpha \\ \beta \end{bmatrix}$
17: else
18: $\text{Q}-\text{bit} \leftarrow \begin{bmatrix} \cos(-\Delta\theta) - \sin(-\Delta\theta) \\ \sin(-\Delta\theta) + \cos(-\Delta\theta) \end{bmatrix} \begin{bmatrix} \alpha \\ \beta \end{bmatrix}$
19: end if
20: end procedure

- Migration: Migration process is used to induce variation of the probabilities in Q-bit individuals. For details of procedures, reader may refer to Han and Kim (2002).

23.3 ADAPTATION OF EA ON IBMQ

23.3.1 Quantum routine

Implementation of QIEA on classical computer discussed in Section 23.2 is available online at (Mohtasham, 2019). In this section, the method for porting EA on IBMQ machine is presented. As mentioned earlier, Q-bits behave like qubits, but operate in a polar plot. However, qubits operate in the Bloch sphere and therefore the angle θ of Q-bit is divided by two to obtain θ for qubit. For porting these qubits on quantum computing, we have considered N operational qubits.

The key process that is implemented on the quantum computer for EA is to observe qubits using a quantum circuit (QC) model. Therefore, the observe (Q) routine in Algorithm 23.2.3.2 is replaced by observing IBM (Q) routine presented in Algorithm 23.3.1.1. Here, the qubits are prepared using rotation gates in a bunch of N qubits at a time and observed on the IBMQ machine.

Algorithm 23.3.1.1: observeIBMQ (Q)

```
1:  Procedure observeIBMQ (Q)
2:          for all qubits in individual q in Q do
3:                  C-bits of p in P ← QuantumCirc (qubits)
4:          end for
5:  end procedure
6:  procedure QuantumCirc (qubits, N=5)
7:          for each set of N qubits do
8:                  C-bits ← qubits observed using rotation and mea-
    surement on IBMQ
9:          end for
10:         return C-bits
11: end procedure
```

23.3.2 Quantum circuit

Quantum computers are generally accessed by using a quantum circuit model which provides an abstraction layer for obfuscation of the underlying physical implementation of the machine. IBM provides access to their quantum system through internet cloud using APIs. The Qiskit (al, n.d.) framework in python language provides functionality to create algorithms for a quantum computer using circuits and allows them to be run on quantum machines.

For the procedure QuantumCirc (*qubits*, N) in Algorithm 23.3.1.1, multiple circuits are prepared of N qubits each. These circuits have rotation gates based on the value of each qubit and measurement gates to capture observed values in classical qubits. A sample circuit to observe five qubits having values corresponding to angles $q0_0=0.01\pi$, $q0_1=0.19\pi$, $q0_2=0.23\pi$, $q0_3=0.55\pi$ and $q0_4=0.77\pi$, is shown in Figure 23.2. Prior to running the circuits, it needs to be transpiled for the specific device. The circuit in Figure 23.2 when transpiled to run *ibmq_bogota* is shown in Figure 23.3. It may be noted from the quantum circuits that since the depth of each circuit is shallow, we expect the minimal effect of noise. Further, this minimal noise is likely to increase diversity in the population by increasing random mutations.

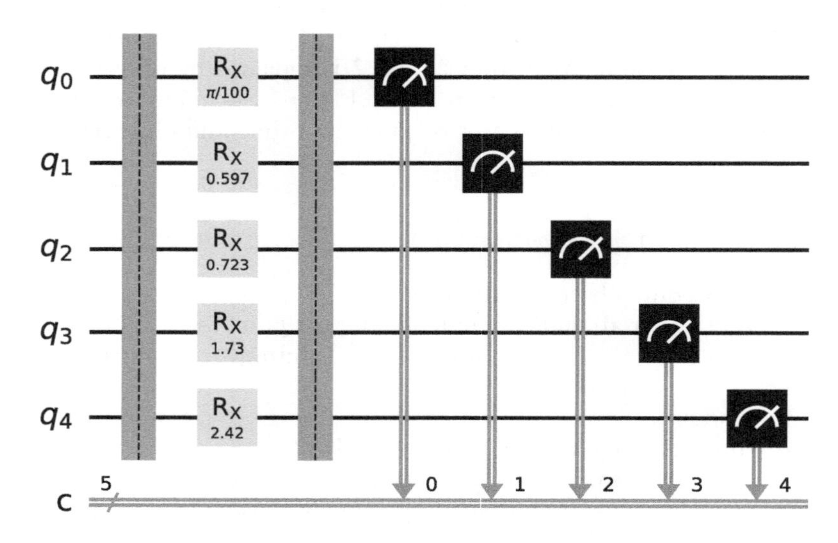

Figure 23.2 Sample circuit to observer 5 qubits.

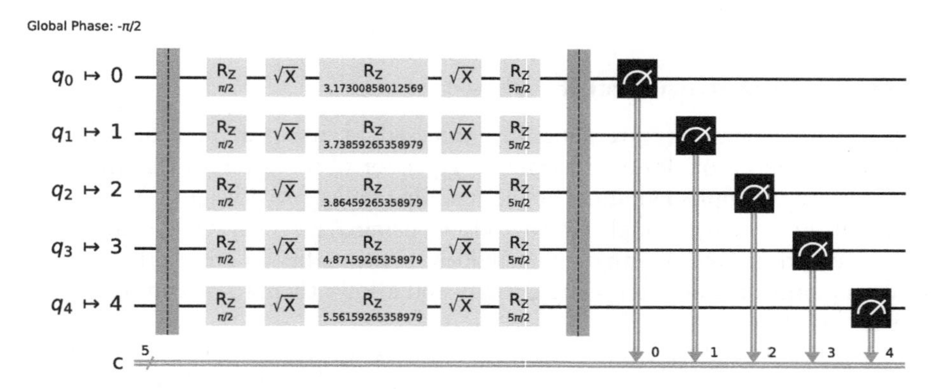

Figure 23.3 Transpiled circuit to observer 5 qubits on "ibmq_bogota".

23.3.3 Proposed methodology

Based on the quantum routine at Algorithm 23.3.1.1, multiple circuits are generated for execution on five qubits IBM NISQ machines. These circuits are packaged together for each generation using *IBMQJobManager* function in Qiskit. This methodology allows solving medium size knapsack problems using 5 qubit quantum machines.

23.4 FRAMEWORK AND EXPERIMENTATION PARAMETERS

23.4.1 Framework

The framework is developed to run EA simultaneously for the same set of problems on classical and quantum computers. Running of EA discussed in Section 23.2 on classical computers is trivial. However, to run EA on the quantum computer using Qiskit on IBMQ machines, the following techniques were used.

- Bunching of circuits: Access to IBMQ machines is provided in *fair-share* mode, where all the circuits submitted are queued to be executed in FIFO mode. To reduce wait time to run each circuit, the *IBMQJobManager* function of Qiskit is used to bunch the circuits together. This allows submission of a large number of circuits to observe hundreds of qubit in one job.
- Selection of IBMQ machine: The load on each IBMQ machine varies based on the number of jobs being submitted. Two methods were used to select a machine based on the access provided at given time:
- Fairshare access: This is the default access provided by IBMQ. Under this access, the *least_busy* function was used to identify the machine with requisite number of qubits and minimum load at a given time.
- Dedicated access: IBM has been kind to provide Amity University 3 hours of dedicated access on three quantum machines every month[1]. During the period of reservation for this access, the reserved machine was used for all the jobs.

23.4.2 Experimentation

Twenty benchmark knapsack problems from Onoue (n.d.) were used to evaluate the performance of QEIA on classical computers and EA on quantum computers. These problems were selected from the four categories of knapsack problems discussed in Section 23.2. The parameters and key variables used during the experimentation are as follows.

- Number of items (n): All knapsack problems were selected with \$n=100\$.
- Number of agents: 25 agents were used for each run of EA.
- Maximum generations (MAX_GEN): Each EA was run for 100 generations.
- Number of qubits (N): Quantum machines were selected with at-least 5 operational qubits and minimum load.

Based on the parameters above, in each generation, 2,500 (25×100) qubits had to be observed. These qubits were submitted as 500 (2,500/5) circuits to IBMQ. Therefore, for each EA with 100 generations, 50,000 circuits were generated. Finally, during the experimentations for 20 problems, each run for 15 times, total circuits submitted to IBM exceeded 1,50,00,000 (50,000×20×15) to observe 7,50,00,000 qubits. The results in the next section are presented based on these 15 runs of EA on each problem. The algorithms for experimentation were implemented in python[2] on Linux OS.

23.5 RESULTS

23.5.1 Readings

The values obtained from experimentation are presented in Table 23.1. The following values are shown in the table, for each problem.

- Classical profit: This is the mean profit obtained for each knapsack problem when solved 15 times using QIEA on a classical computer.
- Quantum profit: This is the mean profit obtained for each knapsack problem when solved 15 times using EA on a quantum computer.
- %Profit diffn in classical and quantum: This is the percentage difference in the mean value of profit for each knapsack problem when solved by a classical computer using QIEA vs solution for same problem using EA on a quantum computer. This value is calculated using Equation (23.7).

$$\left(\left(\text{Classical Profit - Quantum Profit}\right) * 100\right)/\text{Quantum Profit} \tag{23.7}$$

- Stdev in classical profit: This is the standard deviation in profit obtained for each knapsack problem when solved 15 times using QIEA on a classical computer.
- Stdev in quantum profit: This is the standard deviation in profit obtained for each knapsack problem when solved 15 times using EA on a quantum computer.

Table 23.1 Record of EA on classical and quantum computer

Problem	*Classical profit*	*Quantum profit*	*%Profit diffn in classical & quantum*	*Stdev in classical profit*	*Stdev in quantum profit*
ucsw_s090.kp	30,069.92	30,076.58	−0.02	607.39	609.91
ucsw_s091.kp	31,423.14	31,309.43	0.36	446.45	407.80
ucsw_s092.kp	32,685.58	32,576.04	0.34	430.94	489.14
ucsw_s093.kp	29,467.92	29,494.77	−0.09	575.53	461.75
ucsw_s094.kp	28,616.57	29,016.29	−1.38	193.74	429.31
sc_s020.kp	276,190.27	274,992.64	0.44	1,324.12	1,194.49
sc_s021.kp	309,885.89	309,327.79	0.18	844.43	957.42
sc_s022.kp	329,634.29	328,937.57	0.21	661.21	916.47
sc_s023.kp	306,113.44	305,408.89	0.23	486.14	675.95
sc_s024.kp	319,710.33	318,942.50	0.24	637.70	963.98
wc_s090.kp	262,548.22	262,520.91	0.01	879.40	1,011.72
wc_s091.kp	281,825.18	282,383.64	−0.20	653.64	1,217.13
wc_s092.kp	288,869.52	289,107.94	−0.08	942.13	1,505.97
wc_s093.kp	254,028.18	253,787.18	0.09	1,331.54	1,317.29
wc_s094.kp	258,165.40	258,567.73	−0.16	836.22	1,085.37
uc_s000.kp	352,959.06	351,515.75	0.41	6,260.90	6,038.22
uc_s001.kp	314,536.58	313,738.42	0.25	3,380.82	5,551.75
uc_s002.kp	336,070.83	339,215.17	−0.93	8,318.07	7,083.54
uc_s003.kp	324,961.67	326,032.33	−0.33	10,167.89	5,854.34
uc_s004.kp	300,101.21	300,224.00	−0.04	7,959.19	3,046.22

23.5.2 Graphical representation

To get the graphical appreciation, the values of profits obtained for 15 runs on 4 problems for quantum vs classical machines are presented in Figure 23.4. One knapsack problem of each category has been shown in the plot. The values of profit obtained by classical method is indicated a CProfit_<*Problem Name*>, while the best profit value of EA on Quantum Computer is indicated as QProfit<*Problem Name*>.

23.5.3 Time evaluation

Quantum computers with five qubits were used over internet cloud for these experiments. Direct comparison of quantum run times in this environment vis-à-vis implementation of QIEA on classical computers, with billions of bits locally available, may not provide a realistic insight. However, typical operations involved in executing algorithm on a quantum computer with observed run times are listed below.

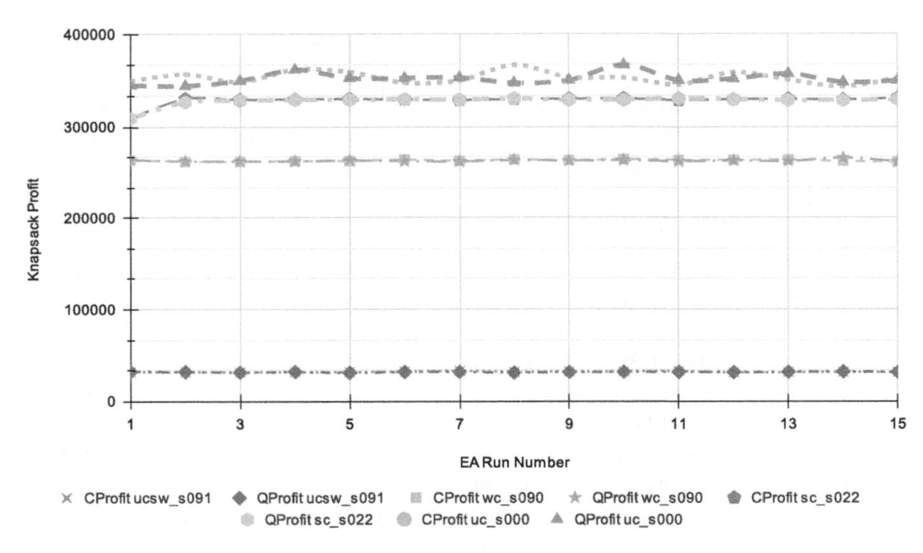

Figure 23.4 Knapsack profit on quantum vs classical machine using EA.

- Time to transpile: On submission of job to IBM cloud, all the quantum circuits in the job are transpiled to the specific device on which circuits are to be implemented. This has been typically observed to take 0.03–0.1 seconds.
- Time to validate: All jobs are validated prior to submission to quantum machine. This operation was observed to take 1–2 seconds.
- In queue time: This is the waiting period for the job to be submitted for running on a quantum machine. It is highly variable depending on the load on the machine and may vary from few seconds to minutes. In some cases, this waiting period may be in hours for specific machines.
- Running time: This is the time which is reported by IBM as the run time for the job on the quantum machine. This time was observed to be 7–20 seconds.
- Time taken: This is the time taken by the job on quantum machine and is a sub part of Running Time. This time is reported by IBM as "*Time Taken*" and typically varied between 5 and 17 seconds for our circuits.
- Get least busy device: This is an auxiliary function for the smooth operation of the algorithm. Using this function, the name of the least busy device with desired number of operational qubits (=5) is obtained from IBM. The time taken by this operation was observed to be 1–10 seconds.

Table 23.2 Pairwise wins–quantum EA vs classical QIEA

Problem	*Win quantum*	*Win classical*	*Total runs*
ucsw_s090.kp	8	7	15
ucsw_s091.kp	7	8	15
ucsw_s092.kp	9	6	15
ucsw_s093.kp	7	8	15
ucsw_s094.kp	6	9	15
sc_s020.kp	9	6	15
sc_s021.kp	8	7	15
sc_s022.kp	6	9	15
sc_s023.kp	8	7	15
sc_s024.kp	10	5	15
wc_s090.kp	7	8	15
wc_s091.kp	6	9	15
wc_s092.kp	7	8	15
wc_s093.kp	8	7	15
wc_s094.kp	8	7	15
uc_s000.kp	7	8	15
uc_s001.kp	7	8	15
uc_s002.kp	8	7	15
uc_s003.kp	7	8	15
uc_s004.kp	9	6	15
Total	152	148	300

23.5.4 Evaluation of results

Perusal of the percentage difference in mean profit values in Table 23.1 and plots of values in Figure 23.4, indicate that comparable values of profit are obtained by QIEA on the classical computer and EA on the quantum computer for all the cases. The following hypothesis is proposed to prove that the quality of results from both the implementations are equivalent.

- H_0: Equivalent quality of results are obtained on knapsack problems using QIEA on classical computers and EA on quantum computers.

To prove the hypothesis we apply "*A simple first-sight procedure: the Sign test*" as proposed in (Derrac, 2011). The test mandates that, if it is assumed in the null hypothesis that two algorithms being compared are equivalent, then each algorithm must win on approximately $n/2$ times out of n problems. Table 23.2 shows the record of wins for EA on quantum machines and QIEA on classical machines, for each run. The values clearly indicate that both procedures win approximately $n/2$ times, thus proving the proposed hypothesis.

23.6 CONCLUSION

In this chapter, we presented adaptation of QIEA on quantum computers. However, the present implementation did not exploit key quantum functions like entanglement and complex quantum gates. The study and proposal in his chapter make the following innovative contributions.

A novel practical implementation of EA on IBMQ machines to solve 100 item knapsack problems.

Proof that the solution for knapsack using QIEA on a classical computer is comparable with the results obtained by EA on a quantum computer.

Successful implementation of EA on present-day NISQ quantum machines to solve medium-sized combinatorial problems paves a way to find new methods to harness nascent quantum computing technology.

NOTES

1 We place on record our gratitude to IBM for providing access to Quantum Machines, without which this research may not have been feasible.
2 Code files are available upon request to author.

BIBLIOGRAPHY

al, H. A. (n.d.). *Qiskit*. Retrieved July 17, 2022, from Qiskit: https://qiskit.org/.

Assi, M. & Haraty, R. A. (2018). A survey of the knapsack problem. *2018 International Arab Conference on Information Technology (ACIT)*, Lebanon, pp. 1–6.

Bowen, J. P. (2016). *Alan Turing: Founder of Computer Science*. School on Engineering Trustworthy Software Systems. Cham: Springer.

Derrac, J., Garcíab, S., Molinac, D., & Herrera, F. (2011). A practical tutorial on the use of nonparametric statistical tests as a methodology for comparing evolutionary and swarm intelligence algorithms. *Swarm and Evolutionary Computation*, 1(1), 3–18.

Deutsch, D. (1985). Quantum theory, the church-turing principle and the universal quantum computer. *Proceedings of the Royal Society of London*, 400(1818), 97–117. Royal Society.

Feynman, R. P. (1986). Quantum mechanical computers. *Foundations of Physics*, 16(6), 507–532.

Frank Arute, K. A. (2019). Quantum supremacy using a programmable superconducting processor. *Nature*, 574(7779), 505–510.

Grover, L. K. (1998). Quantum computers can search rapidly by using almost any transformation. *Physical Review Letters*, 80, 4329–4332.

Han, K. H. & Kim, J. H. (2002). Quantum-inspired evolutionary algorithm for a class of combinatorial optimization. *IEEE Transactions on Evolutionary Computation*, 6, 580–593.

Karp, R. (1974). Reducibility among combinatorial problems. In R. T. Miller, ed. *Complexity of Computer Computations*, pp. 85–103. Bostan: Springer.

Kellerer, H., Pferschy, U., & Pisinger, D. (2004). Exact solution of the knapsack problem. In *Knapsack Problems*, pp. 117–160. Amsterdam: Elsevier.

Mani, N., Srivastava, G., & Mani, A. (2016). Solving combinatorial optimization problems with quantum inspired evolutionary algorithm tuned using a novel heuristic method. *CoRR*.

Mani, N., Srivastava, G., Sinha, A. K., & Mani, A. N. (2014). Effect of population structures on quantum-inspired evolutionary algorithm. *Applied Computational Intelligence and Soft Computing* 2014(24), 24.

Mohtasham, M. B. (2019). Quantum-evolutionary-algorithm-knapsack-python. Retrieved July 17, 2022, from Github: https://github.com/mjBM/Quantum-Evolutionary-Algorithm-Knapsack-Python-.

Narayanan, A., Moore, M. (1996). Quantum-inspired genetic algorithms. *Proceedings of IEEE International Conference on Evolutionary Computation*, Nagoya, pp. 61–66.

Onoue, Y. (n.d.). *likr/kplib*. Retrieved July 17, 2022, from github: https://github.com/likr/kplib.

Platel, M. D., Schliebs, S., & Kasabov, N. (2009). Quantum-inspired evolutionary algorithm: A multimodel EDA. *IEEE Transactions on Evolutionary Computation*, vol. 13, no. 6, pp. 1218–1232.

Preskill, J. (2018). Quantum computing in the NISQ era and beyond. *Quantum* 2, 79.

Ross, O. H. (2020). A review of quantum-inspired metaheuristics: Going from classical computers to real quantum computers. *IEEE Access* 8, 814–838.

Shor, P. W. (1994). Algorithms for quantum computation: Discrete logarithms and factoring. *35th Annual Symposium on Foundations of Computer Science*, Santa Fe, pp. 124–134. IEEE.

Soloviev, V. P. & Bielza, C. (2021). Quantum-inspired estimation of distribution algorithm to solve the travelling salesman problem. *2021 IEEE Congress on Evolutionary Computation (CEC)*, Poland, pp. 416–425.

Vitiello, G. A. (2021). Implementing evolutionary optimization on actual quantum processors. *Information Sciences* 575, 542–562.

Zubairy, M. S. (2020). *Quantum Mechanics for Beginners: With Applications to Quantum Communication and Quantum Computing*. London: Oxford University Press.

Index